GEOLOGIC DIVISION OFFICES

U.S. Geological Survey
National Center
12201 Sunrise Valley Drive
Reston, Virginia 22092
(703) 860-6531

USGS
Federal Center, Bldg. 25
Denver, Colorado 80225
(303) 234-3624

USGS
345 Middlefield Road
Menlo Park, California 94025
(415) 323-2214

USGS
San Juan Observatory
Box 936
Cayey, Puerto Rico 00633
(809) 738-2281

USGS
P.O. Box 5917, Puerta de Tierra Sta.
San Juan, Puerto Rico 00906
(809) 722-3142

USGS
Quissett Campus, Bldg. B
Woods Hole, Massachusetts 02543
(617) 548-8700

USGS
Fredericksburg Geomagnetic Center
Corbin, Virginia 22446
(703) 373-7601

USGS
Branch of Oil & Gas Resources
Fisher Island Station
Miami Beach, Florida 33139
(305) 672-1784

USGS
Rm. 227, Geology Bldg.
University of Indiana
Bloomington, Indiana 47401
(812) 337-7597

USGS
Rm. 222 Science Hall
University of Wisconsin
Madison, Wisconsin 53706
(608) 233-3083

USGS
P.O. Box 6732
Corpus Christi, Texas 78411
(512) 888-3294

USGS
c/o Arkansas Geological Commission
3815 W. Roosevelt Road
Little Rock, Arkansas 72204
(501) 371-1616

USGS
Rm. 508, Post Office Bldg.
Salt Lake City, Utah 84101
(801) 588-5640

USGS
2255 North Gemini Drive
Flagstaff, Arizona 86001
(602) 261-1455

USGS
Albuquerque Seismological Center
Kirtland AFB, East Bldg. 10002
Albuquerque, New Mexico 87115
(505) 264-4637

USGS
Hawaiian Volcano Observatory
Hawaii National Park
Hawaii, 96718
(808) 967-7485

USGS
1107 N.E. 45th Street, Suite 10
Seattle, Washington 98105
(206) 442-1995

USGS
301 E. McLoughlin St., Suite 4
Vancouver, Washington 98660
(206) 696-7693

USGS
Newport Observatory
Rt. 4, Box 56A
Newport, Washington 99156
(509) 456-0111

USGS
U.S. Court House, Rm. 656
W. 920 Riverside Avenue
Spokane, Washington 99201
(509) 456-4677

USGS
1209 Orca Street
Anchorage, Alaska 99501
(907) 271-4150

USGS
Box 80586
College, Alaska 99701
(907) 456-7084

Directory of the Geologic Division, U.S. Geological Survey

with staff listings
and office and branch rosters

Published 1981 by

AMERICAN GEOLOGICAL INSTITUTE
One Skyline Place
5205 Leesburg Pike
Falls Church, Virginia 22041

Printed in the United States of America
ISBN 0-913312-45-2

Chief Geologists

G. K. Gilbert
1889-1892

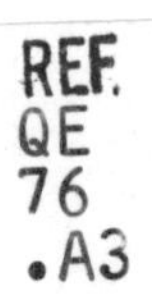

C. D. Walcott
1893-1894

Bailey Willis
1897-1902

C. W. Hayes
1902-1911

Waldemar Lindgren
1911-1912

David White
1912-1922

W.C. Mendenhall
1922-1930

T.W. Stanton
1930-1935

G.F. Loughlin
1935-1944

W.H. Bradley
1944-1959

C.A. Anderson
1959-1964

W.T. Pecora
1964-1965

H.L. James
1965-1971

V.E. McKelvey
1971-1972

R.P. Sheldon
1972-1977

D.L. Peck
1977-

FOREWORD

The U.S. Geological Survey celebrated its Centennial Anniversary on March 3, 1979. During this 100-year span, the growth of the Survey's scientific and engineering research programs have paralleled and contributed to the growth of the Nation. And the emphasis of the Survey's work, still patterned after the spirit of exploration and scientific interest that guided the Lewis and Clark Expedition through the West in 1804–6, is to gain a measure of the land and the resources it contains.

The Survey's wide-ranging programs are divided among four main organizational units—the Geologic Division, the Water Resources Division, the National Mapping Division, and the Conservation Division. Support services are provided by the Administrative and Computer Center Divisions. These Divisions are headed by the Office of the Director.

Headquarters for the Geological Survey is located at the National Center in Reston, Virginia. Regional offices have been established to help plan and manage the Survey's scientific programs. An Eastern Region office is located in Reston, Virginia; a Central Region office is located in Denver, Colorado; and a Western Region office is located in Menlo Park, California. Numerous local offices are scattered throughout the country (see addresses listed on the inside front cover).

The research programs of the Geologic Division are planned and managed by seven main offices and their subordinate branches. The names of the offices and branches show in capsule form the work of the Division. The work ranges from detailed study of active volcanoes on the Island of Hawaii (Hawaiian Volcano Observatory) to the exploration of the Solar System (Branch of Astrogeology), to the preparation of estimates of the country's oil and gas resources (Branch of Oil and Gas Resources). An Office of Scientific Publications provides editorial and technical support to assure the timely publication of the results of the scientific investigations of the entire Division. All Geologic Division offices are headed by the Office of the Chief Geologist.

The investigations of the Geologic Division require the combined efforts of many kinds of scientists, engineers, and technical assistants. Thus the organization chart on the inside back cover plus the Staff listings and the Directory of Personnel shown on the following pages reflect the highly diversified and far-ranging nature of its research programs.

Dallas L. Peck
Chief Geologist
Reston, Virginia

GEOLOGIC DIVISION STAFF

OFFICE OF THE CHIEF GEOLOGIST

Chief Geologist—Dallas L. Peck
Associate Chief Geologist—Gordon P. Eaton
Deputy Associate Chief Geologist—Jerald M. Goldberg
Deputy Chief Geologist, Operations—Penelope M. Hanshaw
Program Analysis Officer—David A. Seyler
Assistant Program Officer—Carl Koteff
Chief Chemist—vacant
Computer System Administrator—R. Michael Gall
Administrative Officer—vacant

OFFICE OF THE REGIONAL GEOLOGISTS

Regional Geologist, Eastern—Avery A. Drake, Jr.
 Deputy Regional Geologist—John D. Peper
Regional Geologist, Central—William R. Keefer
 Deputy Regional Geologist—Donald E. Watson
 Administrative Officer—Joseph S. Pillera
Regional Geologist, Western—Joseph I. Ziony
 Deputy Regional Geologist—Maureen G. Johnson
 Deputy Regional Geologist, Flagstaff—Gordon A. Swann
 Administrative Officer—Linda L. Like

OFFICE OF SCIENTIFIC PUBLICATIONS

Chief—Robert E. Davis
Deputy Chief—Vacant
Staff Scientist—Henry R. Spall
Staff Geologist—Clifford M. Nelson
Chief Librarian—George H. Goodwin, Jr.
Chief, Technical Reports Unit, Eastern—John M. Aaron
Chief, Technical Reports Unit, Central—Arthur C. Tarr
Chief, Technical Reports Unit, Western—John S. Derr
Chief, Geologic Inquiries Group—Helen P. Withers
Geologic Map Editor—vacant
Chairman, Geologic Names Committee—Norman F. Sohl
Administrative Officer—Mary L. Ratliff

OFFICE OF MINERAL RESOURCES

Chief—A. Thomas Ovenshine
Deputy Chief—Donald J. Grybeck
Deputy Chief, Wilderness Program—Gus H. Goudarzi
Deputy Chief, Scientific Programs—vacant
Deputy Chief, BIA/BLM Programs—Helen M. Beikman
Staff Geologist for CUSMAP—Gary C. Curtin
Administrative Officer—Douglas K. Fridrich
Chief, Branch of Alaskan Geology—Thomas P. Miller
Chief, Office of Resource Analysis—Richard F. Meyer
Chief, Branch of Exploration Research—Glenn H. Allcott
Chief, Branch of Eastern Mineral Resources—Donald A. Brobst
Chief, Branch of Central Mineral Resources—Richard B. Taylor
Chief, Branch of Western Mineral Resources—Roger P. Ashley

OFFICE OF INTERNATIONAL GEOLOGY

Chief—John A. Reinemund
Deputy Chief, Operations—Alfred H. Chidester
Staff Scientist for Research—David F. Davidson
Assistant for Program Coordination—Hershell L. Fleming
Assistant for Program Administration—Alvin F. Holzle
Administrative Officer—Helen Willingham
Chief, Branch of Middle Eastern & Asian Geology—Maurice J. Terman
Chief, Branch of Latin American & African Geology—Gene E. Tolbert

OFFICE OF GEOCHEMISTRY & GEOPHYSICS

Chief—Robert I. Tilling
Deputy Chief—Charles J. Zablocki
Deputy Chief, Geothermal Research—Donald W. Klick
Deputy Chief, Program—Jeffrey C. Wynn
Staff Geochemist—John R. Keith
Staff Geologist—Allan N. Kover
Administrative Officer—John D. Wall
Chief, Branch of Experimental Geochemistry & Mineralogy—Philip M. Bethke
Chief, Branch of Field Geochemistry & Petrology—Benjamin A. Morgan, III
Scientist-in-Charge, Hawaiian Volcano Observatory—Robert W. Decker
Chief, Branch of Regional Geophysics—William F. Hanna
Chief, Branch of Isotope Geology—Bruce R. Doe
Chief, Branch of Regional Geochemistry—Richard J. Ebens
Chief, Branch of Electromagnetism & Geomagnetism—Adel A.R. Zohdy
Chief, Branch of Petrophysics & Remote Sensing—Graham R. Hunt
Chief, Branch of Analytical Laboratories—Brent P. Fabbi

OFFICE OF ENERGY RESOURCES

Acting Chief—Linn Hoover
Deputy Chief—vacant
Deputy Chief, Uranium & Thorium—Edwin A. Noble
Deputy Chief, Oil & Gas Resources—Oswald W. Girard, Jr.
Administrative Officer—Lee L. Benton
Chief, Branch of Oil & Gas Resources—Richard F. Mast
Chief, Branch of Coal Resources—Jack H. Medlin
Chief, Branch of Sedimentary Mineral Resources—Marc W. Bodine, Jr.
Chief, Branch of Uranium & Thorium Resources—Terry W. Offield

OFFICE OF ENVIRONMENTAL GEOLOGY

Chief—Douglas M. Morton
Deputy Chief—Richard D. Krushensky
Deputy Chief, Coal Environmental Programs—John O. Maberry
Deputy Chief, Reactor Programs—Robert H. Morris
Staff Geologist for Land Resource Programs—Thomas M. Cronin
Administrative Officer—Scott E. Tilley
Chief, Branch of Eastern Environmental Geology—Douglas W. Rankin
Chief, Branch of Central Environmental Geology—Robert B. Raup, Jr.
Chief, Branch of Western Environmental Geology—Desiree E. Stuart-Alexander
Chief, Branch of Engineering Geology—Donald R. Nichols
Chief, Branch of Special Projects—William S. Twenhofel
Chief, Branch of Astrogeologic Studies—Laurence A. Soderblom
Chief, Branch of Paleontology & Stratigraphy—William V. Sliter

OFFICE OF EARTHQUAKE STUDIES

Acting Chief—John R. Filson
Deputy Chief—vacant
Deputy for Research—Roger M. Stewart
Deputy for Plans & Programs—Mark D. Zoback
Deputy for Research Applications—Walter W. Hays
Administrative Officer—William E. Phelps, Jr.
Chief, Branch of Global Seismology—Eric R. Engdahl
Chief, Branch of Seismology—David P. Hill
Acting Chief, Branch of Seismic Engineering—Roger D. Borcherdt
Chief, Branch of Ground Motion & Faulting—Roger D. Borcherdt
Chief, Branch of Earthquake Tectonics & Risk—Robert C. Bucknam
Chief, Branch of Tectonophysics—James H. Dieterich
Acting Chief, Branch of Network Operations—Jerry P. Eaton

OFFICE OF MARINE GEOLOGY

Chief—N. Terence Edgar
Deputy Chief, Environmental Studies—Robert W. Rowland
Deputy Chief, Programs & Budget—Paul G. Teleki
Chief, Branch of Pacific-Arctic Geology—H. Edward Clifton
Chief, Branch of Atlantic-Gulf of Mexico Geology—David W. Folger
Administrative Officer—Lee L. Benton

AARON, JOHN M. **Barbara**
Geologist
Office of Scientific Publications
Reston **(703) 860-6491**
Franklin & Marshall-BS; Penn State-PhD
North Atlantic environmental studies
Marine geology; sedimentology

ABEYTA, VONA V.
Secretary
Br. of Isotope Geology
Denver **(303) 234-3624**

ABRAMS, GERDA A.
PST
Br. of Electromag. & Geomagnetism
Denver **(303) 234-3624**
U. of Colorado-BA

ACKERMANN, HANS D. **Charlotte**
Geophysicist **(303) 988-4321**
11215 W. Center Avenue
Lakewood, Colorado 80226
Br. of Regional Geophysics
Denver **(303) 234-3141**
Carnegie Inst. of Technology-BS; Penn State-MS
Seismic refraction & reflection data acquisition, processing & interpretation

ADAM, DAVID P. **Wansley**
Geologist **(415) 591-5860**
750 Cedar Street
San Carlos, California 94070
Br. of Western Environmental Geology
Menlo Park **(415) 323-8111 x2669**
Harvard-BA; U. of Arizona-MS, PhD
Upper Quaternary climatic history of California
Palynology; chrysophyte cysts; computer modeling

ADDICOTT, WARREN O. **Suzanne**
Geologist **(415) 493-6352**
957 Los Robles Avenue
Palo Alto, California 94306
Office of International Geology
Menlo Park **(415) 323-8111 x2370**
Pomona Coll.-BA; Stanford U.-MA; U. of California (Berkeley)-PhD
Circum-Pacific map project
Stratigraphy; Cenozoic molluscan paleontology

AFFOLTER, RONALD H.
Geologist
Br. of Coal Resources
Denver **(303) 234-5280**
Augsburg Coll.-BS; Northeastern Illinois U.-MS
Coal geochemistry

AFRA, DONNA W. **Tom**
Secretary
19201 Shubert Drive
Saratoga, California 95070
Br. of Isotope Geology
Menlo Park **(415) 323-8111 x2368**
U. of Minnesota-BA

AGARD, SHERRY S.
Geologist
Br. of Engineering Geology
Denver **(303) 234-3413**
U. of Wyoming-BA; U. of Colorado-MS
Tree-ring dating of recent mass movements
Geomorphology; dendrochronology; environmental geology

AGER, THOMAS A.
Geologist **(703) 536-5548**
1007 N. Sycamore Street
Falls Church, Virginia 22046
Br. of Paleontology & Stratigraphy
Reston **(703) 860-7745**
Wayne State U.-BS; U. of Alaska-MS; Ohio State-PhD
Quaternary climate, southwest Alaska
Late Cenozoic palynology; paleoclimatology; paleoecology

ALBAN, EUGENE H.
Engineering Technician **(301) 243-8150**
1615 Division Avenue
Lutherville, Maryland 21093
Br. of Analytical Laboratories
Reston **(703) 860-7662**

ALBERS, JOHN P. **Eileen**
Geologist
647 Fairmede Avenue
Palo Alto, California 94306
Br. of Western Mineral Resources
Menlo Park **(415) 323-8111 x2004**
Carleton Coll.-BA; U. of Minnesota-MS; Stanford-PhD
Mineral resources of California
Economic geology; tectonic framework; metallogeny

ALBERT, NAIRN R.D. **Elaine**
Geologist **(415) 566-9540**
1562 27th Avenue
San Francisco, California 94122
Br. of Alaskan Geology
Menlo Park **(415) 323-8111 x2026**
San Jose State U.-BA, MS
Remote sensing; Radiolaria

ALBRIGHT, VERNA T.
Secretary
Library
Reston **(703) 860-6618**
James Madison U.

ALDEN, ANDREW L. **Fleur P. Helsingor**
Technical Publications Editor
Office of Scientific Publications
Menlo Park **(415) 323-8111 x2302**
U. of New Hampshire-BA
Geological language; history of science; Quaternary

ALDRICH, THOMAS C. **Peggy**
Physical Scientist **(617) 548-1975**
61 Drumlin Road
Falmouth, Massachusetts 02540
Br. of Atlantic-Gulf of Mex. Geology
Woods Hole **(617) 548-8700**
Park Coll.-BA
Marine Operations administration
Gravimetry; navigation; geophysical systems

ALEINIKOFF, JOHN N. **Debra**
Geologist **(303) 861-2814**
1422 E. 8th Avenue
Denver, Colorado 80218
Br. of Isotope Geology
Denver **(303) 234-3876**
Beloit-BA; Dartmouth-MA, PhD
Geochronology in Alaska, Utah, Saudi Arabia, northeast U.S.
U-Th-Pb geochronology

ALGERMISSEN, SYLVESTER T. **Marta**
Geophysicist **(303) 755-4024**
2868 S. Oakland Circle E
Aurora, Colorado 80014
Br. of Earthquake Tectonics & Risk
Denver **(303) 234-4014**
Missouri Schl. of Mines-BS; Washington U. (St. Louis)-MA, PhD
Regional & national seismic hazard & risk assessment
Earthquake intensity & damage

ALLAN, MARY A.
PST
Br. of Pacific-Arctic Geology
Menlo Park **(415) 856-7084**
Diablo Valley-AA; U. of California (Davis)-BS
X-ray mineralogy

ALLCOTT, GLENN H. **Shirley**
Geologist **(303) 986-9675**
13539 W. Dakota Place
Lakewood, Colorado 80228
Br. of Exploration Research
Denver **(303) 234-5301**
Black Hills State Coll.; South Dakota State U.; U. of Idaho
Geochemical exploration

ALLDREDGE, LEROY, R. **Larita**
Geophysicist **(303) 499-9422**
4475 Chippewa Drive
Boulder, Colorado 80303
Br. of Electromag. & Geomagnetism
Denver **(303) 234-5164**
U. of Arizona-BS, MS; Harvard-MS; U. of Maryland-PhD
Geomagnetism-secular change

ALLEN, REX V. **Mary Elizabeth**
Geophysicist
Br. of Network Operations
Menlo Park **(415) 323-8111 x2240**
U. of California (Berkeley)-BA
Automatic seismic data processing
Instrumentation

ALLEY, LONNIE B.
Geophysicist
Br. of Oil & Gas Resources
Denver **(303) 234-4668**
U. of Utah-BS, BS, MS
Oil & gas energy data system development
Computer data analysis, graphics systems, & data file design/implementation

ALLINGHAM, JAMES M.
Engineering Technician
613 West Holly Avenue
Sterling, Virginia 22170
Br. of Analytical Laboratories
Reston **(703) 860-7442**

ALLISON, ANNIE L.
Administrative Clerk
Br. of Western Environmental Geology
Menlo Park **(415) 323-8111**

ALPHA, TAU R. **Ann**
Cartographer **(415) 856-6126**
2854 Louis Road
Palo Alto, California 94303
Br. of Pacific-Arctic Geology
Menlo Park **(415) 856-7109**
California State U. (Northridge)-BA;
California State U. (Hayward)
Oblique maps; physiographic diagrams; map projections

ALSOP, KAREN S.
PST
Br. of Western Mineral Resources
Menlo Park **(415) 323-8111 x2549**
San Diego State U.-BA
Wilderness evaluations
Graphic arts; land use evaluation

ALTSCHULER, ZALMAN S.
Geologist **(301) 493-6705**
10201 Grosvenor Place
Rockville, Maryland 20852
Br. of Coal Resources
Reston **(703) 860-6649**
Brooklyn Coll.-BA; U. of Cincinnati-MS
The Everglades as a coal basin model
Geochemistry of sediments; phosphates; coal; weathering

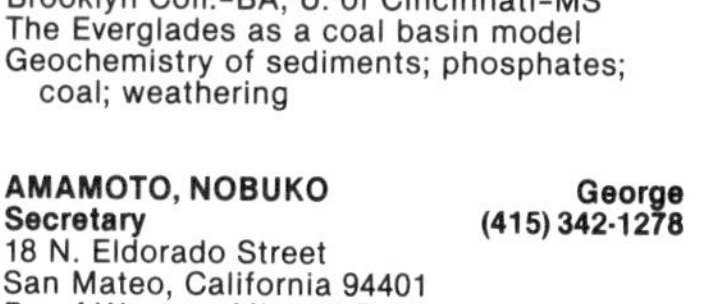

AMAMOTO, NOBUKO **George**
Secretary **(415) 342-1278**
18 N. Eldorado Street
San Mateo, California 94401
Br. of Western Mineral Resources
Menlo Park **(415) 323-8111 x2336**

AMBEAU, ELSIE PINNACE **Nelson**
Secretary **(415) 586-8180**
2630 San Jose Avenue
San Francisco, California 94112
Br. of Oil & Gas Resources
Menlo Park **(415) 856-7026**
City Coll. of San Francisco

AMBUTER, BRUCE P. **Susan Hollins**
Electrical Engineer **(617) 754-2395**
43 Beach Street
Monument Beach, Massachusetts 02553
Br. of Atlantic-Gulf of Mex. Geology
Woods Hole **(617) 548-8700**
Tufts-BS, MS
Ocean bottom seismograph operations
Ocean bottom instrument packages; single & multichannel systems

AMOS, DEWEY H. **Dora**
Geologist **(217) 345-7131**
2003 University Drive
Charleston, Illinois 61920
Br. of Central Environmental Geology
Charleston **(217) 345-7131**
Marietta Coll.-BS; U. of Illinois-MA, PhD
Paducah 1° x 2° sheet
Areal geology; mineral deposits

ANDERS, DONALD E. **Rose**
Chemist **(303) 989-3355**
11473 Briarwood Drive
Lakewood, Colorado 80226
Br. of Oil & Gas Resources
Denver **(303) 234-4744**
Illinois Coll.-BS; U. of Wyoming-MS
Molecular characterization of complex organic mixtures & geochemical applications
Org. geochemistry; analytical chemistry

ANDERSON, BARBARA M.
Botanist **(303) 988-9410**
10071 West Exposition Drive
Lakewood, Colorado 80226
Br. of Regional Geochemistry
Denver **(303) 234-5239**
U. of Denver-BS, BA
Geochemical baselines for *Atriplex* sp.
Environmental geochemistry of western powerplant sites; background chemical composition of native plant species

ANDERSON, KENNETH E.
Supply Technician
360 South Irving Street
Denver, Colorado 80219
Br. of Analytical Laboratories
Denver **(303) 234-6405**

ANDERSON, ROBERT C. **Lynda**
Geophysicist **(303) 989-4434**
12561 West Asbury Place
Lakewood, Colorado 80228
Br. of Oil & Gas Resources
Denver **(303) 234-5008**
Colorado Schl. of Mines-BS, MS
Research on seismic methods of detecting stratigraphic traps
Seismic exploration & acquisition; seismic data processing & interpretation

ANDERSON, R. ERNEST — **Mary**
Geologist — **(303) 423-2739**
3990 Garland
Wheat Ridge, Colorado 80033
Br. of Earthquake Tectonics & Risk
Denver **(303) 234-5109**
Marietta Coll.-BS; U. of Montana-MS; Washington U. (St. Louis)-PhD
Tectonics of southwestern Utah
Tectonics of Basin and Range province

ANDREANI, VERONICA A. — **Frank**
Administrative Officer
7362 Roxbury Avenue
Manassas, Virginia 22110
Office of Mineral Resources
Reston **(703) 860-6571**

ANDREWS, DUDLEY J. — **Judy**
Geophysicist — **(415) 856-0181**
2253 Santa Ana Street
Palo Alto, California 94303
Br. of Ground Motion & Faulting
Menlo Park **(415) 323-8111 x2752**
Tulane-BS; Yale-MS; Washington State U.-PhD
Physics of ground motion source
Physical theory of faulting & ground motion

ANDREWS, GEORGE W.
Geologist
Br. of Paleontology & Stratigraphy
Washington, D.C. **(202) 343-3796**
U. of Wisconsin-BA, MA, PhD
Marine diatom biostratigraphy of eastern U.S.
Marine & nonmarine diatoms

ANDREWS, J. STACEY — **Gary Fuis**
Secretary
745 La Para Avenue
Palo Alto, California 94306
Br. of Sedimentary Mineral Resources
Menlo Park **(415) 323-8111 x2336**
Canada College

ANTWEILER, JOHN C. — **Zairah**
Chemist
8461 S. Blue Creek Road
Evergreen, Colorado 80439
Br. of Exploration Research
Denver **(303) 234-3665**
Colorado State U.-BS; U. of Wyoming; U. of Colorado
Geochemistry in wilderness studies
Geochemistry of gold; geochemical exploration

AQUILINO, LINDA M. — **Daniel**
Clerical Assistant — **(703) 759-3970**
803 Lunenburg Road
Great Falls, Virginia 22066
Br. of Eastern Environmental Geology
Reston **(703) 860-6406**

ARCHULETA, RALPH J. — **Lucy**
Geophysicist
Br. of Ground Motion & Faulting
Menlo Park **(415) 323-8111 x2062**
U. of Wyoming-BS; U. of California (San Diego)-MS, PhD
Strong ground motion prediction in a layered medium
Numerical modeling of earthquake sources; synthetic seismograms

ARMBRUSTMACHER, THEODORE — **Jeannette**
Geologist — **(303) 674-5854**
29741 Fairway Drive
Evergreen, Colorado 80439
Br. of Uranium-Thorium Resources
Denver **(303) 234-3204**
Michigan State U.-BS; Miami U.-MS; U. of Iowa-PhD
Geology & resources of thorium deposits
Geology & geochemistry of mineral deposits; igneous & metamorphic petrology

ARMIN, RICHARD A. — **Shelley Evanne**
Geologist
Br. of Western Mineral Resources
Menlo Park **(415) 323-8111 x2667**
San Jose State U.-BSc, MSc
Geologic & resource studies in Sierra Nevada Mts.
Sedimentology; geology of Sierra Nevada Mts.

ARMSTRONG, AUGUSTUS K. — **Shirley**
Geologist — **(408) 241-1565**
753 Silvertip Way
Sunnyvale, California 94086
Br. of Paleontology & Stratigraphy
Menlo Park **(415) 323-8111**
U. of New Mexico-BS; U. of Cincinnati-MS, PhD
Devonian carbonate rocks
Late Paleozoic biostratigraphy, carbonate deposition-diagenesis

ARMSTRONG, FRANK C. — **Jean Ann**
Geologist — **(509) 448-0527**
2614 E. 40th Avenue
Spokane, Washington 99203
Br. of Uranium-Thorium Resources
Spokane **(509) 456-4677**
Yale-BA; U. of Washington-MS; Stanford-PhD
Uranium resources, Wyoming
Uranium; ore deposits; structural geology

ARNAL, ROBERT E. — **Annabelle**
Geologist — **(408) 659-4567**
170 El Caminito Road
Carmel Valley, California 93924
Br. of Oil & Gas Resources
Menlo Park **(415) 856-7166**
U. of Southern California-PhD
Southern California continental borderland
Micropaleontology; sedimentation; oceanography

ARNDT, HAROLD H. — **Joyce**
Geologist — **(303) 985-4631**
8284 West Woodard Drive
Lakewood, Colorado 80227
Br. of Coal Resources
Denver **(303) 234-3533**
Franklin & Marshall-BS
Coal deposits & Carboniferous stratigraphy

ARNOLD, EDOUARD P.
Geophysicist
Br. of Global Seismology
Denver **(303) 234-3994**
Rensselaer Polytechnic Inst.-BS; U. of Cambridge-PhD
National earthquake information service
Earthquake locations; earthquake economics

ARTH, JOSEPH G. — **Andrea**
Geologist — **(703) 860-9216**
12322 Panama Road
Reston, Virginia 22091
Br. of Isotope Geology
Reston **(703) 860-6593**
SUNY (Stony Brook)-BS, MS, PhD
Geochronology; igneous petrology

ARTHUR, DAVID W.G.
Research Cartographer
210 Zuni Drive
Flagstaff, Arizona 86001
Br. of Astrogeologic Studies
Flagstaff **(602) 779-3311**
Planetary heights
Photogrammetry; selenodesy; planetary cartography

ARTHUR, MICHAEL A.
Geologist — **(303) 988-0997**
184 S. Zang Way, 3-303
Lakewood, Colorado 80228
Br. of Oil & Gas Resources
Denver **(303) 234-4026**
U. of California (Riverside)-BS, MS; Princeton-PhD
Carbon cycle (global)
Geochemistry; stable isotope geochemistry; sedimentology; oceanography

ARUSCAVAGE, PHILIP J. **Janet**
Chemist **(703) 791-5250**
7213 Ridgeway Drive
Manassas, Virginia 22110
Br. of Analytical Laboratories
Reston **(703) 860-6144**
George Washington U.-BS
Trace element analysis; atomic spectroscopy

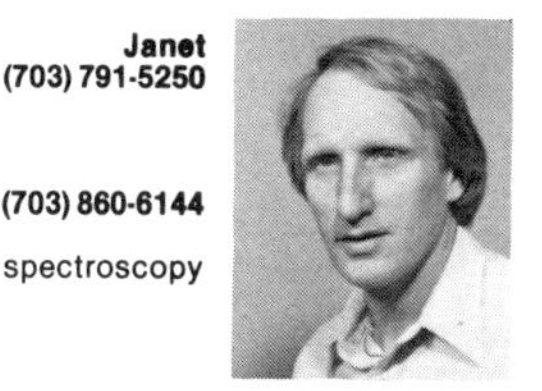

ASHER-BOLINDER, SIGRID **Richard Asher**
Geologist **(303) 744-8408**
586 S. Williams
Denver, Colorado 80209
Br. of Sedimentary Mineral Resources
Denver **(303) 234-5017**
Indiana U.-BS; U. of Montana
Lithium in sedimentary environments
Lithium commodity geology; sedimentation; carbonate petrology

ASHLEY, ROGER P. **Mary**
Geologist **(415) 494-1829**
3114 Cowper Street
Palo Alto, California 94306
Br. of Western Mineral Resources
Menlo Park **(415) 323-8111 x2650**
Carleton Coll.-BA; Stanford U.-PhD
Mineral resources of Spirit Lake quadrangle
Economic geology

ATHERTON, NANCY C. **Bob**
Personnel Technician **(602) 526-2189**
3686 N. Steves Boulevard
Flagstaff, Arizona 86001
Office of Environmental Geology
Flagstaff **(602) 779-3311 x1326**
Mary Mansfield Coll.; Huntington Coll.

ATTANASI, EMIL D. **Diana**
Economist **(703) 281-2367**
605 Thelma Circle
Vienna, Virginia 22180
Office of Resource Analysis
Reston **(703) 860-6455**
Evangal-BA; U. of Missouri-MA, PhD
Petroleum economics; design of resource information systems; methods of resource appraisal

ATWATER, BRIAN F.
Geologist
2849 Fillimore Street, #3
San Francisco, California 94123
Br. of Western Environmental Geology
Menlo Park **(415) 323-8111 x2939**
Stanford U.-BS, MS; U. of Delaware-PhD
Quaternary geology; paleobotany of tidal marshes

ATWOOD, THOMAS J. **Dorothy**
PST **(415) 494-2574**
4257 McKellar Lane
Palo Alto, California 94306
Br. of Pacific-Arctic Geology
Menlo Park **(415) 323-8111 x7100**
U. of Oregon; Stanford U.-BS
Eastern Gulf of Alaska resource evaluation
Marine geology

BABCOCK, RICHARD N. **Elaine**
PST **(303) 985-0781**
13234 W. Utah Circle
Lakewood, Colorado 80228
Br. of Coal Resources
Denver **(303) 234-3519**
Mesa Coll.; U. of Colorado
Subsurface geophysics of coal
Borehole geophysics

BACHERT, LES
Administrative Officer **(303) 494-5038**
4440 Laguna Place, #106
Boulder, Colorado 80303
Office of Earthquake Studies
Denver **(303) 234-5085**

BACHSTEIN, ELIZABETH
Secretary
3207 N. Pine Drive
Flagstaff, Arizona 86001
Br. of Astrogeologic Studies
Flagstaff **(602) 779-3311 x1438**
U. of Leyden, The Netherlands

BACK, JUDITH M.
Geologist **(303) 355-3050**
1538 Harrison Street
Denver, Colorado 80206
Br. of Central Mineral Resources
Denver **(303) 234-3226**
Bryn Mawr Coll.-BA; U. of Oregon-MS
Igneous petrology; sedimentology

BACON, CHARLES R. **Cynthia Dusel-Bacon**
Geologist **(415) 324-8237**
139 Princeton Road
Menlo Park, California 94025
Br. of Field Geochemistry & Petrology
Menlo Park **(415) 323-8111 x2332**
Stanford U.-BS; U. of California (Berkeley)-PhD
Volcanic evolution of Crater Lake region
Igneous petrology; volcanology; geochemistry

BAEDECKER, PHILIP A. **Mary Jo**
Chemist **(703) 860-2577**
11314 Handlebar Road
Reston, Virginia 22091
Br. of Analytical Laboratories
Reston **(703) 860-6853**
U. of Ohio-BS; U. of Kentucky-MS, PhD
Radiochemistry
Trace element geochemistry; activation analysis

BAILEY, EDGAR H. **Gwen**
Geologist **(415) 366-2690**
1835 Edgewood Road
Redwood City, California 94062
Br. of Western Mineral Resources
Menlo Park **(415) 323-8111 x2283**
Stanford U.-PhD
Mercury deposits of Nevada
Mercury deposits; California coast ranges; economic geology

BAILEY, NORMAN G.
Marine Data Manager **(617) 548-4328**
49 Fishermans Cove Road
E. Falmouth, Massachusetts 02536
Br. of Atlantic-Gulf of Mex. Geology
Woods Hole **(617) 548-8700**
Ohio State U.
Data management
Data management; ground-water geology; engineering geology

BAILEY, ROY A **Patrice**
Geologist **(301) 229-8390**
5005 Newport Avenue
Bethesda, Maryland 20016
Br. of Field Geochemistry & Petrology
Reston **(703) 860-7468**
Brown U.-BA; Cornell U.-MS; Johns Hopkins U.-PhD
Geology of Long Valley-Mono Basin geothermal area, California
Volcanology; igneous petrology

BAILEY, SHIRLEY A.
Scientific Illustrator **(415) 949-0839**
49 Showers Drive, #468
Br. of Pacific-Arctic Geology
Menlo Park **(415) 856-7086**
Ohio State

BAKER, JAMES W. **Sandra**
Chemist **(303) 278-9019**
1030 Rogers Street
Golden, Colorado 80401
Br. of Analytical Laboratories
Denver **(303) 234-6401**
U. of Colorado-BA
X-ray spectroscopy
Major and trace element studies by X-ray fluorescence

BAKER, KAY M.
Library Technician **(303) 424-1350**
4025 Kendall St., Apt. #1
Wheat Ridge, Colorado 80033
Library
Denver **(303) 234-2722**
U. of Denver-BA; U. of Illinois-MLS

BAKER, LAWRENCE M.
Computer Programmer **(415) 494-3521**
650 Kendall Avenue
Palo Alto, California 94306
Br. of Ground Motion & Faulting
Menlo Park **(415) 323-8111 x2703**
Stanford U.-BS
National Strong Motion Data Center (Project Chief)

BAKUN, WILLIAM H.
Geophysicist
Br. of Seismology
Menlo Park **(415) 323-8111 x2777**
U. of California (Berkeley)-PhD
Digital signal processing of seismic data
Seismology; tectonophysics

BALANC, MARIJA
PST
Br. of Paleontology & Stratigraphy
Washington, D.C. **(202) 343-2253**
U. of Ljubljana

BALCH, ALFRED, H. **Manie**
Geophysicist
671 E. Davies Avenue
Littleton, Colorado 80122
Br. of Oil & Gas Resources
Denver **(303) 234-5008**
Stanford U.-BS; Colorado Schl. of Mines-DSc
Geophysical properties of stratigraphic oil & gas traps
Seismic exploration; applied seismology; vertical seismic profiling

BALDAUF, JACK G.
PST
Br. of Paleontology & Stratigraphy
Menlo Park **(415) 323-8111 x2801**
San Jose State U.-BA
Cenozoic marine diatoms; biostratigraphy & paleoecology

BALDWIN, FRANCIS W.
PST **(509) 447-4685**
Rt. 4, Box 475
Newport, Washington 99156
Br. of Global Seismology
Newport, Washington **(303) 234-3994**
Canton Agr., & Tech. Inst.-AAS

BALIN, DONNA F.
GFA **(415) 328-2196**
Br. of Alaskan Geology
Menlo Park **(415) 323-8111 x2470**
U. of Texas (Austin)-BS
Devonian clastic rocks, Brooks Range, Alaska
Carbonate petrology; fluvial & deltaic sedimentology

BALL, MAHLON M. **Marilyn**
Geophysicist **(617) 564-4582**
496 Hatchville Road
Hatchville, Massachusetts 02536
Br. of Oil & Gas Resources
Woods Hole **(617) 548-8700**
U. of Kansas-BSc, MSc, PhD; U. of Birmingham (Eng.)-MSc
Petroleum geology of Atlantic-Gulf
Marine geology & geophysics

BALTZ, ELMER H. **Diana**
Geologist
719 South Lee Court
Lakewood, Colorado 80226
Br. of Central Environmental Geology
Denver **(303) 234-3147**
U. of New Mexico-BS, MS, PhD
Structure; stratigraphy

BANKEY, VIKI L.
PST
6175 Habitat
Boulder, Colorado 80301
Br. of Electromag. & Geomagnetism
Denver **(303) 234-6590**
U. of Michigan-BS
Gravity surveys in Montana disturbed belt
Gravity reduction & modeling

BANKS, NORMAN G. **Jane**
Geologist
P.O. Box 14, Hawaii Volcanoes National Park, Hawaii 96718
Br. of Field Geochemistry & Petrology
Hawaiian Volcano Observ. **(808) 967-7328**
New Mexico Inst. of Mining & Tech.-BS; U. of California (San Diego)-MS, PhD
Volcano studies
Petrology; geochemistry; field mapping

BARARI, RACHEL A. **Assad**
Geologist **(605) 624-2848**
RR2, Box 25AA
Vermillion, South Dakota 57069
Office of Resource Analysis
Vermillion, South Dakota **(605) 624-4471**
U. of South Dakota-BA, MA
Computerized Resources Information Bank
Computer studies

BARCUS, LORETTA A. **James**
Mathematician
Br. of Astrogeologic Studies
Flagstaff **(602) 779-3311 x1515**
U. of Oregon-BS
Photogrammetric applications of radar imagery
Photogrammetric applications of extra terrestrial imagery

BARGAR, KEITH E. **Chere**
Geologist **(415) 683-2245**
1505 E. San Martin Avenue
San Martin, California 95046
Br. of Field Geochemistry & Petrology
Menlo Park **(415) 323-8111 x2830**
San Jose State U.-MS
Hydrothermal alteration

BARKER, CHARLES E.
Geologist **(303) 278-1647**
1112 6th St.
Golden, Colorado 80401
Br. of Oil & Gas Resources
Denver **(303) 234-4640**
San Diego State U.-BS; U. of California (Riverside)-MS
Thermal alteration of organic matter
Vitrinite reflectance geothermometry; time-temperature-rank functions

BARKER, FRED
Geologist
Br. of Field Geochemistry & Petrology
Denver **(303) 234-3521**
MIT-BS; Cal Tech-MS, PhD
Granitic rocks
Precambrian rocks; petrology; geochemistry

BARKER, JUDY A.
PST **(415) 964-5539**
1361 Ormonde Way
Mountain View, California 94043
Br. of Western Environmental Geology
Menlo Park **(415) 323-8111 x2098**
Stanford U.-BS
Quaternary geology; geomorphology; fluvial processes

BARKER, RACHEL M.
Geologist **(703) 860-0067**
2100 Golf Course Drive
Reston, Virginia 22091
Br. of Engineering Geology
Reston **(703) 860-6062**
Bennington Coll.-BA; Smith Coll.-MA
Environmental geology of the Bighorn Basin, Wyoming
Engineering geology; glacial geology; surficial geology

BARNARD, JAMES B.
PST **(408) 296-2638**
915 Vermont St., #2
San Jose, California 95126
Br. of Western Mineral Resources
Menlo Park **(415) 323-8111 x2668**
California State U. (Hayward)-BS
Kalmiopsis specific gravity; serpentinization map
Ophiolite complexes; structural analysis; geologic mapping

BARNES, CAROL S. **Lonnie**
Secretary **(415) 225-4687**
5700 Venado Court
San Jose, California 95123
Br. of Western Mineral Resources
Menlo Park **(415) 323-8111 x2650**

BARNES, CONNIE A.
Geologist
Br. of Coal Resources
Denver **(303) 234-3519**
Fort Lewis Coll.-BS
Coal exploratory drilling-drill support group

BARNES, DAVID F. **Ann**
Geophysicist **(415) 593-6531**
107 Northam Avenue
San Carlos, California 94070
Br. of Regional Geophysics
Menlo Park **(415) 323-8111 x2249**
Harvard-BS, MA; U. of California (Berkeley)
Alaskan gravity surveys
Application of geophysics to Arctic Research

BARNES, PETER W.
Geologist
Br. of Pacific-Arctic Geology
Menlo Park **(415) 856-7008**
Antioch Coll.-BS; U of Southern California-PhD
Beaufort Sea environmental studies
Sedimentology; oceanography; arctic processes

BARNETT, CATHERINE A.
Admnistrative Officer **(415) 726-9681**
Route 1, Box 444 T.
Half Moon Bay, California
Br. of Western Environmental Geology
Menlo Park **(415) 323-8111 x2001**
San Jose State U.-BS, MPA

BARNHARD, LYNN M.
PST
9973 W. Canyon Avenue
Littleton, Colorado 80123
Br. of Earthquake Tectonics & Risk
Denver **(303) 234-5604**
Colorado State U.-BS

BARNHART, MERILEE A. **Robert**
Administrative Clerk
Office of Scientific Publications
Menlo Park **(415) 323-8111 x2563**
DeAnza Coll.-AA; San Jose State U.

BARNHART, REBECCA K.
Administrative Officer **(415) 493-5259**
250 Curtner Avenue, #30
Palo Alto, California
Br. of Western Mineral Resources
Menlo Park **(415) 323-2214**
Sheridan Junior Coll.-AA

BARRON, JOHN A.
Geologist **(415) 962-0680**
821 Runningwood Circle
Mountain View, California 94040
Br. of Paleontology & Stratigraphy
Menlo Park **(415) 323-8111 x2806**
UCLA-BS, PhD
Cenozoic marine diatoms & silicoflagellates

BARTA, JIMMIE L.
Cartographic Technician **(512) 937-5054**
3350 Bali
Corpus Christi, Texas 78418
Br. of Atlantic-Gulf of Mex. Geology
Corpus Christi **(512) 888-3294**
Del Mar Tech.-AS; Texas A&I

BARTEL, ARDITH J.
Chemist
410 Gladiola St.
Golden, Colorado 80401
Br. of Analytical Laboratories
Denver **(303) 234-6401**
Kansas State U.-BS
X-Ray fluorescence analysis
Radiochemistry; radioactivation analysis

BARTH, JOSEPH J. **Irene**
Electronics Technician
2590 Pierson St.
Lakewood, Colorado 80215
Br. of Petrophysics & Remote Sens.
Denver **(303) 234-5488**
Colorado State U.-BS; U. of Colorado-MPA
Borehole instrumentation research
Human resources management

BARTON, HARLAN N.
Chemist **(303) 494-0927**
4977 Moorhead Avenue, Apt. 102
Boulder, Colorado 80303
Br. of Exploration Research
Denver **(303) 234-3601**
U. of Colorado-BA, MA
Ajo 2° Quad., Williams Fork Rare II
Emission spectroscopy, laser microprobe, portable x-ray spectrograph

BARTON, PAUL, B., JR. **Martha**
Geologist **(703) 860-3369**
12842 Oxon Road
Herndon, Virginia 22070
Office of Mineral Resources
Reston **(703) 860-6601**
Penn State-BS; Columbia-MA, PhD
Mineral deposits; geochemistry of metals & sulfur; thermodynamics & phase equilibria

BARTOW, J. ALAN
Geologist
280 Waverley Street
Palo Alto, California 94301
Br. of Western Environmental Geology
Menlo Park **(415) 323-8111 x2020**
Fresno State U.-BA; UCLA-MA; Stanford U.-PhD
Cenozoic tectonics, San Joaquin Valley
Sedimentology; stratigraphy

BARTSCH-WINKLER, SUSAN **Gary**
Geologist
Br. of Alaskan Geology
Anchorage
U. of Wisconsin-BS; California State U. (San Jose)-MS
Alaskan coastal environments-Upper Cook Inlet
Recent sedimentary processes; sandstone petrography & diagenesis

BASKERVILLE, CHARLES A. **Susan**
Geologist **(703) 734-1474**
6713 Van Fleet Drive
McLean, Virginia 22101
Br. of Engineering Geology
Reston **(703) 869-6680**
CCNY-BS; New York U.-MS, PhD
Engineering geology of New York City
Engineering geology; astrogeology

BASKETT, SHARON L.
Clerk-Typist
Br. of Pacific-Arctic Geology
Menlo Park **(415) 856-2071**
Coll. of San Mateo

BASLER, JAMES R.
PST
Br. of Pacific-Arctic Geology
Menlo Park **(415) 856-7152**
U. of California (Santa Cruz)-BS
Marine mineral studies
Hydrothermal; analytical; U-Th dating

BATH, GORDON D. **Ruth**
Geophysicist **(303) 985-9455**
13341 West Alaska Place
Lakewood, Colorado 80228
Br. of Special Projects
Denver **(303) 234-2146**
U. of Washington; Colorado Schl. of Mines-BS
Waste isolation studies
Magnetic method of geophysical exploration

BATSON, RAYMOND M. **Rhoda**
Superv. Res. Cartographer **(602) 526-1642**
1476 Wakonda
Flagstaff, Arizona 86001
Br. of Astrogeologic Studies
Flagstaff **(602) 779-3311 x1352**
U. of Colorado-BA
Planetary cartography
Planetary cartography; cartographic image processing; computer graphics

BAWIEC, WALTER J. **Ellen**
Geologist **(703) 476-4388**
2032 Headlands Circle
Reston, Virginia 22091
Office of Resource Analysis
Reston **(703) 860-6446**
Waynesburg Coll.-BS
Resource appraisal; petroleum
Computer graphics; resource estimations

BAYER, KENNETH C. **Vivian**
Geophysicist **(703) 860-3583**
2461 Alsop Court
Reston, Virginia 22070
Br. of Oil & Gas Resources
Reston **(703) 860-6634**
New Mexico Inst. of Mining & Tech.-BS
Seismic (CDP) data acquisition & processing, onshore & offshore
Geophysical interpretation; earthquake seismology

BEARDSLEY, MARY A. **Keith**
Secretary **(703) 938-0571**
9933 Lindel Lane
Vienna, Virginia 22180
Br. of Eastern Environmental Geology
Reston **(703) 860-6404**
Iowa State U.; Bowie State Coll.-BA

BECK, MYRL, E. JR.
Geologist
Br. of Petrophysics & Remote Sens.
Department of Geology, Western Washington U., Bellingham, Washington 98225
(206) 676-3595
Stanford U.-BA, MS; U. of California (Riverside)-PhD
Paleomagnetism & tectonics of the western Cordillera of North America
Paleomagnetism; tectonics

BECKE, DONNA M.
Secretary
Br. of Oil & Gas Resources
Reston **(703) 860-6634**
George Mason U.-BA

BECKER, DAVID G.
Cartographic Technician
3268 Sydenham Street
Fairfax, Virginia 22031
Br. of Oil & Gas Resources
Reston **(703) 860-6634**
Northern Virginia Community Coll.-AA

BECKER, KEIR
Geophysicist
Br. of Regional Geophysics
San Diego **(714) 452-3505**
Harvard-BA
Heat flow in Gulf of California
Heat flow; marine geophysics

BEDETTE, BARBARA A.
PST **(202) 966-0156**
3660 38th Street, N.W.
Washington, D.C. 20016
Br. of Paleontology & Stratigraphy
Washington, D.C. **(202) 343-6682**
Bowling Green State U.-BA; American U.
Cenozoic mollusks

BEEN, JOSH M.
PST **(303) 494-0527**
1435 Kendall Drive
Boulder, Colorado 80303
Br. of Uranium-Thorium Resources
Denver **(303) 234-5531**
U. of Colorado-BA
Developing uranium exploration spectrometer

BEESON, FERN E. **Glen**
Funds Management Analyst **(602) 774-4167**
1126 Azure Drive
Flagstaff, Arizona 86001
Office of Environmental Geology
Flagstaff **(602) 779-3311 x1332**
Northern Arizona U.
Budget & fiscal administration

BEESON, MELVIN, H. **Kay**
Geologist **(415) 792-8958**
38355 Jacaranda Drive
Newark, California 94560
Br. of Field Geochemistry & Petrology
Menlo Park **(415) 323-8111 x2507**
Eastern Oregon Coll.; U. of Oregon-BA, MA; U. of California (Berkeley); U. of California (Santa Cruz)-PhD
Hydrothermal alteration in the Cascades
Igneous petrology; geochemistry; mineralogy

BEHRENDT, ELIZABETH C.
Librarian **(303) 279-5759**
15626 West First Drive
Golden, Colorado 80401
Library
Denver **(303) 234-4133**
Hunter Coll.-BA; U. of Portland-MLS

BEHRENDT, JOHN C.
Geophysicist
Br. of Regional Geophysics
Denver **(303) 234-5917**
U. of Wisconsin-BS, MS, PhD
Charleston earthquake studies; U.S. Atlantic margin; Antarctica
Seismology; magnetics; gravity

BEIKMAN, HELEN H.
Geologist (703) 476-9370
11714 Newbridge Court
Reston, Virginia 22091
Office of Mineral Resources
Reston (703) 860-6567
Indiana U.-BA
Coordinating BIA and BLM programs

BELKIN, HARVEY E.
Geologist (703) 471-5065
11142 Forest Edge Drive
Reston, Virginia 22090
Br. of Exper. Geochem. & Mineralogy
Reston (703) 860-6639
Franklin & Marshall Coll.-BA; George Washington U.
Fluid inclusions
Petrology; mineralogy; structural geology

BELL, EDITH E. Ted
PST
San Carlos, California
Br. of Analytical Laboratories
Menlo Park (415) 323-8111 x2948
Sample control officer

BELL, HENRY, III
Geologist
Br. of Eastern Mineral Resources
Reston (703) 860-6915

BENNETT, GLENN, E.
Computer Programmer (602) 779-1511
21 South Walnut
Flagstaff, Arizona 86001
Office of Environmental Geology
Flagstaff (602) 779-3311 x1367
Northern Arizona U.
Budget modeling system
Scientific data reduction; graphics; system analysis

BENNETT, HUGH F.
Geophysicist
Br. of Seismology
Menlo Park (415) 323-2214
U. of Wisconsin-PhD
Studies in seismic shear wave anisotropy
Seismic wave propagation; anisotropic materials; stress fields

BENNETT, MICHAEL J. Marjorie
PST (408) 259-3828
2551 Abed Court
San Jose, California 95116
Br. of Engineering Geology
Menlo Park (415) 856-7122
California State U. (Northridge)-BS; San Jose State U.-MS
Liquefaction potential
Engineering geology; Quaternary geology

BENNETTI, JOHN B., JR.
Electronics Engineer
Br. of Engineering Geology
Denver (303) 234-5560
San Jose State U.-BSEE
Instrumentation

BENTON, LEE L. Ellie
Administrative Officer (703) 339-6232
5936 River Drive
Lorton, Virginia 22079
Office of Energy Resources
Reston (703) 860-6434
George Washington U.

BERARDUCCI, ALAN M.
PST
Br. of Electromag. & Geomagnetism
Denver (303) 234-2588
U. of Michigan-BS

BERDAN, JEAN M.
Geologist (202) 338-4292
510 21st Street, N.W.
Washington, D.C. 20006
Br. of Paleontology & Stratigraphy
Washington, D.C. (202) 343-3658
Vassar-BA; Yale-MA, PhD
Lower Paleozoic ostracodes; Silurian-Devonian stratigraphy

BERG, HENRY C.
Geologist
Br. of Alaskan Geology
Menlo Park (415) 323-8111 x2266
Harvard-MA
Geotectonics, metallogenesis & resource appraisal of southeastern Alaska
Economic geology; tectonics

BERGER, BYRON R. Sally
Geologist (303) 237-4885
690 Dudley Street
Lakewood, Colorado 80215
Br. of Exploration Research
Denver (303) 234-6161
Occidental Coll.-BA; U. of California (Los Angeles)-MS
Geochem.-Dillon, MT-ID, 2° quad. (CUSMAP)
Economic geology; igneous petrogenesis & mineral deposits

BERGIN, MARION J. JoAnne
Geologist
100 Yeonas Drive, S.W.
Vienna, Virginia 22180
Office of International Geology
Reston (703) 860-6551
U. of Wyoming-BS
Energy resources assessments in developing countries
Geology & resource appraisal of fossil fuels

BERGQUIST, JOEL R. Margaret King
Geologist (415) 325-5305
50 Winchester Drive
Atherton, California 94025
Br. of Western Mineral Resources
Menlo Park (415) 323-8111 x2167
Williams Coll.-BA; U. of Michigan-JD; Stanford U.-PhD

BERGQUIST, WENONAH E. Harlan
Geologist (301) 299-5745
9016 Marseille Drive
Potomac, Maryland 20854
Office of International Geology
Reston (703) 860-6551
U. of Minnesota-BS

BERLAGE, LINDA J.
PSA
2016 Race Street
Denver, Colorado 80205
Br. of Coal Resources
Denver (303) 234-3536

BERMAN, SOL Esther
Chemist (301) 299-6488
11525 Gainsborough Road
Potomac, Maryland 20854
Br. of Analytical Laboratories
Reston (703) 860-7652
Oklahoma A&M-BS
Spectrographic analysis & research

BERRY, ANNE L.
PST **(415) 325-7405**
Br. of Isotope Geology
Menlo Park **(415) 323-8111**
Colorado Coll.-BA; Stanford U.-MS
Rb/Sr dating

BERRYHILL, HENRY L. **Louise**
Geologist
Br. of Atlantic-Gulf of Mex. Geology
Corpus Christi **(512) 888-3294**
U. of North Carolina-BS, MS
Environmental geology, NW Gulf of Mexico shelf and slope
Marine geology and geophysics

BETHKE, PHILIP M. **Jean**
Geologist **(301) 933-6080**
4220 Franklin Street
Kensington, Maryland 20795
Br.of Exper. Geochem. & Mineralogy
Reston **(703) 860-6602**
Amherst Coll.-BA; Columbia U.-MA, PhD
Environment of ore deposition
Ore petrology; ore genesis

BEYER, LARRY A.
Geophysicist
Br. of Oil & Gas Resources
Menlo Park **(415) 856-7069**
U. of California (Riverside)-BA, MA; Stanford U.-PhD
Geophysics & petroleum geology of Pacific margin sedimentary basins
Borehole & marine gravity; physical properties of rocks

BIAGI, CARLO H.
Clerk
1015 Cotton Street
Menlo Park, California 94025
Br. of Pacific-Arctic Geology
Menlo Park **(415) 856-2074**

BICE, TOM **Ann**
Electronics Technician **(303) 279-6580**
17232 Rimrock Drive
Golden, Colorado 80401
Br. of Earthquake Tectonics & Risk
Denver **(303) 234-6352**
Navarro Coll.; Mississippi State U.
Seismic networks

BIER, ROBERT A., JR. **Susan**
Librarian
2357 Southgate Square
Reston, Virginia 22091
Library
Reston **(703) 860-6671**
Middlebury Coll.-BA; U. of Maryland-MLS
Maps

BILLINGS, PATTY
Geologist **(303) 459-3488**
Box 165
Ward, Colorado 80481
Br. of Isotope Geology
Denver **(303) 234-4201**
New Mexico Tech.-BS; Colorado Schl. of Mines-MS
Fission track geochronology

BINGHAM, MICHAEL P.
Geologist **(415) 948-8079**
2680 Fayette Drive, #614
Mountain View, California 94040
Br. of Pacific-Arctic Geology
Menlo Park **(415) 856-7152**
Northeastern U.-BS; U. of Maine-MS
Glacial marine geology, Gulf of Alaska
Glacial marine sedimentation; stratigraphy; glacial geology

BISCHOFF, JAMES L. **Mary**
Geologist **(415) 856-9730**
774 Christine Drive
Palo Alto, California 94303
Br. of Pacific-Arctic Geology
Menlo Park **(415) 856-7162**
Occidental Coll.-BA; U. of California (Berkeley)-PhD
Marine mineral resources
Geochemistry; geothermal processes

BISDORF, ROBERT J. **Carol**
Geophysicist
2408 Cheyenne Drive
Golden, Colorado 80401
Br. of Regional Geophysics
Golden **(303) 234-5466**
Colorado Schl. of Mines-BS
Computer interpretation of resistivity data

BLACK, DOUGLAS F.B. **Eleanor**
Geologist **(703) 860-8174**
Br. of Eastern Environmental Geology
Reston **(703) 860-6657**
Whittier Coll.-BA; UCLA; South Dakota Tech-MS
Structural analysis of central eastern Kentucky
Structural geology; stratigraphy; areal geology

BLACKMON, PAUL D. **Carol**
Geologist **(303) 777-6568**
575 South Race Street
Denver, Colorado 80209
Br. of Central Mineral Resources
Denver **(303) 234-2873**
U. of Buffalo-BA, MA
Sedimentary mineralogy laboratory
Clay mineralogy; sedimentary petrology

BLACKWELDER, BLAKE W. **Diane**
Geologist **(301) 229-7460**
5129 Massachusetts Avenue
Bethesda, Maryland 20016
Br. of Paleontology & Stratigraphy
Washington, D.C. **(202) 343-5488**
Duke U.-BA; George Washington U.-PhD
Atlantic coastal plain Cenozoic shorelines
Mollusks; paleoecology; biostratigraphy

BLACKWOOD, DANN S.
GFA **(617) 540-5413**
434 Davisville Road
E. Falmouth, Massachusetts 02536
Br. of Atlantic-Gulf of Mex. Geology
Woods Hole **(617) 548-8700**
Florida Inst. of Technology; Syracuse U.-BS
Underwater photography; photojournalism

BLAIR, NANCY L.
Librarian **(415) 941-6744**
675 Berry Avenue
Los Altos, California 94022
Library
Menlo Park **(415) 323-8111**
Southern Illinois U.-BA; U. of California (Berkeley)-MLS
Library science

BLAKE, DORSEY
Computer Technician
Br. of Coal Resources
Denver **(303) 234-3624**
Metropolitan State Coll.
National coal resources data system

BLAKE, MILTON C., JR.
Geologist
4394 Miller Avenue
Palo Alto, California 94306
Br. of Western Environmental Geology
Menlo Park **(415) 323-8111 x2221**
U. of California (Berkeley)-BA; Stanford U.-PhD
Coos Bay, Oregon, 2° sheet
Pacific-Mediterranean subduction complexes

BLAKELY, RICHARD J. **Diane**
Geophysicist **(415) 856-9521**
933 Sycamore Drive
Palo Alto, California 94303
Br. of Regional Geophysics
Menlo Park **(415) 323-8111 x2624**
Oregon State U.-BS; Stanford U.-MS, PhD
Wilderness
Potential theory; analysis of magnetic data; paleomagnetism

BLANK, JEANNE N.
Scientific Illustrator **(415) 494-3971**
3374 Alma Street, Apt. 186
Palo Alto, California 94306
Br. of Oil & Gas Resources
Menlo Park **(415) 856-7087**
Layton Schl. of Art-BFA; Milwaukee Area Technical Coll.-AA

BLISS, JAMES D.
Geologist
Office of Resource Analysis
Reston **(703) 860-6451**
S. Dakota Schl. of Mines-BS; Arizona State U.-MS
Navajo resource assessment project
Computer application in geologic problems; economic geology; statistics

BLUM, JOHN E. **Toni-Ann**
Fiscal Officer **(703) 437-4720**
701 Archer Court
Herndon, Virginia 22070
Office of the Chief Geologist
Reston **(703) 860-6537**
Lynchburg Coll.-BA

BLUNT, DAVID J.
Geologist
Br. of Pacific-Arctic Geology
Menlo Park **(415) 856-7149**
California State U. (Hayward)-BS, MS
Amino-acid geochemistry of Pacific Northwest molluscan fauna
Amino-acid geochronology; sedimentology

BODENLOS, ALFRED J. **Mary Jean**
Geologist **(703) 860-1894**
2112 Golf Course Drive
Reston, Virginia 22091
Br. of Coal Resources
Reston **(703) 860-7440**
Columbia Coll.-BA; Columbia U.-MA, PhD
Sulfur in coal
Economic geology; structural geology; stratigraphy

BODINE, MARC W. JR. **Sallie**
Geologist **(303) 674-2527**
30574 Sun Creek Drive
Evergreen, Colorado 80439
Br. of Sedimentary Mineral Resources
Denver **(303) 234-3785**
Princeton-BA; Columbia-MA, PhD
Sedimentary mineral resources
Clay mineralogy; brine geochemistry; zeolitization of acid volcanics

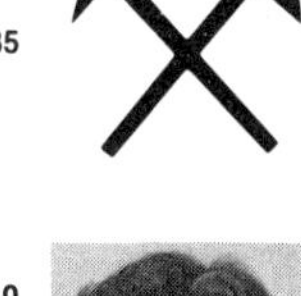

BOERNGEN, JOSEPHINE G.
Computer Specialist **(303) 733-4010**
1825 South Gilpin Street
Denver, Colorado 80210
Br. of Regional Geochemistry
Denver **(303) 234-2806**
Southwestern Coll.; U. of Tennessee; Denver U; U. of Colorado
Statistical analysis of geochemical data

BOGDON, MARIANNE **William**
Clerk-Typist
Br. of Earthquake Tectonics & Risk
Reston **(703) 860-6529**

BOHANNON, ROBERT G. **Susan**
Geologist **(303) 697-9586**
4341 S. Braun Court
Morrison, Colorado 80465
Br. of Central Mineral Resources
Denver **(303) 234-2183**
U. of California (Santa Barbara)-PhD
Tectonic development of Southern Great Basin; regional setting of Cordilleran Thrust Belt

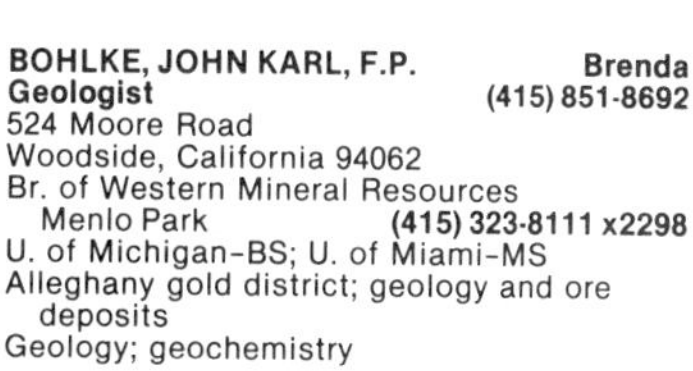

BOHLKE, JOHN KARL, F.P. **Brenda**
Geologist **(415) 851-8692**
524 Moore Road
Woodside, California 94062
Br. of Western Mineral Resources
Menlo Park **(415) 323-8111 x2298**
U. of Michigan-BS; U. of Miami-MS
Alleghany gold district; geology and ore deposits
Geology; geochemistry

BOLCHUNOS, MARILYN C. **John**
Clerk-Typist **(303) 422-6387**
6895 Nelson Street
Arvada, Colorado 80004
Br. of Global Seismology
Denver **(303) 234-4041**
U. of Colorado-BA

BOND, KEVIN R.
Geologist **(703) 471-4668**
1420 Northgate Square, #11B
Reston, Virginia 22090
Br. of Regional Geophysics
Reston **(703) 860-7233**
U. of Maryland-BS
National magnetic anomaly map

BONHAM, SELMA M. **Larry**
Geologist **(301) 229-8864**
4856 Park Avenue
Bethesda, Maryland 20016
Br. of Eastern Environmental Geology
Reston **(703) 860-7256**
Penn State-BS; Stanford U.-MS
Military geology project
Geology of USSR

BONILLA, MANUEL G. **Ruth**
Geologist **(415) 493-8474**
4127 Wilkie Court
Palo Alto, California 94306
Br. of Ground Motion & Faulting
Menlo Park **(415) 323-8111**
U. of California (Berkeley)-BA; Stanford U.-MS
Surface faulting
Engineering geology

BOORE, DAVID M. **Judy**
Geophysicist
Br. of Ground Motion & Faulting
Menlo Park **(415) 323-8111 x2698**
Stanford U.-BS, MS; MIT-PhD
Estimates of ground motion near earthquake faults
Seismology; wave propagation

BOORE, SARA A.
Scientific Illustrator
Office of Scientific Publications
Menlo Park **(415) 323-8111**
San Jose State U.-BA

BOOTH, JAMES S. **Sandy**
Geologist
Br. of Atlantic-Gulf of Mex. Geology
Woods Hole **(617) 548-8700**
U. of Wisconsin-BS; U. of S. California-MS, PhD
Marine geotechnical investigations
Marine geology; sedimentology; geotechnology

BORCHERDT, ROGER D. Judy
Geophysicist
234 Flynn Avenue
Mountain View, California 94043
Br. of Ground Motion & Faulting
Menlo Park **(415) 323-8111 x2755**
U. of Colorado-BA; U. of Wisconsin-MA; U. of California (Berkeley)-MS, PhD
Strong motion seismology
Seismic zonation; viscoelastic wave propagation

BORDERS, GLENN H.
Librarian
Library
Reston **(703) 860-6615**
George Washington U.-BA; McGill U.-BLS

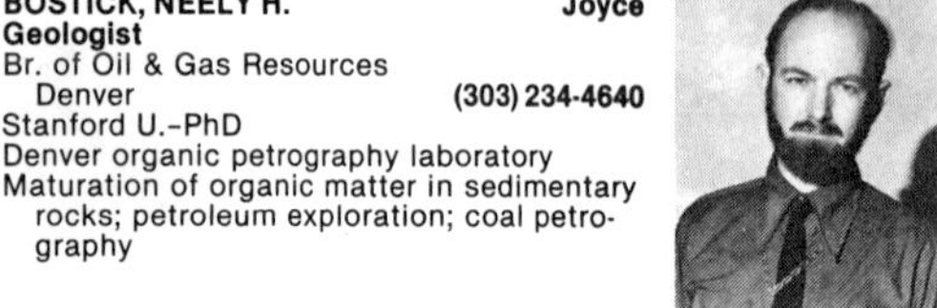

BOSTICK, NEELY H. Joyce
Geologist
Br. of Oil & Gas Resources
Denver **(303) 234-4640**
Stanford U.-PhD
Denver organic petrography laboratory
Maturation of organic matter in sedimentary rocks; petroleum exploration; coal petrography

BOSTWICK, LARRY G.
Electronics Technician **(303) 426-4097**
341 Elbert Way
Denver, Colorado 80221
Br. of Engineering Geology
Denver **(303) 234-4477**
Seismic monitoring in coal mines

BOTBOL, JOSEPH M. Sandra
Geologist **(617) 564-4419**
9 Inkberry Lane
N. Falmouth, Massachusetts 02556
Br. of Atlantic-Gulf of Mex. Geology
Woods Hole **(617) 548-8700**
St. Lawrence U.-BS; U. of Utah-MS, PhD
Computer applications section
Computer applications; economic geology

BOTHNER, MICHAEL H.
Oceanographer
Br. of Atlantic-Gulf of Mex. Geology
Woods Hole **(617) 548-8700**
Bowdoin Coll.-BA; Dartmouth-MA; U. of Washington-PhD
Geochemistry of continental shelf sediments
Trace metal geochemistry; radioactive dating techniques; modern sediment processes

BOTINELLY, THEODORE
Geologist
Br. of Exploration Research
Denver **(303) 234-6158**
Tufts Coll.; Harvard
Trace elements in sulfides in ore deposits

BOUCHER, GARY W. Susan
Geologist **(415) 856-9245**
3838 La Selva Drive
Palo Alto, California 94025
Br. of Pacific-Arctic Geology
Menlo Park **(415) 856-7018**
Colorado Coll.-BSc; Columbia-PhD
Study of physical properties of marine sediments in situ by acoustics
Exploration seismology; acoustics; earthquake seismology

BOUDETTE, EUGENE L.
Geologist
Br. of Uranium-Thorium Resources
Denver **(303) 234-3697**
U. of New Hampshire-BS; Dartmouth-MA, PhD
Uranium in plutonic rocks of New England
Mineral resources in orogenic belts; orogenic evolution

BOUDREAU, RAMONA L.
Photographer **(602) 526-3777**
4633 Northwood Way
Flagstaff, Arizona 86001
Br. of Astrogeologic Studies
Flagstaff **(602) 779-3311 x1318**

BOUMA, ARNOLD H. Mechelina
Geologist **(512) 937-6251**
1631 Graham Road
Corpus Christi, Texas 78418
Br. of Atlantic-Gulf of Mex. Geology
Corpus Christi **(512) 888-3294**
U. of Utrecht-MS, PhD
Continental slope, Gulf of Mexico
Submarine fans; geological hazards; seismic stratigraphy

BOWEN, ROGER W. Judy
Mathematician **(703) 777-8218**
132 Ayrlee Avenue
Leesburg, Virginia 22075
Office of Resource Analysis
Reston **(703) 860-6451**
San Jose State U.-BA; Penn State U.-MS
GRASP
Database management systems; computer applications

BOWKER, PAUL C.
PST
Br. of Atlantic-Gulf of Mex. Geology
Woods Hole **(617) 548-8700**
Boston State Coll.-BS
Geochemistry & hydrochemistry of marine resources
D.C. plasma emission spectroscopy; elemental analysis of marine and earth materials

BOWLES, ROBERT M. Yvette
Gravity/Navigation Specialist **(617) 563-3349**
P.O. Box 465
Falmouth, Massachusetts 02541
Br. of Atlantic-Gulf of Mex. Geology
Woods Hole **(617) 548-8700**
U. of New Hampshire-BA
Integrated navigation-gravity systems
Gravity; navigation; electronics

BOWMAN, ETHEL G.
Clerk-Typist
Br. of Global Seismology
Denver **(303) 234-4041**

BOWN, THOMAS M. Mary J. Kraus
Geologist **(303) 499-3542**
1255 Ithaca Drive
Boulder, Colorado 80303
Br. of Paleontology & Stratigraphy
Denver **(303) 234-6720**
Iowa State U.-BSc; U. of Wyoming-PhD
Tertiary nonmarine basins
Vertebrate paleontology; fluvial sedimentology; taphonomy

BOYER, ROBERT W.
PST **(415) 321-7451**
397 College Avenue, #D
Palo Alto, California 94306
Office of the Chief Geologist
Menlo Park **(415) 323-8111 x2218**
Bowling Green State U.-BS
Community labs

BRABB, EARL E. Gisela
Geologist **(415) 494-7039**
3262 Ross Road
Palo Alto, California 94303
Br. of Western Environmental Geology
Menlo Park **(415) 323-8111 x2203**
Dartmouth-BA; U. of Michigan-MS; Stanford U.-PhD
Seismic zonation, San Francisco Bay region
Environmental geology; Tertiary stratigraphy; seismic zonation

BRADBURY, J. PLATT
Geologist **(303) 278-1481**
4611 Pine Ridge Road
Golden, Colorado 80401
Br. of Paleontology & Stratigraphy
Denver **(303) 234-5863**
U. of New Mexico-PhD
Paleoclimate
Micropaleontology

BRADLEY, JERRY A.
Electronics Technician
Br. of Electromag. & Geomagnetism
Denver **(303) 234-3624**

BRADLEY, LEON A. **Helen**
PST **(303) 371-6423**
14920 E. 55th Avenue
Denver, Colorado 80239
Br. of Analytical Laboratories
Denver **(303) 234-3624**
Texas Southern U.; Rust Coll.-BS

BRADLEY, ROBIN
PST
319 Kingsley Street
Palo Alto, California 94301
Br. of Western Mineral Resources
Menlo Park **(415) 323-8111 x2169**
Stanford U.-BS
Oregon CRIB file; Ajo-Papago Indian Reservations Project

BRAGG, LINDA J. **Eric**
PST **(703) 437-1430**
701 Gentle Breeze Court
Herndon, Virginia 22070
Br. of Coal Resources
Reston **(703) 860-7440**
U. of West Virginia-BS
Geochemistry of trace elements in coal

BRAMSOE, ERIK **Renee**
Electronics Technician
2232 S. Yank Way
Lakewood, Colorado 80228
Br. of Electromag. & Geomagnetism
Denver **(303) 234-2589**

BRANCH, BENJAMIN H., JR. **Marjorie Lee**
Librarian
Library
Reston **(703) 860-6617**
Guilford Coll.; Drexel U.; U. of Illinois

BRANDT, ELAINE L. **Willard**
Chemist **(303) 841-2401**
8267 N. Silo Road
Parker, Colorado 80134
Br. of Analytical Laboratories
Denver **(303) 234-6401**
U. of Denver-BS

BREED, CAROL S.
Geologist
Br. of Astrogeologic Studies
Flagstaff **(602) 779-3311 x1463**
Smith Coll.-BA; Brown U.-MS
Inventory of wind-formed features on Mars (Eolian Atlas)
Desert geomorphology; eolian processes; comparative planetology

BREGER, IRVING A. **Ruth**
Geochemist **(301) 593-8460**
212 Hillsboro Drive
Silver Spring, Maryland 20902
Br. of Coal Resources
Reston **(703) 860-6643**
Worcester Polytech. Inst.-BS; MIT-MS, PhD
Geochemistry of fossil fuels (coal, oil, gas, kerogen)
Organic geochemistry; general geochemistry

BRESSLER, STEPHEN L. **Helen**
Geologist **(602) 526-0506**
1707 Deer Crossing Road
Flagstaff, Arizona 86001
Br. of Petrophysics & Remote Sens.
Flagstaff **(602) 779-3311 x1544**
Northern Arizona U.-BS; U. of Arizona-MS
Paleomagnetism—western United States
Paleomagnetism; stratigraphy

BREW, DAVID A. **Sally**
Geologist **(415) 941-6485**
164 Doud Drive
Los Altos, California 94022
Br. of Alaskan Geology
Menlo Park **(415) 323-8111**
Dartmouth-BA; U. of Vienna; Stanford U.-PhD
Geologic mapping; structural geology; batholithic complexes

BRIDGES, NANCY J. **Denney**
Mathematician **(303) 973-1123**
11201 W. Swarthmore Place
Littleton, Colorado 80123
Office of Resource Analysis
Denver **(303) 234-6284**
Northern Arizona U.-BS
International phosphate resource data base
Computer applications using geologic data

BRIDGES, PATRICIA M. **Al**
Cartographer (Photogrammetry) **(602) 526-2032**
3206 N. Steves
Flagstaff, Arizona 86001
Br. of Astrogeologic Studies
Flagstaff **(602) 261-1536**
Washington U.-BFA
Planetary cartography
Airbrush cartography; planetary photo-interpretation

BRIGGS, MELROSE M.
Supervisory Coordinator **(602) 526-1643**
3225 North Tindle Boulevard
Flagstaff, Arizona 86001
Br. of Astrogeologic Studies
Flagstaff **(602) 779-3311 x1455**
New Mexico Western Coll.

BRIGGS, NANCY D.
Geologist **(303) 985-5575**
1531 S. Owens, #23
Lakewood, Colorado, 80226
Br. of Isotope Geology
Denver **(303) 234-4201**
U. of Arizona-BS; Victoria U. (New Zealand)-PhD
Fission-track dating

BRIGGS, PAUL H.
Chemist
Br. of Analytical Laboratories
Denver **(303) 234-6401**
Adams State Coll.-BA; U. of Colorado; Colorado Schl. of Mines
Atomic absorption spectroscopy

BRIGIDA, MIRIAM J. **Vito**
Technical Publications Editor **(703) 860-1139**
2511 Fowlers Lane
Reston, Virginia 22091
Office of Scientific Publications
Reston **(703) 860-6494**
Augustana Coll., U. of Illinois-BA

BRISKEY, JOSEPH A., JR.
Geologist
Br. of Western Mineral Resources
Menlo Park (415) 323-8111 x2177
Oregon State U.-PhD
Pb-Zn resources
Ore deposit geology

BRITTON, O.J. Janice
Geophysicist (505) 298-6165
10300 Santa Paula
Albuquerque, New Mexico 87111
Br. of Global Seismology
Albuquerque (505) 844-4637
Stephen F. Austin Coll.-BS
Seismological instrumentation; administration

BROBST, DONALD A. Marie
Geologist (703) 860-4759
2268 Wheelwright Court
Reston, Virginia 22091
Br. of Eastern Mineral Resources
Reston (703) 860-6913
Muhlenberg Coll.-BA; U. of Minnesota-PhD
Economic geology; barite

BROCKMAN, STANLEY R. Barbara
Geophysicist (303) 237-4501
11715 W. 33rd Place
Wheat Ridge, Colorado 80033
Br. of Earthquake Tectonics & Risk
Denver (303) 234-5081
Kansas State U.-BS
Nuclear reactor site studies

BRODES, BETTY B. Donald
Library Technician
9544 Barkwood Court
Fairfax, Virginia 22092
Library
Reston (703) 860-6617
U. of Maryland-BA; U. of Virginia

BROKAW, JAMES A.
Administrative Officer
Office of Scientific Publications
Denver (303) 234-3229
Coe Coll.; U. of New Mexico-BA

BROKER, MICHAEL M. Karen
PST
Lakewood, Colorado 80215
Br. of Electromag. & Geomagnetism
Denver (303) 234-2950
Florida State U.-BA
Waste disposal

BROSGÉ, WILLIAM P. Mary
Geologist (415) 851-0772
4 Cedar Lane
Woodside, California 94062
Br. of Alaskan Geology
Menlo Park (415) 323-8111 x2316
Columbia U.-BA
Brooks Range Devonian clastic rocks
Stratigraphy; geologic mapping

BROWN, C. ERVIN Martha
Geologist (703) 759-3708
10608 Good Spring Avenue
Great Falls, Virginia 22066
Br. of Eastern Mineral Resources
Reston (703) 860-6913
Franklin & Marshall-BS; Harvard
Resources & geologic correlations in Precambrian rocks of N.Y.
Mineral resources; talc commodity geologist

BROWN, FLOYD W.
Chemist
2115 Freda Drive
Vienna, Virginia 22180
Br. of Analytical Laboratories
Reston (703) 860-6628
American U.-BS

BROWN, GLEN F. Helen
Geologist (703) 860-2057
2031 Royal Fern Court, Apt. 21C
Reston, Virginia 22091
Office of International Geology
Reston (703) 860-6555
New Mexico Schl. of Mines-BS; Northwestern U.-MS, PhD
Western Arabia
Precambrian geology; economic geology; ground water

BROWN, ROBERT D. Betty
Geologist (415) 854-0265
260 Erica Way
Portola Valley, California 94025
Office of Earthquake Studies
Menlo Park (415) 323-8111 x2461
U. of Oregon-BS, MS
Coordinator, earthquake hazards program
Structural geology; tectonics

BROWNFIELD, ISABELLE K. Michael
PST
Br. of Uranium-Thorium Resources
Denver (303) 234-5819
U. of Oregon
Thorium resources
Thorium, rare earth & carbonatite mineralogy

BRUMLEY, EDITH B. Drexel
Administrative Technician
1778 Robb Street
Lakewood, Colorado 80215
Br. of Sedimentary Mineral Resources
Denver (303) 234-3800

BRUNS, TERRY R.
Geophysicist (415) 366-7877
1771 Maryland Street
Redwood City, California 94061
Br. of Oil & Gas Resources
Menlo Park (415) 467-7106
MIT-MS
Eastern Gulf of Alaska continental margin
Petroleum exploration; marine geology & geophysics; tectonics

BRYANT, BRUCE H. Sandy
Geologist (303) 526-0234
26553 Columbine Glen
Golden, Colorado 80401
Br. of Central Environmental Geology
Denver (303) 234-5113
Dartmouth-BA; U. of Washington-PhD
Wasatch-Uinta tectonics
Regional geology; structure; petrology

BRYDEN, CYNTHIA G.
Oceanographer
39 Marvin Circle
Falmouth, Massachusetts 02540
Br. of Atlantic-Gulf of Mex. Geology
Woods Hole (617) 548-8700
Mount Holyoke Coll.-BA; Boston U.-MA
Continental shelf environmental assessment program
Bottom boundary layer-continental shelf, Georges Bank; bottom sediments

BUCHANAN-BANKS, JANE M. Norman
Geologist
P.O. Box 14
Hawaii National Park, Hawaii 96718
Br. of Engineering Geology
Hawaiian Volcano Observ. (808) 967-7328
San Francisco State U.-BA
Seismic hazards of the Hilo quadrangle, Hawaii
Geologic mapping; geologic hazards & land-use planning

BUCK, CAROLYN E.
Special Assistant to the China Program
6800 Fleetwood Road, McLean House
McLean, Virginia 22101 **(703) 734-0043**
Office of International Geology
Reston **(703) 860-6555**
Ohio Wesleyan U.-BA; MIT
USGS-PRC relations

BUCKNAM, ROBERT C. **Maria**
Geologist
Br. of Earthquake Tectonics & Risk
Denver **(303) 234-5089**
Colorado Schl. of Mines-GE; U. of Colorado-PhD
Late Quaternary surface faulting
Geology of earthquakes; Quaternary geology; tectonics

BUDAHN, JAMES R. **Amy**
Chemist **(303) 420-5654**
10993 W. 38th Place
Wheat Ridge, Colorado 80033
Br. of Analytical Laboratories
Denver **(303) 234-4201**
Southwest (Minn.) State U.-BA; Oregon State U.-MS
Radiochemistry, Denver
Geochemistry of igneous rocks; INAA technique development

BUFE, CHARLES G. **Jacquelyn**
Geophysicist
Br. of Seismology
Reston **(703) 928-6581**
Michigan Tech.-BS, MS; U. of Michigan-PhD
Liaison to DOE
Induced seismicity; geothermal exploration & assessment, earthquake prediction

BUFFA, ELIZABETH A. **Blaine**
Clerk-Typist
Office of Scientific Publications
Reston **(703) 860-6511**

BUFFINGTON, EDWIN C. **Peggy**
Geologist **(415) 325-5585**
213 Lexington Drive
Menlo Park, California 94025
Br. of Pacific-Arctic Geology
Menlo Park **(415) 856-7038**
Carleton Coll.-BA; California Tech-MS; U. of S. California-PhD
Marine Geology

BUHR, GROVER S.
PST
Br. of Seismology
Menlo Park **(415) 323-8111 x2010**
San Francisco State U.-BA
Parkfield prediction experiment
Real time microseismic networks; space-time microseismic patterns

BUKRY, JOHN D.
Geologist
675 S. Sierra Avenue
Solana Beach, California 92075
Br. of Pacific-Arctic Geology
La Jolla **(714) 452-2924**
Johns Hopkins-BA; Princeton-MA; U. of Illinois
Micropaleontology; electron microscopy

BUNKER, CARL M.
Geophysicist **(303) 922-1422**
5894 W. Mexico Avenue
Lakewood, Colorado 80226
Br. of Isotope Geology
Denver **(303) 234-4201**
U. of Dayton-BS
Natural radioelement distribution
Gamma-ray spectrometry

BUNTENBAH, CAROLE J. **Thomas**
Clerk-Typist **(303) 424-4942**
6583 Urban Street
Arvada, Colorado 80004
Br. of Sedimentary Mineral Resources
Denver **(303) 234-3714**
U. of Rochester

BURFORD, ROBERT O. **Gretchen**
Geologist **(415) 493-7239**
947 Ilima Way
Palo Alto, California 94306
Br. of Earthquake Tectonics & Risk
Menlo Park **(415) 323-8111 x2574**
Colorado Coll.-BS; Stanford U.-MS, PhD
Vertical tectonics project;
Crustal movements and deformation

BURKE, DENNIS B.
Geologist
Br. of Ground Motion & Faulting
Menlo Park **(415) 323-8111 x2048**
Cornell U.-BA; Stanford U.-MS, PhD
Quaternary stratigraphy & structure in south-central California
Tectonics; stratigraphy

BURKE, JANET B.
Data Management Clerk
Br. of Atlantic-Gulf of Mex. Geology
Woods Hole **(617) 548-8700**
Cape Cod Community Coll.
Marine data management

BURKHOLDER, ROBERT E. **Clara Jane**
PST **(303) 287-5465**
9371 Ellen Court
Thornton, Colorado 80229
Br. of Paleontology & Stratigraphy
Denver **(303) 234-5920**
U. of Pittsburgh-BS
Upper Cretaceous ammonites

BURNHAM, ROBYN J.
Museum Technician
Br. of Paleontology & Stratigraphy
Menlo Park **(415) 323-8111 x2758**
U. of California-BA
Paleobotany

BURR, GEORGE S.
PST **(301) 926-2707**
13008 Meadow View Drive
Gaithersburg, Maryland 20760
Br. of Isotope Geology
Reston **(703) 860-6113**
U. of Colorado-BA; George Washington U.-MS
Geochemistry

BURROW, GEORGE T. **Virginia**
Chemist **(303) 935-6053**
2670 W. First Avenue
Denver, Colorado 80219
Br. of Analytical Laboratories
Denver **(303) 234-3624**
U. of Alabama-BS

BUSH, ALFRED, L. **Jo Ann**
Geologist **(303) 985-7413**
10445 W. Kentucky Drive
Lakewood, Colorado 80226
Br. of Central Mineral Resources
Denver **(303) 234-2694**
U. of Rochester-BA, MS
U.S. lightweight aggregate resources
Economic geology of nonmetallics; mineral resource appraisal

BUSH, CHARLES A. **Shelley**
Geophysicist **(303) 985-3620**
476 S. Cole Court
Lakewood, Colorado 80228
Br. of Isotope Geology
Denver **(303) 234-4201**
Colorado Schl. of Mines-GE
Natural radioelement distribution
Gamma-ray spectrometry

BUTLER, FREDERICK L.
Electronics Technician
37811 Fremont Blvd, #9
Fremont, California 94536
Br. of Network Operations
Menlo Park **(415) 323-8111 x2920**
City Coll. of San Francisco-AA

BUTLER, HOWELL M. **Carol**
Geophysicist **(505) 294-3095**
8201 Evangeline Court, NE
Albuquerque, New Mexico 87109
Br. of Global Seismology
Albuquerque **(505) 844-4637**
Louisiana Tech U.-BS
Global seismograph networks
Seismology; seismological instrumentation; observatory facilities

BUTLER, LAURIE L.
PST **(303) 425-1375**
8731-B West 54th Place
Arvada, Colorado 80002
Br. of Oil & Gas Resources
Denver **(303) 234-6287**
Hanover Coll.-BA
Tight gas sands project

BUTLER, WILLIAM C. **Nancy**
Geologist
31951 Lodgepole Drive
Evergreen, Colorado 80439
Office of Energy Resources
Denver **(303) 234-4750**
U. of Illinois-BS; U. of Arizona-MS, PhD
Petroleum geology; Permian conodonts; sedimentary petrology

BUTMAN, BRADFORD
Oceanographer
Br. of Atlantic-Gulf of Mex. Geology
Woods Hole **(617) 548-8700**
Cornell U.-BA; MIT-Woods Hole Oceanographic Inst.-PhD
Continental shelf current & sediment dynamics

BUTTERFIELD, WILLARD
Electronic Technician
Br. of Engineering Geology
Denver **(303) 234-4477**
Instrumentation

BYBELL, LAUREL M. **David L. Govoni**
Geologist **(703) 791-5603**
12722 Bristow Road
Nokesville, Virginia 22123
Br. of Paleontology & Stratigraphy
Reston **(703) 860-7745**
Vassar-BA; U.of Miami-MS
Cenozoic calcareous nannofossils

BYERLEE, JAMES D.
Geophysicist **(415) 321-3258**
1540 Oak Creek Drive
Palo Alto, California 94304
Br. of Tectonophysics
Menlo Park **(415) 323-8111**
U. of Queensland (Australia)-BSc; MIT-PhD
Experimental rock mechanics
Rock friction; permeability; acoustic emission

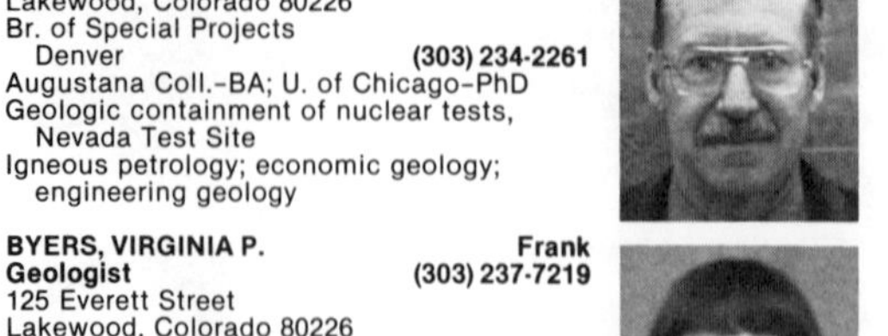

BYERS, FRANK M., JR. **Virginia**
Geologist **(303) 237-7219**
125 Everett Street
Lakewood, Colorado 80226
Br. of Special Projects
Denver **(303) 234-2261**
Augustana Coll.-BA; U. of Chicago-PhD
Geologic containment of nuclear tests, Nevada Test Site
Igneous petrology; economic geology; engineering geology

BYERS, VIRGINIA P. **Frank**
Geologist **(303) 237-7219**
125 Everett Street
Lakewood, Colorado 80226
Br. of Uranium-Thorium Resources
Denver **(303) 234-5148**
U. of Denver-BA
Geologic mapping Az-NM; uranium resources
Mineral deposits & economic geology; stratigraphy

CADIGAN, GERALDINE C. **Bob**
Foreign Participant Asst. **(303) 237-1781**
9125 West Second Avenue
Lakewood, Colorado 80226
Office of International Geology
Denver **(303) 234-3708**
International geology-participant programs

CADIGAN, ROBERT A. **Geraldine**
Geologist **(303) 237-1781**
9125 West Second Avenue
Lakewood, Colorado 80226
Br. of Uranium-Thorium Resources
Denver **(303) 234-2974**
U. of Puget Sound-AB; Penn State-MS
Radium springs & geochemical techniques
Exploration geochemistry; hydrogeochemistry

CADY, JOHN W. **Marith**
Geophysicist **(303) 277-1843**
3955 Douglas Mountain Drive
Golden, Colorado 80401
Br. of Petrophysics & Remote Sens.
Denver **(303) 234-5021**
Harvard-AB; Stanford U.-MS, PhD
Northern Alaska geophysics
Exploration geophysics; tectonics; field geology

CADY, WALLACE M. **Helen**
Geologist **(303) 985-1990**
348 South Moore Street
Lakewood, Colorado 80226
Br. of Central Environmental Geology
Denver **(303) 234-2827**
Middlebury Coll.-BS; Northwestern-MS; Columbia U.-PhD
Archean geology of southwestern Montana
Geologic mapping; regional structural geology; tectonic history

CAIN, JOSEPH C. **Shirley**
Geophysicist **(303) 234-0186**
1080 Zinnia Street
Golden, Colorado 80401
Br. of Electromag. & Geomagnetism
Denver **(303) 234-5166**
U. of Alaska-PhD
Global geomagnetism
Geomagnetic field structure and variations; computer-assisted numerical analysis; satellite geophysical experiments

CALDWELL, DELIA M.
Library Technician
Library
Reston **(703) 860-6613**

CALDWELL, JILL E.
PST
14476 E. Warren Place
Aurora, Colorado 80014
Br. of Electromag. & Geomagnetism
Denver **(303) 234-5506**
Colorado State U.-BS
Process data from U.S. magnetic observatories & repeat stations

CALK, LEWIS C. **Elaine**
Geologist **(415) 948-3351**
609 Arboleda Drive
Los Altos, California 94022
Br. of Field Geochemistry & Petrology
Menlo Park **(415) 323-8111 x2158**
U. of California (Berkeley)-BA; San Jose State U.-MS
Electron microprobe
Electron microprobe analysis

CALKINS, JAMES A. **Betty**
Geologist **(703) 534-5272**
2124 Powhatan Street
Falls Church, Virginia 22043
Office of Resource Analysis
Reston **(703) 860-6455**
Penn State-PhD
CRIB (Computerized Resources Information Bank)
Mineral resources

CALLAHAN, BETTY L. **Larry**
Secretary **(303) 466-0236**
236 Coral Way
Broomfield, Colorado 80020
Br. of Analytical Laboratories
Denver **(303) 234-2521**

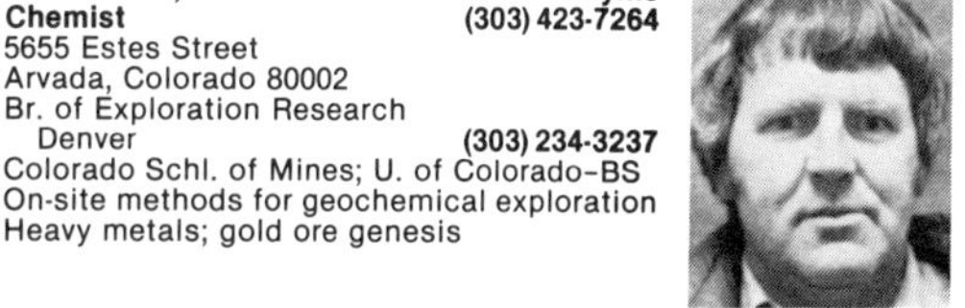

CAMERON, CORNELIA C.
Geologist
Br. of Eastern Minerals Resources
Reston **(703) 860-6916**
U. of Iowa-PhD
Geology of peat
Geomorphology; economic geology

CAMPBELL, CATHERINE C.
Geologist **(415) 776-1992**
1333 Jones Street
San Francisco, California 94109
Br. of Western Environmental Geology
Menlo Park **(415) 323-8111 x2145**
Oberlin Coll.-AB, AM; Radcliffe-PhD
San Francisco Bay region project
Environmental geology

CAMPBELL, DAVID L. **Katherine**
Geophysicist **(303) 986-4353**
1321 S. Jellison Street
Lakewood, Colorado 80226
Br. of Petrophysics & Remote Sens.
Denver **(303) 234-4594**
U. of California (Berkeley)-PhD
Uranium geophysics
Exploration geophysics; seismology; rock deformation

CAMPBELL, DONALD M. **Carol**
Geophysicist **(303) 278-9585**
14551 W. Archer Avenue
Golden, Colorado 80401
Br. of Electromag. & Geomagnetism
Denver **(303) 234-5182**
U. of Texas-BA
Geomagnetic observatory-digital processing

CAMPBELL, ESMA
Chemist
Br. of Analytical Laboratories
Reston **(703) 860-6845**
Western Reserve U.-BA
Analytical techniques

CAMPBELL, JOHN A. **Pat**
Geologist **(303) 674-0251**
27837 Whirlaway Trail
Evergreen, Colorado 80439
Br. of Uranium-Thorium Resources
Denver **(303) 234-5666**
U. of Tulsa-BG; U.of Colorado-MS, PhD
Uranium potential in Permian rocks, SW United States
Sedimentary petrology; sedimentology; stratigraphy

CAMPBELL, RUSSELL H. **Marilee**
Geologist **(703) 476-8924**
12007 Turf Lane
Reston, Virginia 22091
Br. of Engineering Geology
Reston **(703) 860-6062**
U. of California (Berkeley)-AB
Los Padres Wilderness (RARE II), southern California
Landslides; stratigraphy; structural geology

CAMPBELL, WALLACE H. **Leatrice**
Geophysicist **(303) 494-2379**
2405 Cragmoor Road
Boulder, Colorado 80303
Br. of Regional Geophysics
Denver **(303) 234-5481**
Louisiana State U.-BS; Vanderbilt-MA; UCLA-PhD
Geomagnetism
Natural sources of field variations

CAMPBELL, WESLEY L. **Phyllis**
Chemist **(303) 423-7264**
5655 Estes Street
Arvada, Colorado 80002
Br. of Exploration Research
Denver **(303) 234-3237**
Colorado Schl. of Mines; U. of Colorado-BS
On-site methods for geochemical exploration
Heavy metals; gold ore genesis

CANNEY, FRANK C. **Isabel**
Geologist **(303) 237-9296**
2954 Routt Circle
Lakewood, Colorado 80215
Br. of Exploration Research
Denver **(303) 234-4440**
M.I.T.-SB, PhD
Sherbrook-Lewiston 1° x 2° sheet; remote sensing in geochemical exploration
Exploration geochemistry; remote sensing of geochemical anomalies

CANNON, WILLIAM F. **Suzanne**
Geologist **(703) 860-0979**
11797 Indian Ridge Road
Reston, Virginia 22091
Br. of Eastern Mineral Resources
Reston **(703) 860-6914**
Syracuse U.-AB, PhD; Miami U.-MS
Iron River CUSMAP
Geology of Lake Superior region; manganese resources

CARAS, GEOFFREY B.
GFA **(408) 423-1849**
2120 N. Pacific Avenue
Santa Cruz, California 95060
Br. of Pacific-Arctic Geology
Menlo Park **(415) 856-7082**
U. of California (Santa Cruz)
Bar formation
Marine processes

CAREY, MARY ALICE
Geologist
Br. of Coal Resources
Denver **(303) 234-6308**
U. of Maryland-BS; U. of Utah-MS
National Coal Resources Data System
Computer applications in geology

CARGILL, SIMON M. **Madalyne**
Geologist
Office of Resource Analysis
Reston **(703) 860-6455**
George Washington U.-MS
Mineral depletion model

CARLSON, CHRISTINE
Geologist **(415) 321-3906**
2275 Amherst
Palo Alto, California 94306
Br. of Alaskan Geology
Menlo Park **(415) 323-8111 x2485**
U. of Washington-BS; Stanford U.-MS
Provenance & petrofacies of selected Mesozoic basins, Alaska Range
Sedimentary petrology; basin analysis; plate tectonics

CARLSON, KURT H.
Geologist
Golden Gate Canyon
Golden, Colorado 80401
Br. of Oil & Gas Resources
Denver **(303) 234-3435**
Colorado State U.-BS; U. of Colorado; Colorado Schl. of Mines
Oil and Gas Resource Appraisal Methods
Petroleum geology; mathematical and statistical modeling

CARLSON, PAUL R. **Mary**
Marine Geologist **(415) 494-1028**
3174 Bryant
Palo Alto, California 94306
Br. of Pacific-Arctic Geology
Menlo Park **(415) 856-7021**
Gustavus Adolphus Coll.-BA; Iowa State-MS; Oregon State-PhD
Navarin Basin seafloor hazards
Marine geology; sedimentology

CARLSON, ROBERT R. **Wendy**
Geologist **(303) 279-6258**
3115 Isabelle
Golden, Colorado 80401
Br. of Exploration Research
Denver **(303) 234-6154**
Colorado Coll.-BA; U. of Colorado; Colorado Schl. of Mines
Stillwater Complex, Montana; wilderness projects
Pt-group metals

CARPENTER, CARRIE E.
Cruise Materials Coordinator
1113 Los Altos Avenue
Los Altos, California 94022
Br. of Pacific-Arctic Geology
Menlo Park **(415) 856-7039**

CARR, DAVID R.
PST
4621 Kipling Street, Apt. 42
Wheat Ridge, Colorado 80033
Br. of Oil & Gas Resources
Denver **(303) 234-6471**
San Diego State U.-BS
Tight gas sands project

CARR, JOSEPH F.
PST
1909 Chula Vista Drive
Belmont, California 94002
Br. of Analytical Laboratories
Menlo Park **(415) 323-8111 x2950**
U. of California (Santa Cruz)-BS
X-ray fluorescence laboratory

CARR, MICHAEL D.
Geologist
Br. of Western Environmental Geology
Menlo Park **(415) 323-8111 x2956**
Franklin & Marshall-BA, MA; Rice-PhD
Structural geology; regional geology; stratigraphy

CARR, MICHAEL H. **Rachel**
Geologist **(415) 851-0258**
1389 Canada Road
Woodside, California 94062
Br. of Astrogeologic Studies
Menlo Park **(415) 323-8111**
U. of London-BS; Yale-PhD
Planetary studies

CARR, WILFRED J. **Janet**
Geologist **(303) 526-0819**
Box 730, Rt. 5
Golden, Colorado 80401
Br. of Special Projects
Denver **(303) 234-2365**
Princeton-BA; UCLA
Tectonics, seismicity & volcanism of the Great Basin
Structure; volcanism

CARRARA, PAUL E.
Geologist
Br. of Central Environmental Geology
Denver **(303) 234-5123**
San Francisco State Coll.-BA; U. of Colorado-MSc
Quaternary geology, Glacier National Park, Montana
Quaternary geology; geomorphology; dendrochronology

CARROLL, LINDA J. **Roger**
Planetary Data Clerk **(602) 526-9671**
3800 N. Swiss Road
Br. of Astrogeologic Studies
Flagstaff **(602) 779-3311 x1505**

CARROLL, RODERICK D.
Geophysicist
1029 E. 8th
Denver, Colorado 80218
Br. of Special Projects
Denver **(303) 234-2371**
U. of Missouri (Rolla)-BS, MS; U. of Colorado-BS
Geophysical studies of nuclear test sites
Engineering geophysics

CARSON, SCOTT E.
PST
Br. of Western Environmental Geology
Menlo Park **(415) 323-8111 x2939**
U. of California (Santa Barbara)-BA
Quaternary geology; geomorphology; environmental geology

CARTER, CLAIRE
Geologist
288 Waverly
Palo Alto, California 94301
Br. of Alaskan Geology
Menlo Park **(415) 323-8111 x2683**
U. of Oregon-BA; Stanford-MS
Graptolites of Alaska
Graptolites

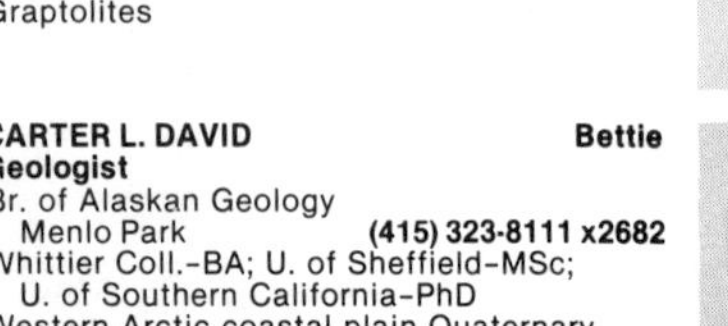

CARTER L. DAVID **Bettie**
Geologist
Br. of Alaskan Geology
Menlo Park **(415) 323-8111 x2682**
Whittier Coll.-BA; U. of Sheffield-MSc; U. of Southern California-PhD
Western Arctic coastal plain Quaternary
Quaternary geology; geomorphology; paleoenvironmental interpretation

CARTER, LORNA M.
Technical Editor
981 S. Pierson Way
Lakewood, Colorado 80226
Office of Scientific Publications
Denver **(303) 234-2866**
U. of Nebraska (Lincoln)-BA, MA
Innovations in publications techniques; scenic geologic features of energy lands areas

CARTER, M. DEVEREUX
Geologist **(703) 821-8875**
6865 Melrose Drive
McLean, Virginia 22101
Br. of Coal Resources
Reston **(703) 928-7464**
George Washington U.-BA
National coal resources data system
Coal resource assessment; coal commodity geologist

CARVER, DAVID L.
Geophysicist
301 Columbine
Golden, Colorado 80401
Br. of Earthquake Tectonics & Risk
Denver **(303) 234-4041**
U. of South Dakota-BA
Post-earthquake investigations
Aftershocks; tectonics; seismicity

CASADEVALL, TOM J.
Geologist
Box 106, Hawaii Volcanoes National Park
Hawaii 96718
Br. of Field Geochemistry & Petrology
Hawaiian Volcano Observ. **(808) 967-7328**
Beloit-BA; Penn State-MA, PhD
Geochemistry of volcanic gases
Volcanology; geology of volcanically related ore deposits; stable isotope geochemistry

CASE, JAMES E. **Irma**
Geophysicist **(415) 854-7440**
2101 Harkins Avenue
Menlo Park, California 94025
Br. of Regional Geophysics
Menlo Park **(415) 323-8111 x2496**
U. of Arkansas-BS, MS; U. of California (Berkeley)-PhD
Geophysical expression of mafic and ultramafic belts
Regional geology; gravity; magnetics

CASHION, WILLIAM B. **Elaine**
Geologist
2190 S. Gray
Denver, Colorado 80227
Br. of Sedimentary Mineral Resources
Denver **(303) 234-4869**
U. of New Mexico-BS
Stratigraphic studies, Unita Basin
Oil shale; tar sand; low-permeability gas sands

CASHMAN, KATHERINE V.
Geologist **(617) 548-4772**
P.O. Box 358
Woods Hole, Massachusetts 02543
Br. of Atlantic-Gulf of Mex. Geology
Woods Hole **(617) 548-8700 x193**
Middlebury Coll.-BA; Victoria U. of Wellington-MS
Side-scan sonograph interpretation-S. Atlantic
Vulcanology; marine geology

CASSIDY, ROBERT J. **Elizabeth**
Computer Programmer **(303) 426-1335**
7963 Greenwood Boulevard
Denver, Colorado 80221
Br. of Oil & Gas Resources
Denver **(303) 234-3435**
Duluth Area Inst. of Tech.; Metropolitan State Coll., Denver
Accounting applications

CATHCART, JAMES B. **Doris**
Geologist **(303) 279-4247**
17225 W. 16th Place
Golden, Colorado 80401
Br. of Sedimentary Mineral Resources
Denver **(303) 234-2881**
U. of California (Berkeley)-BA
Phosphate, southeast United States
Economic geology of sedimentary deposits; phosphate mineralogy; stratigraphy

CATHCART, J. DANIEL
PST
6430-B W. 80th Drive
Arvada, Colorado 80003
Br. of Coal Resources
Denver **(303) 234-4230**

CATHRALL, JOHN B. **Judy**
Geologist **(303) 237-4006**
7000 W. 26th Avenue
Lakewood, Colorado 80215
Br. of Exploration Research
Denver **(303) 234-4813**
Colgate-BS; U. of Oklahoma-MS
Mineral resources appraisal SE Alaska & Brooks Range, Alaska
Geochemistry; resource appraisal

CECIL, C. BLAINE
Geologist **(703) 777-6288**
12 Governors Drive
Leesburg, Virginia 22075
Br. of Coal Resources
Reston **(703) 860-7734**
U. of West Virginia-BS, MS, PhD
Origin of mineral matter in coal
Sedimentary geology & sedimentary geochemistry

CEDER, HERBERT W. **Jean**
Photographer
11706 E. Dakota Avenue
Aurora, Colorado 80012
Br. of Petrophysics & Remote Sens.
Denver **(303) 234-5470**
Photographer & darkroom technician

CHAFFEE, MAURICE A. **Annette**
Geologist
Br. of Exploration Research
Denver **(303) 234-6157**
Colorado Schl. of Mines-GE; U. of Arizona-MS, PhD
Walker Lake 2° quadrangle; wilderness and RARE II areas; SE Arizona porphyry Cu
Geochemical zoning, ore deposits; regional geochemical zoning; biogeochemistry related to ore search

CHAMPION, DUANE E.
Geologist
Br. of Petrophysics & Remote Sens.
Menlo Park **(415) 323-8111 x2226**
SUNY (Buffalo)-BA, MA; Cal Tech-PhD
Holocene geomagnetic secular variation
Paleomagnetism; volcanology; tectonics

CHAO, EDWARD C.T. **Vera**
Geologist **(703) 533-1094**
2200 N. Roosevelt Street
Arlington, Virginia 22205
Br. of Coal Resources
Reston **(703) 860-6787**
U. of Chicago-PhD
Petrochemistry of coal & related investigations
Petrology; geochemistry; geology

CHAO, TSUN TIEN **Ru Kwa**
Chemist **(303) 986-6533**
9770 West Ohio Avenue
Lakewood, Colorado 80226
Br. of Exploration Research
Denver **(303) 234-6167**
National Central (China)-BS; Cornell; Oregon State-MS, PhD
Chemical methodology for geochemical exploration
Analytical chemistry & geochemistry

CHAPMAN, NANCY L. **John**
Clerk-Typist **(415) 782-2322**
1850 Egret Lane
Hayward, California 94545
Office of Scientific Publications
Menlo Park **(415) 323-8111**

CHAPMAN, ROBERT M. **Jo**
Geologist **(415) 323-0750**
155 Pineview Lane
Menlo Park, California 94025
Br. of Alaskan Geology
Menlo Park **(415) 323-8111 x2670**
Carleton Coll.; Northwestern-BS; Syracuse
Stratigraphy & structure of central Yukon River area
Regional geology & mineral deposits of interior Alaska

CHAPPELL, BARBARA A. **Gregg**
Librarian
3720 Camelot Drive
Annandale, Virginia 22003
Library
Reston **(703) 860-6675**
U. of Texas, Austin-BS, MLS

CHARPENTIER, RONALD R.
Geologist **(303) 988-7098**
484 S. Wright Street, Apt. 109
Lakewood, Colorado 80228
Br. of Oil & Gas Resources
Denver **(303) 234-5235**
U. of Kansas-BS, MS; U. of Wisconsin (Madison)-PhD
Oil and gas resource appraisal group
Geomathematics; petroleum geology; paleontology

CHASE, BARBARA R. **George**
Secretary **(703) 670-5811**
14621 Danville Road
Woodbridge, Virginia 22193
Office of Energy Resources
Reston **(703) 860-6431**
Indiana U.-BA; U. of Montana

CHASE, THOMAS E. **Marlene**
Oceanographer
1249 Marilla Avenue
San Jose, California 95129
Br. of Pacific-Arctic Geology
Menlo Park **(415) 856-7132**
San Diego State U.-BA
Data synthesis
Marine bathymetry; seismic data flowage

CHEN, ALBERT T.F. **Kathryn**
Civil Engineer
Br. of Engineering Geology
Menlo Park **(415) 323-8111 x2605**
Taiwan U.-BS; U. of Massachusetts-MS; Rensselaer-PhD
Dynamic soil behavior
Earthquake response of soils; consolidation theory; computer application

CHESSON, SHARON ANN **Robert**
Geologist **(303) 279-8229**
1810 Ford Street
Golden, Colorado 80401
Office of Resource Analysis
Denver **(303) 234-6377**
Georgia State-BS
Wallace 2° sheet
Economic geology; paleontology

CHIDESTER, ALFRED H. **Louise**
Geologist **(703) 860-9179**
2427 Alsop Court
Reston, Virginia 22091
Office of International Geology
Reston **(703) 860-6418**
Augustana Coll.-AB; U. of Chicago-PhD
Ultramafic rocks; igneous & metamorphic petrology

CHILDRESS, ANNE E.
Chemist
Br. of Analytical Laboratories
Reston **(703) 860-6638**
U. of Arkansas-BA

CHIN, JOHN L.
PST **(415) 321-4503**
1136 Fulton
Palo Alto, California 94301
Br. of Pacific-Arctic Geology
Menlo Park **(415) 856-7081**
U. of Texas-BA; San Jose State U.
Coastal processes
Marine-coastal geology

CHLEBORAD, ALAN F. **Anita**
Geologist
Br. of Engineering Geology
Denver **(303) 234-3360**
U. of Colorado-BA
Engineering geology of Cook Inlet Basin coal lands, Alaska
Regional engineering geology; topical studies

CHOPKO, GLENDA I.
Clerk-Typist
7135 W. Vassar Avenue
Lakewood, Colorado 80227
Office of Scientific Publications
Denver **(303) 234-2445**
U. of Nebraska

CHOU, I-MING **Chiu-Jung**
Geologist **(703) 860-1142**
2516 Congreve Court
Herndon, Virginia 22070
Br. of Exper. Geochem. & Mineralogy
Reston **(703) 860-6654**
National Taiwan U.-BS; Johns Hopkins-PhD
Radioactive waste, rock-fluid interactions
Mineral stabilities at various pressures, temperatures, and fluid compositions

CHOY, GEORGE L.
Geophysicist
Br. of Global Seismology
Denver **(303) 234-4041**
Cooper Union-BS; Columbia-M.Phil, PhD
Wave propagation; source mechanism; Earth structure

CHRISTIAN, RALPH P.
Geologist **(303) 674-3606**
29896 Woods Drive
Evergreen, Colorado 80439
Br. of Central Mineral Resources
Denver **(303) 234-3544**
U. of Colorado-BA; George Washington U.-MS
Electron microprobe
Electron microprobe & X-ray energy dispersive analysis; SEM

CHRISTIANSEN, ANN COE **Charles**
Geologist **(303) 526-1665**
Box 338
Georgetown, Colorado 80444
Br. of Central Environmental Geology
Denver **(303) 234-5114**
U. of Michigan-BA
Geologic map of Wyoming
Compilation of geologic data; geologic maps & illustrations

CHRISTIANSEN, ROBERT L. **Susan**
Geologist
1118 Harker Avenue
Palo Alto, California 94301
Br. of Field Geochemistry & Petrology
Menlo Park **(415) 467-2680**
Stanford-BS, MS, PhD
Igneous petrology; volcanology

CHRISTOPHER, RAYMOND A.
Geologist
Br. of Paleontology & Stratigraphy
Reston **(703) 860-6523**
U. of Rhode Island-BS, MS; Louisiana State U.-PhD
Palynology of Cretaceous rocks of the Cape Fear region
Cretaceous palynology; quantitative approaches to biostratigraphy

CHRISTOPHERSON, KAREN R.
Geologist **(303) 499-1008**
6079 Simmons Drive
Boulder, Colorado 80303
Br. of Electromag. & Geomagnetism
Denver **(303) 234-5157**
U. of Colorado-BS, MS
Geothermal
Audiomagnetotellurics; telluric profiling; geologic interpretation

CHURCH, STANLEY E.
Geologist
16095 Ridge Tee Drive
Morrison, Colorado 80465
Br. of Exploration Research
Denver **(303) 234-3665**
Kansas U.-BS, MS; U. of California (Santa Barbara)-PhD
Glacier Peak Wilderness
Trace element geochemistry; isotope geochemistry

CHURKIN, MICHAEL, JR. **Carol**
Geologist
Br. of Alaskan Geology
Menlo Park **(415) 323-8111 x2256**
U. of California (Berkeley)-BA, MS; Northwestern-PhD; Columbia
Tectonostratigraphic studies
Stratigraphy; paleontology; tectonics

CIRCE, RON C. **Sharon**
PST **(512) 855-0532**
4317 Brentwood
Corpus Christi, Texas 78415
Br. of Atlantic-Gulf of Mex. Geology
Corpus Christi **(512) 888-3241**
Texas A&I-BS; Corpus Christi State U.-MS
Automatic core liner cutter
Design & fabrication of sampling & testing devices

CLEMENSEN, MARGARET A. **Berle**
Editorial Assistant **(303) 986-9851**
2287 S. Beech Way
Lakewood, Colorado 80228
Br. of Central Mineral Resources
Denver **(303) 234-3836**

CLAGUE, DAVID A.
Geologist
1180 Grant Road
Los Altos, California 94022
Br. of Pacific-Arctic Geology
Menlo Park **(415) 856-7133**
Scripps Institution of Oceanography-PhD
Linear island chains
Petrology; marine geology

CLIFTON, H. EDWARD **Ann**
Geologist **(415) 961-6122**
1933 Fallen Leaf Lane
Los Altos, California 94022
Br. of Pacific-Arctic Geology
Menlo Park **(415) 856-7073**
Ohio State-BSc; Johns Hopkins-PhD
Coastal sedimentology; petrology of ancient coastal deposits.

CLARK, ALLEN L.
Geologist **(703) 860-0389**
12124 Quorn Lane
Reston, Virginia 22091
Office of International Geology
Reston **(703) 860-6555**
Iowa State-BS; U. of Idaho-MS, PhD
Economic geology; resource analysis; computer applications

CLOW, GARY D.
Geophysicist
Br. of Astrogeologic Studies
Menlo Park **(415) 323-8111 x2361**
U. of California (Berkeley)-AB
Structure & evolution of planetary bodies

CLARK, HAROLD E., JR. **Betty**
Electronic Engineer **(505) 298-8811**
4905 Northridge Place, N.E.
Albuquerque, New Mexico 87111
Br. of Global Seismology
Albuquerque **(505) 844-4637**
New Mexico State U.-BSEE; U. of New Mexico-MSEE
Microprocessor-based seismic systems
Electronic engineering; computer programming

CLYNNE, MICHAEL A.
Geologist
148 El Bosque Avenue
San Jose, California 95134
Br. of Exper. Geochem. & Mineralogy
Menlo Park **(415) 323-8111 x2124**
U. of Santa Cruz-BS
Southern Cascades geothermal evaluation & water-rock interaction
Field geology; volcanic terranes; brines

CLARK, JOSEPH C. **Helen**
Geologist **(814) 845-7621**
Glen Campbell, Pennsylvania 15742
Br. of Western Environmental Geology
Indiana University of Pennsylvania
Indiana, Pa. 15705 **(412) 357-2379**
U. of Texas-BS, MA; Stanford-PhD
Seismic zonation, San Francisco Bay region
Tertiary stratigraphy; sedimentary petrology

COAKLEY, JOHN M.
PST **(415) 547-5522**
501 Scenic Avenue
Piedmont, California 94611
Br. of Network Operations
Menlo Park **(415) 323-8111 x2517**
U. of California (Berkeley)-BA

CLARK, MALCOLM M.
Geologist
Br. of Ground Motion & Faulting
Menlo Park **(415) 323-8111 x2591**
Stanford-PhD
Quaternary faulting in S. California
Geology of active faults; glacial geology

COATES, DONALD A. **Mary-Margaret**
Geologist
Br. of Central Environmental Geology
Denver **(303) 234-3343**
U. of Colorado-BA, MS; UCLA-PhD
Origin & structure of clinker in western coal basins
Stratigraphy; surficial geology

CLARK, SANDRA H.B.
Geologist
Br. of Eastern Mineral Resources
Reston **(703) 860-6914**
Iowa State U.; U. of Idaho-PhD

COATS, ROBERT R. **Elizabeth**
Geologist **(415) 493-8795**
3836 La Selva Drive
Palo Alto, California 94306
Br. of Western Mineral Resources
Menlo Park **(415) 323-8111 x2373**
U. of Washington-BS, MS; U. of California (Berkeley)-PhD
Elko County, Nevada
Areal mapping; tectonics; volcanology

CLARKE, JAMES W.
Geologist **(703) 759-4487**
914 Leigh Mill Road
Great Falls, Virginia 22066
Br. of Eastern Environmental Geology
Reston **(703) 860-6595**
Emery U.-BA; Yale-PhD
Field mapping, Greenville 2° sheet; oil fields of U.S.S.R.
Tectonics; igneous and metamorphic petrology; petroleum geology

COBB, EDWARD H. **Ruth Twitch**
Geologist **(415) 324-3465**
1140 Cotton Street
Menlo Park, California 94025
Br. of Alaskan Geology
Menlo Park **(415) 323-8111 x2483**
Yale-BS, MS
Mineral resources of Alaska
Synthesis of mineral-resources information

CLAYPOOL, GEORGE E. **Carol**
Geochemist **(303) 237-8273**
8910 West 2nd Avenue
Lakewood, Colorado 80226
Br. of Oil & Gas Resources
Denver **(303) 234-3561**
Colorado State U.-BS; UCLA-PhD
Origin of petroleum
Sedimentary, organic, isotope geochemistry

COBBAN, WILLIAM A. **Ruth**
Geologist **(303) 233-6337**
70 Estes Street
Lakewood, Colorado 80226
Br. of Paleontology & Stratigraphy
Denver **(303) 234-5860**
U. of Montana-BA; Johns Hopkins-PhD
Chronostratigraphy of Cretaceous hydrocarbon source rocks
Upper Cretaceous ammonites (systematics, paleoecology)

COCHRANE, GUY R.
PST **(415) 582-5561**
3094 Horseshoe Court
Hayward, California 94541
Br. of Pacific-Arctic Geology
Menlo Park **(415) 856-7100**
Humboldt State U.-BA, BS
Oil & gas, western Gulf of Alaska

COIT, THERESA A.
Secretary
Mountain View, California 94043
Br. of Oil & Gas Resources
Menlo Park **(415) 856-7033**

COLE, DAVID
PST
Br. of Petrophysics & Remote Sens.
Denver **(303) 234-4897**
Northeastern U.-BS; Colorado Schl. of Mines
Remote sensing; mineral exploration

COLE, GRACE L. **John**
Administrative Assistant
Br. of Paleontology & Stratigraphy
Reston **(703) 860-7745**
Highland Park Jr. Coll.-AS

COLE, JAMES C. **Estella**
Geologist
U.S.G.S. c/o American Embassy
A.P.O. New York 09697
Br. of Latin American & African Geology
Jiddah, Saudi Arabia
U. of California (Santa Barbara)-BA; U. of Colorado-PhD
Precambrian shield, Arabia
Igneous & metamorphic petrology; structural geology

COLEMAN, ANNETTE L.
Administrative Clerk **(415) 327-6982**
750 Coleman Avenue, #9
Menlo Park, California 94025
Br. of Western Environmental Geology
Menlo Park **(415) 323-8111 x2001**
U. of California (Davis)-BS

COLEMAN, ROBERT G. **Cathryn**
Geologist **(415) 854-3641**
2025 Camino Al Lago
Menlo Park, California 99025
Br. of Field Geochemistry & Petrology
Menlo Park **(415) 323-8111**
Oregon State U.-BS, MS; Stanford-PhD
Oman ophiolite
Petrology; mineralogy; field geology

COLEMAN, S. LYNN
Geologist **(703) 437-1299**
1639 Parkcrest Circle, #300
Reston, Virginia 22090
Br. of Coal Resources
Reston **(703) 860-7734**
West Georgia Coll.-BA; Bryn Mawr
Collection, chemical analysis & evaluation of coal samples from eastern U.S.
Coal geochemistry; trace elements in health & disease

COLLINS, DONLEY S.
Geologist
Br. of Engineering Geology
Denver **(303) 234-3433**
Colorado State U.-BS; U. of Colorado-BA
Kimberlite occurrences; coal mine subsidence
Uranium deposits

COLLINS, IRENE L.
Administrative Clerk
45763 Cayuga Court
Fremont, California 94538
Br. of Alaskan Geology
Menlo Park **(415) 323-8111**

COLMAN, STEVEN M. **Marian**
Geologist
Br. of Central Environmental Geology
Denver **(303) 234-5170**
Notre Dame-BS; Penn State-MS; U. of Colorado-PhD
Quaternary dating & neotectonics
Quaternary geology; rock weathering; paleoclimatology

COLTON, ROGER B. **Eve**
Geologist **(303) 279-4633**
1949 Mt. Zion Drive
Golden, Colorado 80401
Br. of Central Environmental Geology
Denver **(303) 234-3624**
Yale-BA, MS
Williston Basin environmental geology

COMMEAU, JUDITH A.
Geologist
52 Sippewissett Road
Falmouth, Massachusetts 02540
Br. of Atlantic-Gulf of Mex. Geology
Woods Hole **(617) 548-8700**
Case-Western Reserve-BA
Laboratory support group
X-ray fluorescence spectroscopy; trace metals; specialized techniques for ultrafine particles

COMMEAU, ROBERT F.
Geologist **(617) 540-2189**
52 Sippewissett Road
Falmouth, Massachusetts 02540
Br. of Atlantic-Gulf of Mex. Geology
Woods Hole **(617) 548-8700**
U. of Maine-BA
Electron microscopy; X-ray microanalysis

COMPTON, ELLEN E.
PST
9906 Longford Court
Vienna, Virginia 22180
Br. of Paleontology & Stratigraphy
Reston **(703) 860-7745**
Radford Coll.-BS

CONDIT, CHRISTOPHER D.
Geologist/Pilot
Dept. of Geology, Univ. of New Mexico
Albuquerque, New Mexico 87131
Br. of Central Environmental Geology
Flagstaff **(602) 261-1556**
Coll. of William & Mary-BA; Northern Arizona U.-MS
Basalt stratigraphy; White Mountains volcanic field, east-central Arizona
Colorado Plateau-Basin & Range transition

CONDON, STEVEN M.
PST **(303) 279-0412**
812 Cheyenne
Golden, Colorado 80401
Br. of Uranium-Thorium Resources
Denver **(303) 234-5564**
Iowa State-BA
Subsurface study of Grants mineral belt
Stratigraphy; sedimentation; geomorphology

CONDROTTE, CHARLES G.
GFA **(707) 763-7850**
1871 Mountain View Avenue
Petaluma, California 94952
Br. of Ground Motion & Faulting
Menlo Park **(415) 323-8111 x2514**
College of Marin-AS; Sonoma State U.; U. of California (Davis)
Alaska seismic project
Seismic telemetry; strong motion instrumentation

CONKLIN, NANCY M.
Spectrographer
Br. of Analytical Laboratories
Denver **(303) 234-6405**
Colorado Woman's Coll.; U. of Colorado; U. of Denver
Quantitative spectrographic analysis
Rare earth minerals; ilmenites; spectrographic standards

CONLEY, EDWARD
Electronics Technician **(602) 526-1570**
Route 3, 146 Campbell Avenue
Flagstaff, Arizona 86001
Br. of Astrogeologic Studies
Flagstaff **(602) 779-3311 x1515**
Grossmont Jr. College

CONLEY, SANDRA J.
Geologist **(617) 540-1905**
94 Turner Road
Hatchville, Massachusetts 02536
Br. of Atlantic-Gulf of Mex. Geology
Woods Hole **(617) 837-4155**
Northeastern-BS

CONNOR, CAROL WAITE
Geologist
Br. of Coal Resources
Denver **(303) 234-3689**
Washington U. (St. Louis)-BA; Louisiana State U.-MS
Sedimentology; coal geology

CONNOR, JON J. **Carol**
Geologist
Br. of Regional Geochemistry
Denver **(303) 234-2924**
Ohio State-BSc; U. of Colorado-PhD
Geochemistry of belt rocks
Geochemistry; geostatistics; petrology

CONSAGRA, KATHLEEN M.
Clerical Assistant
Br. of Exper. Geochem. & Mineralogy
Reston **(703) 860-6692**

COOK, AMY E.
PST
2649 Alma
Palo Alto, California 94306
Br. of Field Geochemistry & Petrology
Menlo Park **(415) 323-8111 x2398**
U. of California (Santa Cruz)-BS

COOK, HARRY E **Judy**
Geologist
#3 Porto Marino
San Carlos, California 94070
Br. of Oil & Gas Resources
Menlo Park **(415) 856-7032**
U. of California (Santa Barbara)-BA; U. of California (Berkeley)
Petroleum geology of continental margins
Submarine mass-transport processes; continental slopes; submarine fans

COOK, JENNIFER L.
Geologist
1180 Daleview Drive
McLean, Virginia 22102
Office of International Geology
Reston **(703) 860-6555**
Washington State U.-BS
Thailand potash & oil shale assessment
Computer applications in geology

COOKE, JAMES E.
Electronics Engineer
Br. of Electromag. & Geomagnetism
Denver **(303) 234-2589**
New Mexico State U.-BS
Portable tensor AMT unit
Geophysical instrumentation

COOKRO, THERESA M.
Geologist
Office of Resource Analysis
Denver **(303) 234-3386**
Kent State-BS; New Mexico Inst. of Mining & Tech.-MS
Potential mineral resource analysis, Challis quad., Idaho
Resource analysis; economic geology; petrology

COOLEY, ELMO F. **Deloris**
Chemist
1005 Maple Drive
Denver, Colorado
Br. of Exploration Research
Denver **(303) 234-6190**
NE Missouri State Teacher Coll.-BS
Pt metals analysis
Spectrographic analysis of geologic materials

COONRAD, WARREN L. **Doris**
Geologist **(415) 851-7483**
235 Highland Terrace
Woodside, California 94062
Br. of Alaskan Geology
Menlo Park **(415) 323-2608**
Colorado Coll.-BA; U. of California (Berkeley)
Mineral deposits; eng. & photogeology

COOPER, ALAN K.
Geophysicist **(415) 493-2685**
264 N. Whisman, #20
Mountain View, California 94043
Br. of Pacific-Arctic Geology
Menlo Park **(415) 856-7094**
U. of Rochester-BS; San Jose State-MS; Stanford-MS, PhD
Tectonics of Bering Sea region
Marine geophysics; crustal structure

COOPER, MARGARET
Geologist **(202) 737-7351**
532 20th Street, NW, Apt. 601
Washington, D.C. 20006
Br. of Eastern Mineral Resources
Washington, D.C. **(202) 343-7753**
Hunter Coll.-BA; Columbia-MA
Uranium & onshore energy resources on federal lands
Economic geology; bibliographic research; environmental impact studies

CORCHARY, GEORGE S. **Phyllis**
Geologist **(303) 237-8353**
8725 West Second Avenue
Lakewood, Colorado 80926
Br. of Special Projects
Denver **(303) 234-2365**
U. of New Hampshire-BS; U. of Illinois-MS
Geologic data center, Nevada test site
Surficial materials; engineering geology

CORNWALL, HENRY R. **Mimi**
Geologist **(415) 325-7544**
1701 Oak Avenue
Menlo Park, California 94025
Br. of Western Mineral Resources
Menlo Park **(415) 323-8111**
Princeton-AB; PhD
Nickel & chromite deposits-NW California, SW Oregon
Ni & Cu deposits

CORREIA, GEORGE A.
PST
3100 Racine Street
Aurora, Colorado 80011
Br. of Coal Resources
Denver **(303) 234-3519**
Drill Support-coal resources evaluation

CORWIN, GILBERT **Ginny**
Geologist **(703) 525-5725**
2362 N. Nelson Street
Arlington, Virginia 22207
Office of Marine Geology
Reston **(703) 860-7539**
Harvard-BS; U. of Minnesota-PhD
Western Pacific Islands; geology, volcanic petrology

COSTELLO, JOYCE A. **Rick**
Secretary
21 Nansemond Street
Leesburg, Virginia 22072
Office of Earthquake Studies
Reston **(703) 860-6471**

COUBROUGH, MARY A.
Secretary **(303) 986-8051**
10555 W. Jewell Avneue, #3-302,
Lakewood, Colorado 80226
Br. of Regional Geochemistry
Denver **(303) 234-3715**

COURY, ANNY B. **Glenn**
Geologist **(303) 234-1644**
6600 West 13th Avenue
Lakewood, Colorado 80214
Br. of Oil & Gas Resources
Denver **(303) 234-3435**
U. of Texas-BS
Estimation of undiscovered petroleum resources
Sedimentary basins, worldwide; subsurface exploration for hydrocarbons

COUSINS, CAROLYN G.
Administrative Officer
Br. of Atlantic-Gulf of Mex. Geology
Woods Hole **(617) 548-8700**

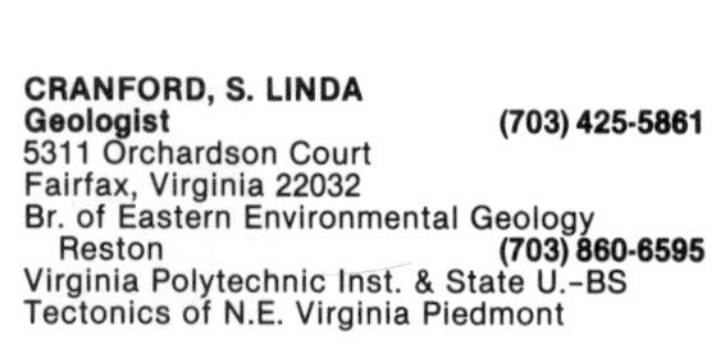

COVINGTON, PAMELA A. **Harry**
Geophysicist
9014 S. U.S. Highway 285
Morrison, Colorado 80465
Br. of Ground Motion & Faulting
Denver **(303) 234-4268**
Colorado State U.; U. of Colorado-BA
Seismic zonation, Los Angeles Basin
Computer programming; ground motion studies; engineering seismology

COWARD, ELIZABETH L.
PST **(617) 540-5292**
P.O. Box 10
West Falmouth, Massachusetts 02543
Br. of Altantic-Gulf of Mex. Geology
Woods Hole **(617) 548-8700 x193**
Mt. Holyoke-BA; Boston Coll.
Environmental hazards & stratigraphy, S. Atlantic
Marine geology & geophysics

COX, DENNIS P.
Geologist
3918 Grove Avenue
Palo Alto, California 94303
Br. of Western Mineral Resources
Menlo Park **(415) 323-8111 x2310**
Stanford-BS, PhD
Ore forming processes, Ambler District, AK.
Copper resources; ore deposits studies

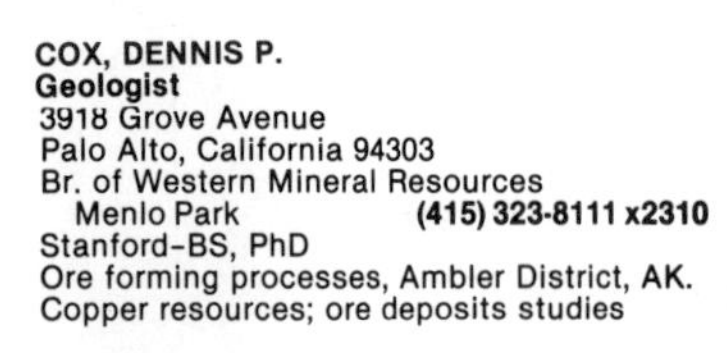

COX, LESLIE J.
Geologist
Br. of Central Mineral Resources
Denver **(303) 234-3454**
Ohio State-BS; Virginia Poly. Inst. & State U.-MS
Massive sulfides; ore microscopy; field mapping

COXE, BERTON W.
PST
Br. of Central Mineral Resources
Denver **(303) 234-6727**
Hamilton Coll.; U. of Colorado-BA
Gospel Hump Wilderness Project, Idaho
Geochemical exploration; granite systems

CRAIG, LAWRENCE C. **Rachel**
Geologist **(303) 237-8294**
1801 Balsam Street
Lakewood, Colorado 80215
Br. of Uranium-Thorium Resources
Denver **(303) 234-4756**
Swarthmore-BA; Columbia-MS, PhD
Uranium potential, Uinta & Piceance Basins, Colorado-Utah
Stratigraphy; economic geology

CRANDELL, DWIGHT R. **Marion**
Geologist **(303) 233-1360**
7101 West 9th Place
Lakewood, Colorado 80215
Br. of Engineering Geology
Denver **(303) 234-2834**
Knox Coll.-BA; Yale-MS, PhD
Volcanic hazards
Late Quaternary eruptive histories of volcanoes; volcanic hazards assessments

CRANDELL, WILLIAM B. **Martha**
PST **(703) 437-1565**
411 Patrick Lane
Herndon, Virginia 22070
Br. of Analytical Laboratories
Reston **(703) 860-7463**
Optical spectroscopy project

CRANFORD, S. LINDA
Geologist **(703) 425-5861**
5311 Orchardson Court
Fairfax, Virginia 22032
Br. of Eastern Environmental Geology
Reston **(703) 860-6595**
Virginia Polytechnic Inst. & State U.-BS
Tectonics of N.E. Virginia Piedmont

CRANSWICK, EDWARD
Geophysicist **(415) 326-6453**
2881 Alma Street
Palo Alto, California 94306
Br. of Ground Motion & Faulting
Menlo Park **(415) 323-8111 x2154**
Columbia Coll.-BA; Columbia U.
Strong ground motion prediction
Data processing.

CRAVOTTA, CHARLES A., III
Technical Reports Editor **(703) 280-1794**
3187 Eakin Park Court
Fairfax, Virginia 22031
Office of Scientific Publications
Reston **(703) 860-6492**
U. of Virginia-BA
Technical editor
Environmental geology

CRAWFORD, JANIE M.
Clerk-Typist **(202) 882-8069**
5720 13th Street, NW
Washington, D.C. 20011
Br. of Paleontology & Stratigraphy
Washington, D.C. **(202) 343-3406**
National Academy of Music

CREMER, MARCELYN J.
Chemist **(415) 339-1173**
6885 Thornhill Drive
Oakland, California 94611
Br. of Analytical Laboratories
Menlo Park **(415) 323-8111 x2942**
U. of Minnesota-BA; U. of Southern California-MS
Geochemical analysis

CRESSMAN, EARLE R. **Rhondda**
Geologist **(303) 279-2239**
1030 N. Columbine Street
Golden, Colorado 80401
Br. of Central Environmental Geology
Denver **(303) 279-2239**
Penn State-BS
Prichard Formation, Belt Supergroup
Stratigraphy; sedimentary petrology

CRIM, WILLIAM D.
PST
Br. of Exploration Research
Denver **(303) 234-6170**
U. of Wyoming; U. of Colorado
Alaskan mineral resource assessment program (Circle Quadrangle)
Regional geochemical exploration

CRITTENDEN, MAX D., JR. **Mabel**
Geologist **(415) 851-0704**
50 Alhambra Court
Portola Valley, California 94025
Br. of Western Environmental Geology
Menlo Park **(415) 323-8111 x2317**
San Jose State U.-BA; U. of California (Berkeley)-PhD
Tectonics of eastern Great Basin; Precambrian geology

CROCK, JAMES G. **Jahn**
Chemist **(303) 422-6320**
12332 West 65th Avenue
Arvada, Colorado 80004
Br. of Analytical Laboratories
Denver **(303) 234-6401**
Mount Union Coll.-BS; Penn State-MS
Automated chemistries; spectroscopy

CROCKETT, REUBEN D.
Computer Operator
Br. of Pacific-Arctic Geology
Menlo Park **(415) 323-2074**

CRONAN, MARY
Editorial Assistant
Br. of Paleontology & Stratigraphy
Washington, D.C. **(202) 343-3206**

CRONE, ANTHONY J.
Geologist
2065 Norwood Avenue
Boulder, Colorado 80302
Br. of Earthquake Tectonics & Risk
Denver **(303) 234-6885**
Clark U. (Mass.)-BA; U. of Colorado-PhD
Seismotectonics of the Mississippi Valley
Sedimentary petrology; stratigraphy; Quaternary geology

CRONIN, THOMAS M. **Margarita**
Geologist **(703) 860-9184**
11986 Barrel Cooper Court
Reston, Virginia 22091
Office of Environmental Geology
Reston **(703) 860-6411**
Colgate-BA; Harvard-MA, PhD
Quaternary sea level; climate; micropaleontology of Atlantic Coast
Climates; micropaleontology; sea level tectonics

CROVELLI, ROBERT A. **Nancy**
Mathematical Statistician **(303) 989-6884**
724 S. Braun Street
Lakewood, Colorado 80228
Br. of Oil & Gas Resources
Denver **(303) 234-5235**
Bucknell U.-BS, MS; Michigan State U.-MS; Colorado State U.-PhD
Resource appraisal methodology
Probability & statistics

CROWLEY, SHARON S. **Jim**
Physical Scientist **(703) 437-0364**
903 Long View Court
Herndon, Virginia 22070
Br. of Coal Resources
Reston **(703) 860-7734**
U. of Virginia-BA

CROWNOVER, LINDA M.
Computer Technician **(303) 756-2203**
4600 E. Kentucky, #106
Denver, Colorado 80222
Br. of Petrophysics & Remote Sens.
Denver **(303) 234-4897**
El Paso Community Coll.-AA; Metropolitan State Coll.
Image processing

CSEJTEY, BELA, JR. **Karen**
Geologist **(415) 494-1842**
3805 Carlson, Circle
Palo Alto, California 94306
Br. of Alaskan Geology
Menlo Park **(415) 323-8111 x2613**
Eotvos Science U. (Budapest); Princeton-PhD
Geologic investigations in the Healy Quadrangle, Alaska
Tectonics; petrology; field geology

CULBERTSON, WILLIAM C. **Maryjane**
Geologist **(303) 233-4049**
28 S. Chase Drive
Lakewood, Colorado 80226
Br. of Coal Resources
Denver **(303) 234-5634**
Cal Tech.-BS
Coal geology of Birney 1° quadrangle, Montana
Coal; oil shale; trona

CUNNINGHAM, CHARLES G. **Cher**
Geologist **(303) 674-6936**
31192 Cree Drive
Evergreen, Colorado 80439
Br. of Central Mineral Resources
Denver **(303) 234-4855**
Amherst-BA; U. of Colorado-MS; Stanford-PhD
Mineral resources of the Richfield, Utah 1° x 2° quadrangle
Fluid inclusion geothermometry

CUNNINGHAM, DAVID R. **Connie**
Electronics Technician **(303) 429-4151**
7898 Stuart Place
Westminster, Colorado 80030
Br. of Special Projects
Denver **(303) 234-3624**
Missouri Valley Coll.; National Electronics Inst.
Geophysical studies
Geophysical research & development

CUSHING, GRANT W.
GFA **(415) 326-7862**
859 Lytton Avenue
Palo Alto, California 94301
Br. of Alaskan Geology
Menlo Park **(415) 323-8111 x2331**
Middlebury Coll.-BA
Alaska mineral resource assessment project

CZAMANSKI, GERALD K. **Beverly**
Geologist **(415) 328-7153**
750 W. Greenwich Place
Palo Alto, California 94303
Br. of Exper. Geochem. & Mineralogy
Menlo Park **(415) 323-8111 x2502**
Sulfides in mafic and ultramafic rocks
Sulfide petrology; mineral chemistry

DADISMAN, SHAWN V.
PST
Br. of Pacific-Arctic Geology
Menlo Park **(415) 856-7020**
San Jose State U.-BS
Eastern Gulf of Alaska
Sedimentology; computers; oceanography

D'AGOSTINO, CATHERINE M.
Geologic Inquiries Asst. (703) 620-3786
11651 Stoneview Square, #12C
Reston, Virginia 22091
Office of Scientific Publications
Reston (703) 860-6517

D'AGOSTINO, JOHN P. Sue
Geologist
One Shetland Court
Rockville, Maryland 20851
Br. of Eastern Mineral Resources
Reston (703) 860-7359
U. of Buffalo-BA, MA
Charlotte 2° sheet; mineral deposits
Migration of metallic suites; engineering geology; photogeology

DAHL, ALFRED G.
Engineering Technician (512) 991-6146
1913 Tara
Corpus Christi, Texas 78412
Br. of Atlantic-Gulf of Mex. Geology
Corpus Christi (512) 888-3241
U. of Washington-BS

DALECHEK, MARJORIE E. Edward
Library Technician (303) 424-6145
8777 Yukon Street
Arvada, Colorado 80005
Library
Denver (303) 234-4004
Kearney State Coll.

DALLIN, MADELYN
Clerk-Typist
Br. of Sedimentary Mineral Resources
Denver (303) 234-3494
A&T Coll. (Farmingdale)-AAS

DALRYMPLE, G. BRENT
Geologist
Br. of Isotope Geology
Menlo Park (415) 323-8111 x2369
Occidental Coll.-AB; U. of California (Berkeley)-PhD
K-Ar dating, oceanic volcanism

DALZIEL, MARY C.
Geologist
404 Marine Street
Boulder, Colorado 80302
Br. of Oil & Gas Resources
Denver (303) 234-4586
Monmouth Coll.-BA; Rosenstiel Schl. of Marine & Atmospheric Sci., U. of Miami-MS
Remote detection of oil & gas deposits
Biogeochemical prospecting for oil & gas deposits

d'ANGELO, WILLIAM M. Teresa
Chemist (301) 258-0767
16 Nina Court
Gaithersburg, Maryland 20760
Br. of Analytical Laboratories
Reston (703) 860-6638
U. of Maryland-BS
Optical spectroscopy
Analytical chemistry

DANIELS, DAVID L. Carol
Geologist (703) 860-4986
2311 Whitetail Court
Reston, Virginia 22091
Br. of Regional Geophysics
Reston (703) 860-7233
Antioch Coll.-BS; U. of California (Berkeley)-MA
Eastern overthrust
Magnetic & gravity interpretation

DANIELS, JEFFREY J.
Geophysicist (303) 278-1755
523 N. Jackson
Golden, Colorado 80401
Br. of Petrophysics & Remote Sens.
Denver (303) 234-4743
Michigan State U.-BS, MS; Colorado Schl. of Mines-PhD
Borehole geophysics; nuclear waste isolation; coal exploration
Borehole geophysics; mining geophysics

DANSEREAU, DANNY A. Valerie
Cartographic Technician (303) 989-3091
1568 South Robb Court
Lakewood, Colorado 80226
Br. of Regional Geophysics
Denver (303) 234-5472
Rick's Coll.; U. of Texas, El Paso
Seismic refraction-reflection
Refraction-reflection computer graphics

DART, RICHARD L. Lora
PST (303) 278-1563
512 Cheyenne Street
Golden, Colorado 80401
Br. of Earthquake Tectonics & Risk
Denver (303) 234-4041
U. of South Alabama-BS
Puerto Rico seismic program
Processing seismic data

DATES, MERID D.
PST
Br. of Pacific-Arctic Geology
Menlo Park (415) 856-7041
Principia Coll.-BS

DAUL, WILLIAM B.
Computer Specialist
P.O. Box 2181
Stanford, California 94305
Br. of Tectonophysics
Menlo Park (415) 323-8111 x2933
U. of Montana
GEOLAB; computerized geophysical laboratory

DAVIDSON, DAVID F. Claire
Geologist (703) 620-2015
2605 Mt. Laurel Place
Reston, Virginia 22091
Office of International Geology
Reston (703) 860-6418
LeHigh U.-BS; U. of Oklahoma; U. of Birmingham (U.K.)
Geology of phosphate deposits; sedimentary ore deposits

DAVIES, WILLIAM E. Geraldine
Geologist (703) 532-7588
125 W. Greenway Boulevard
Falls Church, Virginia 22046
Br. of Engineering Geology
Reston (703) 860-6421
M.I.T.-BS; Michigan State Coll.-MS
Appalachian landslides
Engineering geology; karst; glacial geology; polar areas

DAVIS, LEO E. Josefina
Geophysicist (602) 792-6420
7290 East Tanque Verde Road
Tucson, Arizona 85715
Br. of Electromag. & Geomagnetism
Tucson (602) 792-6420
U. of Florida-BS
Tucson magnetic observatory

DAVIS, RAYMOND E.
Mechanical Engineer
Br. of Atlantic-Gulf of Mex. Geology
Woods Hole (617) 548-8700
Worcester Polytechnical Inst.-BS
Design & development of ocean bottom instrumentation
Electro-mechanical instrumentation; oceanographic hardware

DAVIS, ROBERT E. — **Phyllis**
Geologist — **(703) 437-7256**
1420 Yellowwood Court
Reston, Virginia 22090
Office of Scientific Publications
Reston — **(703) 860-6575**
U. of Arizona-BS, MS
Publications; engineering geology

DAVIS, SUSAN L.
Scientific Illustrator — **(602) 526-2029**
10 Blackhorse Road
Flagstaff, Arizona 86001
Br. of Astrogeologic Studies
Flagstaff — **(602) 779-3311 x1535**
Northern Arizona U.-BS
Planetary shaded relief mapping
Airbrush techniques; photo interpretation

DAWSON, MARGARET P. — **Phillip**
Clerk-Typist — **(703) 860-0072**
13016 Thompson Road
Fairfax, Virginia 22030
Br. of Analytical Laboratories
Reston — **(703) 860-7543**

DAY, JOHN H.
Physicist
Br. of Isotope Geology
Reston — **(703) 860-7662**
Bethune-Cookman Coll.-BS; Howard U.-MS
Borehole analysis of uranium & thorium
Gamma-ray spectroscopy

DEADMON, CHARLES E. — **Marianne**
PST — **(512) 991-0193**
4826 Elmhurst Lane
Corpus Christi, Texas 78413
Br. of Atlantic-Gulf of Mex. Geology
Corpus Christi — **(512) 888-3294**
Central State U.
Marine operations
Observatory geophysics; marine operations

DEAN, WALTER E.
Geologist — **(303) 674-8069**
30107 Carriage Loop Drive
Evergreen, Colorado 80439
Br. of Regional Geochemistry
Denver — **(303) 234-2310**
Syracuse U.-AB; U. of New Mexico-MS, PhD
Sedimentology; geochemistry; paleolimnology

DeCILLIS, MARIA C. — **Jack**
Clerical Assistant
704 Nutley Street, S.W.
Vienna, Virginia 22180
Br. of Eastern Environmental Geology
Reston — **(703) 860-6503**
Salter Secretarial School

DECKER, JOHN E.
Geologist — **(415) 321-3914**
555 Lytton Avenue, #2
Palo Alto, California 94301
Br. of Alaskan Geology
Menlo Park — **(415) 323-8111**
U. of California (Berkeley)-AB; U. of Alaska-MS; Stanford-PhD
Cretaceous subduction complex, southern Alaska
Sedimentology; tectonics; paleomagnetism

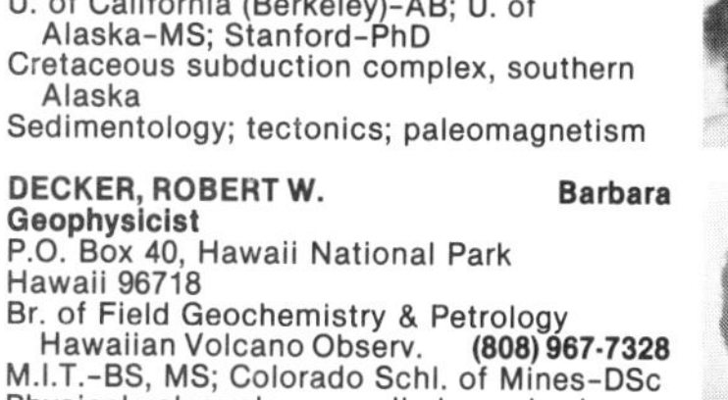

DECKER, ROBERT W. — **Barbara**
Geophysicist
P.O. Box 40, Hawaii National Park
Hawaii 96718
Br. of Field Geochemistry & Petrology
Hawaiian Volcano Observ. — **(808) 967-7328**
M.I.T.-BS, MS; Colorado Schl. of Mines-DSc
Physical volcanology; applied geophysics; tectonics

DeDONTNEY, DOROTHY M.
Administrative Technician
1385 Ranchita Drive
Los Altos, California 94022
Office of Geochemistry & Geophysics
Menlo Park — **(415) 323-2625**

DeJONGE, CHERYL D. — **Duane**
Secretary
4963 Harlan Street
Wheat Ridge, Colorado 80033
Br. of Electromag. & Geomagnetism
Denver — **(303) 234-2588**

DELEVAUX, MARYSE H.
Chemist — **(303) 989-4244**
10492-B West Florida Avenue
Lakewood, Colorado 80226
Br. of Isotope Geology
Denver — **(303) 234-3876**
George Washington U.-BS
Analysis of Pb-U-Th
Analytical chemistry & mass spectrometry

DeMARINIS, SUSAN K.
PST
1669 Nelson Road
Scotts Valley, California 95066
Br. of Oil & Gas Resources
Menlo Park — **(415) 856-7036**
U. of California (Santa Cruz)-BS
Southern California borderland project

DEMATTEO, RONALD E. — **Mary Ann**
Administrative Officer
U.S.G.S., c/o American Embassy
APO New York 09697
Br. of Latin American & African Geology
Jidda, Saudi Arabia — **674188**
Notre Dame-BS

DEMPSEY, WILLIAM J. — **Marianne**
Geophysicist — **(301) 587-0733**
820 Rowen Road
Silver Spring, Maryland 20910
Br. of Eastern Environmental Geology
Reston — **(703) 860-7256**
Catholic U.-BS, MS
Military geology
Magnetics; gravity

DENNIS, PHYLLIS R. — **Roger**
Library Technician — **(303) 278-8171**
18769 W. 60th Avenue
Golden, Colorado 80401
Library
Denver — **(303) 234-4004**
Northern Michigan U.-BA; U. of Denver

DENSON, NORMAN M.
Geologist — **(303) 238-0686**
1160 Pierce
Lakewood, Colorado 80214
Br. of Uranium-Thorium Resources
Denver — **(303) 234-2912**
Montana State U.-BA, MA; Princeton-PhD
Middle & Late Tertiary stratigraphy & structure of northern Rocky Mts. & northern Great Plains

DERICK, JAMES L. — **Glendyne**
Cartographic Technician — **(303) 986-1672**
12443 W. Louisiana
Lakewood, Colorado 80228
Br. of Central Environmental Geology
Denver — **(303) 234-2878**
Colorado Schl. of Mines; U. of Denver
Plotter lab

DERR, JOHN S.
Geologist
Office of Scientific Publications
Menlo Park **(415) 323-8111 x2300**
Amherst-BA; U. of California (Berkeley)-PhD
Earthquake lights; earthquake prediction; free oscillations

DESBOROUGH, GEORGE A.
Geologist
2164 Zang Street
Golden, Colorado 80401
Br. of Central Mineral Resources
Denver **(303) 234-5103**
Southern Illinois U.-BS, MS; U. of Wisconsin-PhD

DETTERMAN, JANIS S. **Robert**
Geologist **(408) 245-0730**
1725 Chitamook Court
Sunnyvale, California 94087
Office of Scientific Publications
Menlo Park **(415) 323-8111**
Coll. of Wooster-BA

DETTERMAN, ROBERT L. **Janis**
Geologist **(408) 245-0730**
1725 Chitamook Court
Sunnyvale, California 94087
Br. of Alaskan Geology
Menlo Park **(415) 323-8111**
Miami U.-BA
Ugashik-Karlok quadrangles, Alaska
Stratigraphy; sedimentation

DEVINE, JAMES F. **Rite**
Geophysicist **(703) 620-3710**
2455 Freetown Drive
Reston, Virginia 22091
Office of Earthquake Studies
Reston **(703) 860-6473**
West Virginia U.-BS
Seismic safety of critical structures
Seismicity; structural response; structural geology

DEWEY, JAMES W. **Kathie**
Geophysicist **(303) 494-8624**
2840 Dover Drive
Boulder, Colorado 80303
Br. of Global Seismology
Denver **(303) 234-4041**
U. of California (Berkeley)-AB, MA, PhD
U.S. seismicity; seismicity precursors
Seismotectonics; earthquake prediction

deWITT, WALLACE, JR. **Jean**
Geologist **(301) 489-4896**
15474 Roxbury Road
Glenwood, Maryland 21738
Br. of Oil & Gas Resources
Reston **(703) 860-6634**
U. of North Carolina-BS
Devonian black shale study program, Appalachian basin
Stratigraphy; geology of petroleum; paleogeography

DeYOUNG, JOHN H., JR. **Sally**
Mineral Economist **(703) 860-2374**
12677 Magna Carta Road
Herndon, Virginia 22071
Office of Resource Analysis
Reston **(703) 860-7356**
Princeton-BSE; U. of Michigan-MS; Penn State-PhD
CUSMAP; mineral resource assessment
Mineral economics; mineral resource assessment; economic geology

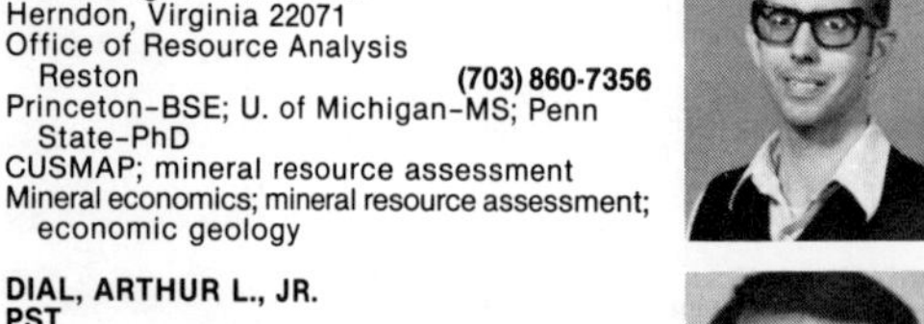

DIAL, ARTHUR L., JR.
PST
P.O. Box 1790
Flagstaff, Arizona 86002
Br. of Astrogeologic Studies
Flagstaff **(602) 779-3311 x1428**
Northern Arizona U.-BS

DICK, PATRICIA A. **Frederick**
Secretary **(703) 754-4441**
2500 Youngs Drive
Haymarket, Virginia 22069
Br. of Exper. Geochem. & Mineralogy
Reston **(703) 860-6659**

DICKEY, DAYTON D. **Jane**
Geologist
1680 S. Iris Way
Lakewood, Colorado 80226
Br. of Engineering Geology
Denver **(303) 234-3892**
Iowa State-BS
Engineering geology of Kemmerer, Wyo., area
Engineering geology; stress in Earth's crust; earthquakes

DICKINSON, KENDELL A. **Shirley**
Geologist **(303) 986-3304**
1749 South Van Gordon Court
Lakewood, Colorado 80228
Br. of Uranium-Thorium Resources
Denver **(303) 234-5667**
U. of Minnesota-BA, MS, PhD
Great Plains & Alaska uranium
Sedimentology; marine geology; mineralogy

DICKSON, JANE, J.
PST
Br. of Field Geochemistry & Petrology
Denver **(303) 234-2836**
Penn State-BS
Volcanic ash project

DIEHL, SHARON F.
PST
835 38th Street
Boulder, Colorado 80303
Br. of Engineering Geology
Denver **(303) 234-3416**
U. of Colorado-BA
Petrography

DIETERICH, JAMES H. **Susan**
Geophysicist
Br. of Tectonophysics
Menlo Park **(415) 323-8111 x2573**
U. of Washington-BS; Yale-PhD
Fault mechanics
Experimental/theoretical rock mechanics; earthquake prediction

DIETERICH, KATHRYN V.
PST
7502 West 10th Avenue
Lakewood, Colorado 80215
Br. of Paleontology & Stratigraphy
Denver **(303) 234-2743**
U. of Illinois-BS
Non-marine diatoms; pollen & spores

DIETRICH, JOHN A. **Joyce**
Electronic Technician **(303) 428-6242**
7978 Raleigh Place
Westminster, Colorado 80030
Br. of Exploration Research
Denver **(303) 234-6182**
National Electronic Institute; Community College
Optical emission and ultraviolet spectroscopy systems

DIETZ, BARBARA D. **Marshall**
Secretary
12601 Magna Carta Road
Herndon, Virginia 22070
Office of Geochemistry & Geophysics
Reston **(703) 860-6581**
Ferris State Coll.-AAS

DIGGLES, MICHAEL F. — **Barbara**
PST — **(415) 323-1649**
2170 Emerson Street
Palo Alto, California 94301
Br. of Alaskan Geology
Menlo Park — **(415) 323-8111 x2176**
Humboldt State U.-BA; San Jose State U.
RARE II, eastern Sierra, Inyo Mts.
Metallic mineral resources; radiometric age dates

DILLON, JOHN J. — **Louise**
Chemist — **(512) 991-5539**
Br. of Atlantic-Gulf of Mex. Geology
Corpus Christi — **(512) 888-3241**
U. of St. Thomas (Houston)-BA
Clays
XRD; trace metals; classical analytical methods

DILLON, WILLIAM P. — **Cynthia**
Geologist — **(617) 548-9439**
Acorn Drive (Box 134)
West Falmouth, Massachusetts 02574
Br. of Atlantic-Gulf of Mex. Geology
Woods Hole — **(617) 548-8700**
Bates-BS; Rensselaer-MS; U. of Rhode Island-PhD
Resource assessment of eastern continental margin south of Cape Hatteras
Continental margins

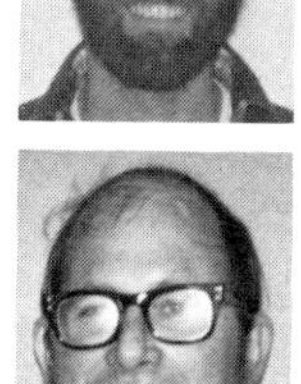

DIMENT, WILLAM H. — **Evelyn**
Geophysicist — **(703) 860-4532**
2040 Headlands Circle
Reston, Virginia 22091
Br. of Earthquake Tectonics & Risk
Reston — **(703) 860-6520**
Williams Coll.-BA; Harvard-PhD
Seismotectonics of northeastern United States
Geophysics

DINGLER, JOHN R. — **Janet**
Oceanographer — **(415) 493-0841**
730 Josina Avenue
Palo Alto, California 94306
Br. of Pacific-Arctic Geology
Menlo Park — **(415) 856-7071**
M.I.T.-SB, SM; Scripps Inst. of Oceanography-PhD
Coastal processes
Wave-sediment interactions

DINTER, DAVID A.
Geophysicist
Br. of Pacific-Arctic Geology
Menlo Park — **(415) 323-8111 x2417**
Stanford-BS, MS
Arctic environmental geology
Structural geology; marine geophysics; stratigraphy

DIXON, GARY L.
Geologist
Br. of Special Projects
Denver — **(303) 234-2391**
Western State Coll.-BA
Nevada nuclear waste storage investigations
Geologic reconnaissance mapping; volcanism in the Great Basin

DIXON, H. ROBERTA
Geologist — **(303) 526-0095**
3011 Rainbow Hills
Golden, Colorado 80401
Br. of Central Environmental Geology
Denver — **(303) 234-5528**
Carleton Coll.-BA; U. of California (Berkeley)-MA; Harvard-PhD
Precambrian of central Wyoming
Metamorphic petrology; structure

DOBROVOLNY, ERNEST E. — **Elsie**
Geologist — **(303) 777-5946**
2210 S. Corona
Denver, Colorado 80210
Br. of Engineering Geology
Denver — **(303) 234-3471**
Kansas State U.-BS; U. of Michigan-MS
Geologic mapping, Powder River, Wyo.; foreign geology
Engineering & environmental geology; stratigraphy

DOBSON, STEVEN W. — **Anita**
Geologist — **(303) 985-8137**
1050 S. Miller Way
Lakewood, Colorado 80226
Br. of Isotope Geology
Denver — **(303) 234-5531**
U. of Georgia-BS
Radiometric age data bank

DODD, JAMES E.
Electronics Engineer
Br. of Atlantic-Gulf of Mex. Geology
Woods Hole — **(617) 548-8700**
Bucknell U.-BS, BA
Integrated navigation-gravity system
Navigation, gravity, and biological instrumentation

DODGE, FRANKLIN C.W. — **Janice**
Geologist
440 Churchill Avenue
Palo Alto, California 94301
Br. of Field Geochemistry & Petrology
Menlo Park — **(415) 323-8111**
U. of California (Berkeley)-AA, BA; Stanford-MS, PhD
Geochemistry of Sierra Nevada batholith
Igneous petrology & geochemistry

DODGE, HARRY W., JR. — **Arden**
Geologist — **(303) 237-7433**
130 Everett Street
Lakewood, Colorado 80226
Br. of Uranium-Thorium Resources
Denver — **(303) 234-5039**
Princeton-BA; U. of Kansas-MS; UCLA
Analysis of Cretaceous sedimentary basins, northern Rocky Mts.
Paleoenvironments of fluvial & marginal-marine Cretaceous rocks

DOE, BRUCE R. — **Nellija**
Geologist — **(303) 238-9217**
815 Estes Street
Lakewood, Colorado 80215
Br. of Isotope Geology
Denver — **(303) 234-4003**
U. of Minnesota-BS, BGE; Missouri Schl. of Mines-MS; Cal Tech-PhD
Lead isotopes & ore deposits
Isotope geology; economic geology; tectonics

DOERING, WILLIS P.
Physicist
85 Dover Street
Lakewood, Colorado 80226
Br. of Isotope Geology
Denver — **(303) 234-5531**
Columbia Union Coll.-BA
Earthquake prediction
Stable isotope geology; mass spectrometer analysis

DOHER, L. IMOGENE
PST — **(303) 422-1797**
8680 Calvin Drive
Arvada, Colorado 80002
Br. of Paleontology & Stratigraphy
Denver — **(303) 234-5855**
Findlay Coll.-BS; U. of Colorado-MS
Palynology of uranium-bearing Triassic rocks, Colorado Plateau
Mesozoic palynology

DOHRENWEND, JOHN C. — **Valerie**
Geologist — **(415) 591-2542**
9 Christian Court
Belmont, California 94022
Br. of Western Mineral Resources
Menlo Park — **(415) 323-8111 x2580**
Rensselaer-BS; Purdue-MS; Stanford-MS, PhD
Quaternary geology; Walker Lake 2° sheet
Quaternary geology; geomorphology; environmental geology

DOLTON, GORDON L.
Geologist
4730 Harrison Avenue
Boulder, Colorado 80303
Br. of Oil & Gas Resources
Denver — **(303) 234-5235**
Pomona-BA; Claremont-MA
Resource appraisal group

DOMENICO, JAMES A. **Burma**
Chemist **(303) 423-8918**
8750 Independence Way
Arvada, Colorado 80005
Br. of Exploration Research
Denver **(303) 234-4440**
Regis Coll.-BS
Wallace 2° sheet (CUSMAP)
Research in exploration spectrochemistry; emission spectroscopy

DONATO, MARY M.
Geologist
1143 Pine Street
Menlo Park, California 94025
Br. of Field Geochemistry & Petrology
Menlo Park **(415) 323-8111 x2058**
Mount Holyoke-BA; U. of Oregon-MS
Igneous & metamorphic petrology; geology of Klamath Mts.

DONNELL, JOHN R. **Hazel**
Geologist **(303) 798-1168**
6035 S. Milwaukee Way
Littleton, Colorado 80121
Br. of Sedimentary Mineral Resources
Denver **(303) 234-3714**
U. of Alabama-BS; Stanford U.
Geology of Sagebrush Hill and Calamity Ridge, Quadrangles
Oil, shale, coal

DONNELLY, CYRIL A.
Administrative Officer
1105 Iron Ridge Court
Herndon, Virginia 22070
Office of Energy Resources
Reston **(703) 860-6434**
Villanova-BA

DONNELLY-NOLAN, JULIE M. **Michael**
Geologist
Br. of Field Geochemistry & Petrology
Menlo Park **(415) 323-8111 x2334**
U. of Washington; U. of California (Berkeley)-BA, PhD
Geology of Medicine Lake Volcano, California
Volcanology; igneous petrology; thermal waters

DONOVAN, TERRENCE J. **Sharon**
Geologist **(602) 526-9233**
2696 N. Oakmont Drive
Flagstaff, Arizona 86001
Br. of Oil & Gas Resources
Flagstaff **(602) 261-1350**
Midwestern U.-BS; U. of California (Riverside)-MA; UCLA-PhD
Remote sensing for oil & gas
Petroleum geology & petroleum exploration research

DOOLEY, JOHN R., JR. **Rosemary**
Physicist **(303) 422-3918**
6579 Lewis Street
Arvada, Colorado 80004
Br. of Isotope Geology
Denver **(303) 234-5531**
Regis Coll.-BS; U. of Denver-MS; U. of Colorado
Lithium & uranium autoradiography
Natural radioactive disequilibrium; fission tracks; U-Pb dating

DOREY, SUSAN E. **Clarence**
Clerk-Typist **(617) 394-6317**
77 North Dennis Road
South Yarmouth, Massachusetts 02664
Br. of Atlantic-Gulf of Mex. Geology
Woods Hole **(617) 548-8700**

DORRZAPF, ANTHONY F., JR. **Mary**
Chemist **(703) 620-3420**
12833 Oxon Road
Herndon, Virginia 22070
Br. of Analytical Laboratories
Reston **(703) 860-7654**
St. Peters Coll.-BS; U. of Maryland-MS
Optical emission spectroscopy
Computer-based emission spectrographic analysis; inductively coupled plasma emission spectroscopy

DOUGLASS, RAYMOND C. **Priscilla**
Geologist **(703) 938-2354**
1636 Irvin Street
Vienna, Virginia 22180
Br. of Paleontology & Stratigraphy
Washington, D.C. **(202) 343-3424**
Stanford-BS, PhD; U. of Nebraska-MS
Penn. coal bains, Appalach.-Illinois, Pennsylvanian/Permian, Western U.S.
Late Paleozoic biostratigraphy; fusulinid foraminifers

DOUKAS, MICHAEL P.
PST
Br. of Western Environmental Geology
Menlo Park **(415) 323-8111**
U. of California (Santa Cruz); California State U. (San Jose)-BS, MS
Sacramento-foothill project

DOW, VIRGINIA L.
Administrative Technician
Br. of Engineering Geology
Denver **(303) 234-2431**

DOYLE, BRIEN F.
PST
3061 S. Glencoe
Denver, Colorado 80222
Br. of Central Mineral Resources
Denver **(303) 234-6724**
Metropolitan State Coll.-BS
Mineral resource appaisal

DRAKE, AVERY A., JR.
Geologist
Office of the Chief Geologist
Reston **(703) 860-6631**

DREW, LAWRENCE J. **Sheila**
Geologist **(703) 476-5242**
12663 Magna Carta Road
Herndon, Virginia 22070
Office of Resource Analysis
Reston **(703) 860-6448**
U. of New Hampshire-BS; Penn State-MS, PhD
Petroleum resources appraisal
Petroleum resources analysis; discovery rate forecasting; statistical analysis

DUBE, MARCEL J. **Cecile**
Electronics Technician
411 Colorado Boulevard
Denver, Colorado 80206
Br. of Isotope Geology
Denver **(303) 234-3876**

DUBIEL, RUSSELL F.
Geologist **(303) 449-1829**
2000 Canyon Boulevard, Apt. 5B
Boulder, Colorado 80302
Br. of Uranium-Thorium Resources
Denver **(303) 234-5611**
Johns Hopkins-BA; U. of Maine-MS
Lake sediment geochemistry; urarnium geochemical exploration

duBRAY, EDWARD A. **Elizabeth**
Geologist
U.S.G.S. c/o American Embassy
APO New York 09697
Br. of Latin American & African Geology
Jiddah, Saudi Arabia **691152**
Stanford-BS, MS
Acid plutonic rocks of Arabian Shield
Petrology, geochemistry & economic geology of acid plutonic rocks

DUFFIELD, WENDELL A. **Anne**
Geologist
13030 La Paloma Road
Los Altos Hills, California 94022
Br. of Field Geochemistry & Petrology
Menlo Park **(415) 323-8111 x2680**
Carleton Coll.-BA; Stanford-MS, PhD
Coordinator, geothermal research program
Volcanology; field geology; geothermal energy

DUFFY, MARY T.
Mgmt. Support Clerk
Br. of Central Mineral Resources
Denver **(303) 234-4842**
Metropolitan State Coll.-BS

DULONG, FRANK T.
Geologist
1615 Valencia Way
Reston, Virginia 22090
Br. of Coal Resources
Reston **(703) 860-7734**
George Washington U.-BS, MS
Contaminants in coal
Mineral matter in coal; X-ray diffraction

DUMOULIN, JULIE A.
Geologist **(907) 278-9762**
801 East 11th Avenue, #16
Anchorage, Alaska 99501
Br. of Alaskan Geology
Anchorage **(907) 271-4150**
U. of Wisconsin-BS, MS
Sedimentology; paleoecology; stratigraphy

DUNCAN, KAREN M.
Editorial Assistant
6145 West 38th Avenue
Wheat Ridge, Colorado 80033
Br. of Exploration Research
Denver **(303) 234-4440**

DUNPHY, GERALD J. **Catherine**
Geophysicist **(303) 494-0948**
2385 Kenwood Drive
Boulder, Colorado 80303
Br. of Global Seismology
Denver **(303) 234-3994**
UCLA-BA
Seismic data systems
Seismicity

DUNRUD, C. RICHARD **Toni**
Geologist **(303) 674-5638**
24568 Giant Gulch Road
Evergreen, Colorado 80439
Br. of Engineering Geology
Denver **(303) 234-3573**
U. of Wyoming-BA, BS, MS
Coal mine deformation studies
Mine subsidence; engineering geology; structural geology

DUPRÉ, WILLIAM R. **Elaine**
Geologist **(713) 723-8490**
12110 Ashcroft
Houston, Texas 77035
Br. of Western Environmental Geology
Houston **(713) 749-3710**
U. of Texas (Austin)-BS, MA; Stanford-MS, PhD
Seismic zonation, San Francisco Bay region
Coastal geomorphology & sedimentation

DURHAM, DAVID L. **Nancy**
Geologist **(415) 948-8783**
306 Alta Vista Avenue
Los Altos, California 94022
Br. of Sedimentary Mineral Resources
Menlo Park **(415) 323-8111 x2227**
Cal Tech-BS
Uranium in Tertiary sedimentary rocks of California
Areal geology; diatomaceous rocks; uranium

DUSEL-BACON, CYNTHIA **Charles**
Geologist **(415) 324-8237**
139 Princeton Road
Menlo Park, California 94025
Br. of Alaskan Geology
Menlo Park **(415) 323-8111 x2202**
U. of California (Santa Barbara)-BA; San Jose State U.-BA
Metamorphic studies, Yukon-Tanana upland, Alaska
Metamorphic petrology; petrography

DUTRO, J. THOMAS, JR. **Nancy**
Geologist **(202) 363-0480**
5173 Fulton Street, NW
Washington, D.C. 20016
Br. of Paleontology & Stratigraphy
Washington, D.C. **(202) 343-3222**
Oberlin-BA; Yale-MS, PhD
Paleozoic biostratigraphy of Alaska
Brachiopoda; late Paleozoic stratigraphy of Alaska, western U.S.

DUTY, DENNIS W.
Drill Operator **(703) 471-4409**
13348 Shea Place
Herndon, Virginia 22070
Br. of Eastern Environmental Geology
Reston **(703) 860-7689**
Old Dominion U.-BS

DUVAL, JOSEPH S. **Ana**
Geophysicist
12899 West Idaho Drive
Lakewood, Colorado 80228
Br. of Petrophysics & Remote Sens.
Denver **(303) 234-5492**
Stephen F. Austin Coll.; Rice U.-BA, MA, PhD
Gamma-ray spectrometry in uranium exploration
Gamma-ray spectrometry; computer modeling; gamma-ray theory

DWORNIK, DEBORAH
Photographer
Br. of Field Geochemistry & Petrology
Reston **(703) 860-6406**
U. of Maryland-BA; Northern Virginia Community Coll.
Scientific & technical photography

DWORNIK, EDWARD J. **Dolores**
Geologist
3117 63rd Place
Cheverly, Maryland 20785
Br. of Analytical Laboratories
Reston **(703) 860-7543**
U. of Buffalo-BA, MA
X-ray spectroscopy
Electron microscopy & analysis of fine-grained minerals

DYMAN, THADDEUS S.
Geologist
924 Cheyenne Street
Golden, Colorado 80401
Br. of Oil & Gas Resources
Denver **(303) 234-6115**
South Dakota Schl. of Mines; Northern Illinois U.-BS, MS; Syracuse U.
Oil & gas energy data systems
Computer applications to geologic problems

DZURISIN, DANIEL **Linda**
Geologist **(808) 966-9851**
P.O. Box 7
Hawaii National Park, Hawaii 96718
Br. of Field Geochemistry & Petrology
Hawaiian Volcano Observ. **(808) 967-7328**
U. of Notre Dame-BS; Cal Tech-MS, PhD
Deformation of Hawaiian volcanoes
Volcano geophysics; stratigraphy of Kilauea caldera; eruption forecast

EATON, GORDON P. **Virginia**
Geologist **(703) 476-5047**
11612 Hunters Green Court
Reston, Virginia 22091
Office of the Chief Geologist
Reston **(703) 860-6532**
Wesleyan U.-BA; Cal. Tech.-MS, PhD
Administration
Regional geological & geophysical synthesis; geological interpretation of geophysical data

EATON, JERRY P. — **Nancy**
Geophysicist — **(415) 941-4530**
220 E. Edith
Los Altos, California 94022
Br. of Network Operations
Menlo Park **(415) 323-8111 x2575**
U. of California (Berkeley)-PhD
California seismic network
Local earthquake seismology; volcano geophysics

EBENS, RICHARD J. — **Beverly**
Geologist — **(303) 985-7254**
1424 S. Ward Street
Lakewood, Colorado 80228
Br. of Regional Geochemistry
Denver **(303) 234-3715**
Beloit Coll.-BS; U. of Wyoming-MA, PhD
Sedimentary geochemistry; geostatistics

EBERLEIN, G. DONALD — **Marilyn**
Geologist — **(408) 354-2058**
P.O. Box 1064
Los Gatos, California 95030
Br. of Alaskan Geology
Menlo Park **(415) 323-8111 x2210**
Yale-BS; Claremont Coll.; Stanford
Alaskan mineral resources; regional mapping, southeastern Alaska
Optical crystallography; igneous & metamorphic petrology; economic geology

ECHERT, ELISE D.
GFA
10671 Holland Street
Broomfield, Colorado 80020
Br. of Global Seismology
Denver **(303) 234-3994**
Denver Community Coll.; U. of Colorado

ECKEL, EDWIN B. — **LaCharles**
Geologist — **(303) 733-0982**
1109 S. High Street
Denver, Colorado 80210
Br. of Central Mineral Resources
Denver **(303) 234-2452**
Lafayette Coll.-BS; U. of Arizona-MS
Indian lands program
Engineering geology; ore deposits; technical writing

EDGAR, N. TERENCE — **Joyce**
Geologist — **(703) 759-3888**
10102 Spring Hollow Lane
Great Falls, Virginia 22066
Office of Marine Geology
Reston **(703) 860-7241**
Middlebury Coll.-BA; Florida State U.-MSC; Columbia U.-PhD

EDWARDS, KATHLEEN L. — **Steve**
Cartographer — **(602) 526-2628**
R.R.I. Box 1060
Flagstaff, Arizona 86001
Br. of Astrogeologic Studies
Flagstaff **(602) 779-3311 x1313**
Northern Arizona U.-BS
Graphics; image processing

EDWARDS, LUCY E.
Geologist — **(703) 968-7976**
4700 Spruce
Fairfax, Virginia 22030
Br. of Paleontology & Stratigraphy
Reston **(703) 860-7745**
U. of Oregon-BA; U. of California (Riverside)-PhD
Micropaleontology (dinoflagellates); biostratigraphy

EGE, JOHN R.
Geologist
Br. of Engineering Geology
Denver **(303) 234-3425**
Michigan State U.-BA, BS; U. of Montana-MS; Colorado Schl. of Mines-PhD
Subsidence & ground failure
Engineering geology; geomechanics; geotechnical mapping

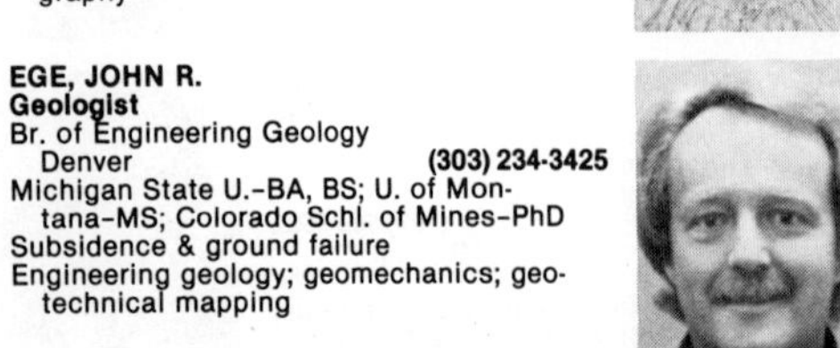

EGELSON, DAVID C.
Geophysicist — **(617) 540-3645**
78 Teaticket Path
East Falmouth, Massachusetts 02536
Br. of Atlantic-Gulf of Mex. Geology
Woods Hole **(617) 548-8700 x128**
Resource assessment
Marine multichannel reflection seismology

EGGLETON, RICHARD E.
Geologist — **(303) 986-3327**
465 S. Wright Street, Apt. 105
Lakewood, Colorado 80228
Br. of Central Environmental Geology
Denver **(303) 234-5152**
U. of Michigan-BSE, MS; U. of Arizona-PhD
Environmental geology of Williston Basin, MT, ND.
Surficial geology; tectonics; planetology

EICHLER, HELEN E. — **Harry**
Education Prgm Administrator — **(303) 422-4830**
6613 Moore Street
Arvada, Colorado 80004
Office of the Chief Geologist
Denver **(303) 234-4790**
Barnes Schl. of Commerce

EISNER, SARA B.
Chemist — **(703) 435-4870**
Br. of Atlantic-Gulf of Mex. Geology
Reston **(703) 860-6965**
Bryn Mawr Coll.-BA
Marine organic geochemistry; resource & environmental assessment
Geochemistry of organic substances

EISTER, MARGARET F.
Technical Information Specialist
Reston, Virginia 22091
Office of Scientific Publications
Reston **(703) 860-6517**
Radcliffe-BA; Penn State; George Washington U.
Geologic map index of the United States

EITTREIM, STEPHEN L. — **Carole-Ann**
Geologist — **(415) 856-6977**
1975 Ivy Lane
Palo Alto, California 94303
Br. of Pacific-Arctic Geology
Menlo Park **(415) 856-7156**
Colby Coll.-BA; Columbia U.-PhD
Abyssal sediment transport & diffusion
Suspended particulate matter; sediment dynamics; marine seismic reflection

EKREN, E. BARTLETT — **Doris**
Geologist
Evergreen, Colorado 80439
Br. of Central Mineral Resources
Denver **(303) 234-2854**
U. of North Dakota-BS; U. of Colorado
Mapping Challis volcanic rocks, Challis 2° CUSMAP
Volcanology; structural geology

ELLEN, STEPHENSON D. — **Carol**
Geologist
Harris Star Route
Garberville, California 95440
Br. of Western Environmental Geology
Menlo Park **(415) 323-8111 x2311**
Amherst Coll.-BA; Stanford U.-MS, PhD
San Francisco Bay region hillside materials
Engineering geology; slope stability

ELLERSIECK, INYO F.
Geologist — **(415) 328-7683**
1143 Pine Street, #4
Menlo Park, California 94025
Br. of Alaskan Geology
Menlo Park **(415) 323-8111 x2147**
U. of California (Berkeley)-BA
Brooks Range stratigraphy
Rock-in-the-box field mapping

ELLICK, DONA M. James
Administrative Officer (415) 851-0399
1180 Westridge Drive
Portola Valley, California 94025
Office of Geochemistry & Geophysics
Menlo Park (415) 323-8111 x2625
U. of California (Berkeley); Menlo Coll.

ELLIOTT, JAMES E. Susan
Geologist (303) 425-7539
3553 Simms Street
Wheat Ridge, Colorado 80033
Br. of Central Mineral Resources
Denver (303) 234-6807
U. of Washington-BS; Stanford-PhD
Butte 2° CUSMAP
Economic geology; geochemistry; petrology

ELLIOTT, RAYMOND L.
Geologist
Br. of Alaskan Geology
Menlo Park (415) 323-8111
Stanford-BS
Bradfield Canal Quad.-AMRAP
S.E. Alaska geol. mapping; igneous & metamorphic petrology

ELLIS, JAMES D. Pat
Electronics Technician (408) 275-8707
1421 Sunshine Court
San Jose, California 95122
Br. of Networks Operations
Menlo Park (415) 323-8111 x2528
Multiple microprocessor on-line seismic data analyzer
Digital circuit design

ELLIS, WILLIAM L. Jeanine
Mining Engineer (303) 985-2218
1352 S. Cape Way
Lakewood, Colorado 80226
Br. of Special Projects
Denver (303) 234-2371
Colorado Schl. of Mines-BSME
Stress investigations; rock mechanics

ELLSWORTH, WILLIAM L.
Geophysicist
Br. of Seismology
Menlo Park (415) 323-8111 x2778
Stanford-BS, MS; MIT-PhD
Earthquake seismology; lithospheric structure

ELSHEIMER, H. NEIL Joanne
Chemist
1130 Plum Avenue
Sunnyvale, California 94087
Br. of Analytical Laboratories
Menlo Park (415) 323-8111 x2950
Wheaton Coll.-BS; Indiana U.-MA; U. of Illinois; U. of Colorado
X-ray fluorescence spectrometry
XRF; flame chemistry; solution chemistry

ELSTON, DONALD P. Shirley
Geologist (602) 526-2641
6300 Country Club Drive
Flagstaff, Arizona 86001
Br. of Petrophysics & Remote Sens.
Flagstaff (602) 779-3311 x1544
Syracuse U.-BA, MS; U. of Arizona-PhD
Precambrian & Tertiary paleomagnetism; dry valley (Antarctic) magnetostratigraphy
Magnetostratigraphy; remote sensing; meteoritics

EMSING, SANDRA L.
Clerk
Br. of Isotope Geology
Denver (303) 234-3876

ENGDAHL, ERIC R. Eileen
Geophysicist (303) 494-0865
255 Inca Parkway
Boulder, Colorado 80303
Br. of Global Seismology
Denver (303) 234-5084
Rensselaer-BS; St. Louis U.-PhD
Earth structure; plate tectonics; earthquake prediction

ENGLEMAN, EDYTHE E. Keith
Physical Scientist
2407 S. Tennyson
Denver, Colorado 80219
Br. of Analytical Laboratories
Denver (303) 234-6401
U. of Nebraska; Nebraska Wesleyan-BA

ENGLISH, KATHY S.
Clerk-Typist
12692 E. Hickman Place
Denver, Colorado 80239
Br. of Central Mineral Resources
Denver (303) 234-3830

ENGLISH, THOMAS T.
Computer Specialist (808) 967-7264
P.O. Box 66
Volcano, Hawaii 96785
Br. of Field Geochemistry & Petrology
Hawaiian Volcano Observ. (808) 967-7328
U. of San Francisco-BS; U. of Hawaii-MBA

ENGLUND, KENNETH J. Virginia
Geologist (703) 327-4765
Route 1, Box 144B
Aldie, Virginia 22001
Br. of Coal Resources
Reston (703) 860-7465
U. of Wisconsin-BA, MA
Appalachian Basin
Coal geology; Carboniferous stratigraphy

EPSTEIN, JACK B.
Geologist (703) 860-2821
2104 Thomas View Road
Reston, Virginia 22091
Br. of Eastern Environmental Geology
Reston (703) 860-6421
Brooklyn Coll.-BS; U. of Wyoming-MA; Ohio State-PhD
National environmental overview program
Environmental geology; mapping, eastern Pennsylvania; stratigraphy

ERD, RICHARD C. Helen
Geologist (415) 493-8185
702 Josina Avenue
Palo Alto, California 94306
Br. of Exper. Geochem. & Mineralogy
Menlo Park (415) 323-8111 x2376
Indiana U.-BA, MA
Mineralogy; crystallography; boron minerals

ERDMAN, JAMES A. Mardi
Botanist (303) 232-5455
2125 Estes Street
Lakewood, Colorado 80215
Br. of Regional Geochemistry
Denver (303) 234-5240
North Central Coll.-BA; U. of Colorado-MA, PhD
Environmental & baseline biogeochemistry
Biogeochemical & geobotantical prospecting

ERICKSEN, GEORGE E. Kelly
Geologist (703) 860-3877
11903-105 Winterthur Lane
Reston, Virginia 22091
Br. of Eastern Mineral Resources
Reston (703) 860-6913
U. of Montana-BA; U. of Indiana-MA; Columbia U.-PhD
Ore deposits of South America
Mineral deposits; geochemistry; engineering geology

ERICKSON, GEORGE S. — **Maxine**
Civil Engineering Technician — **(303) 567-2604**
Route 1, Box 530
Idaho Springs, Colorado 80452
Br. of Engineering Geology
Denver — **(303) 234-2528**
U. of Colorado
Engineering geology laboratory
Geotechnical testing

ERICKSON, M. SUZANNE — **Ralph**
Spectrographer — **(303) 278-9023**
268 Gardenia Court
Golden, Colorado 80401
Br. of Exploration Research
Denver — **(303) 234-6190**
U. of Northern Colorado; U. of Colorado; UCLA
Spectrographic analyses & methods in geochemical exploration

ERICKSON, RALPH L. — **Suzi**
Geologist — **(303) 278-9023**
268 Gardenia Court
Golden, Colorado 80401
Br. of Exploration Research
Denver — **(303) 234-3891**
Miami U. (Ohio)-BA; Michigan State U.-MS; U. of Minnesota-PhD
Geochemical halos (midcontinent region)
Applied geochemistry; geology of ore deposits

ERNST, WALLACE G. — **Charlotte**
Geologist — **(213) 454-2477**
16939 Lirorno Drive
Pacific Palisades, California 90272
Br. of Field Geochemistry & Petrology
Menlo Park — **(415) 323-2214**
Carleton Coll.-BA; U. of Minnesota-MS; Johns Hopkins-PhD
Marble Mountain wilderness project
Petrology & plate pushing

ESCOWITZ, EDWARD C.
Oceanographer
Scientists Cliffs Road
Port Republic, Maryland 20676
Br. of Atlantic-Gulf of Mex. Geology
Reston — **(703) 860-6489**
Rockland C. Coll.-AAS; SUNY (Stony Brook)-BS; New York U.-MS, PhD
Physical oceanography; marine climatology

ESPINOSA, ALVARO F.
Geophysicist
Br. of Earthquake Tectonics & Risk
Denver — **(303) 234-5077**
St. Louis U.; Columbia U.
Seismic wave attenuation in conterminous United States
Seismology; engineering seismology; strong-motion seismology

ESTABROOK, JAMES R. — **Kathy**
Geologist — **(703) 860-0699**
11627 Stoneview Square
Reston, Virginia 22091
Office of Scientific Publications
Reston — **(703) 860-6492**
Carleton Coll.-BA
Map editor

EVANS, DAVID H. — **Melva**
Geophysicist — **(907) 579-3252**
P.O. Box 5250, Adak, Alaska
FPO Seattle, Washington 98791 (Adak, AK)
Br. of Global Seismology
Denver — **(907) 234-3994**
George Washington U.; Colorado Schl. of Mines; U. of Alaska
Adak Seismological Observatory

EVANS, HOWARD T., JR. — **Grace**
Physicist — **(301) 229-6722**
6307 Bannockburn Drive
Bethesda, Maryland 20034
Br. of Exper. Geochemistry & Mineralogy
Reston — **(703) 860-6666**
M.I.T.-BS, PhD
Crystal structure analysis
Mineral structures; crystal chemistry; inorganic chemistry

EVANS, JOHN R.
Geophysicist
c/o 172 Kensington Way
Los Gatos, California 95030
Br. of Seismology
Menlo Park — **(415) 323-8111 x2923**
U. of California (Santa Cruz)-BS; Princeton-MA
Yellowstone-Snake River plain deep velocity structure
Seismology

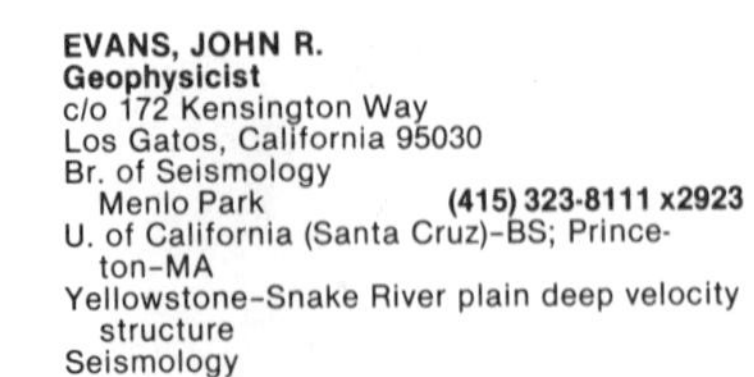

EVARTS, RUSSELL C. — **Alice**
Geologist — **(415) 793-2257**
36379 Spruce Street
Newark, California 94560
Br. of Western Mineral Resources
Menlo Park — **(415) 323-8111 x2666**
Franklin & Marshall-BA; Stanford-PhD
Spirit Lake Quad., Washington
Petrology; mineral deposits; field geology

EVENDEN, GERALD I.
Geophysicist
Br. of Atlantic-Gulf of Mex. Geology
Woods Hole — **(617) 548-8700**
Digital computer systems & application programs

FABBI, BRENT P. — **D. Yolanda**
Physicial Scientist
9810 Meadow Knoll Court
Vienna, Virginia 22180
Br. of Analytical Laboratories
Reston — **(703) 860-7246**
U. of Nevada-BS
X-ray spectroscopy; major & trace analysis

FABIANO, EUGENE B. — **Lutgarde**
Geophysicist
1800 South Pratt Parkway
Longmont, Colorado 80501
Br. of Electromag. & Geomagnetism
Denver — **(303) 234-5180**
U. of Pittsburgh-BS
Magnetic charts

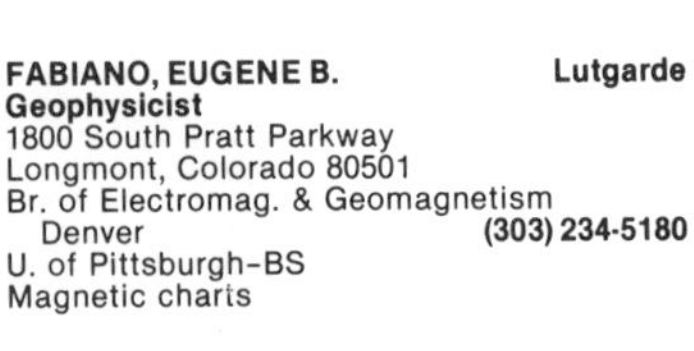

FARROW, RICHARD A. — **Joan**
Geologist — **(512) 937-4957**
305 Caribbean
Corpus Christi, Texas 78418
Br. of Engineering Geology
Corpus Christi — **(512) 888-3241**
U. of Colorado-BA
Marine geotechnical laboratory
Marine geotechnics, engineering geology, soil rock mechanics

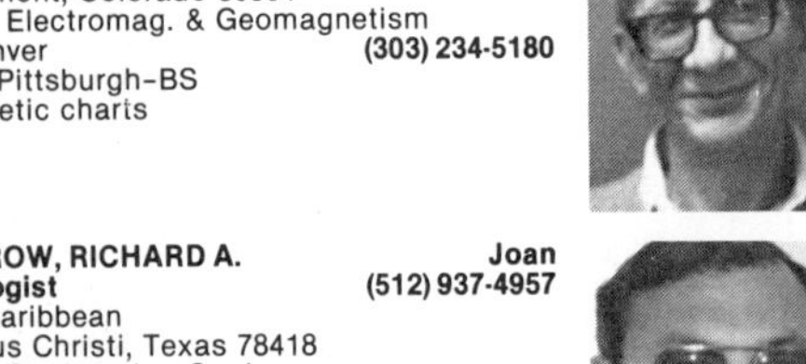

FARY, RAYMOND W., JR. — **Doris**
Geologist — **(301) 593-0238**
304 Penwood Road
Silver Spring, Maryland 20901
Office of International Geology
Reston — **(703) 860-6551**
McKendree Coll.; U. of Washington-AB, MA
Petroleum geology; military geology; applications of remote sensing

FATZ, LORRIE P.
Clerk-Stenographer
Office of Environmental Geology
Reston — **(703) 860-6417**

FEENEY, DIANE D. — **John**
Administrative Clerk — **(703) 437-9038**
1239 Springtide Place
Herndon, Virginia 22070
Office of Mineral Resources
Reston — **(703) 860-6571**

FELMLEE, J. KAREN
Geologist
Br. of Uranium-Thorium Resources
Denver **(303) 234-5213**
Earlham Coll.-BA; U. of Wisconsin-MA
Uranium & daughter products in springs & ground water
Geochemistry; metamorphic geology

FENICLE, KAREN A.
Clerk-Typist
Br. of Isotope Geology
Denver **(303) 234-3876**

FERGUSON, HOLLY M. **Joseph**
Cartographer **(602) 779-2183**
Br. of Astrogeologic Studies
Flagstaff **(602) 779-3311 x1531**
U. of New Mexico-BS
Martian mass wasting; Martian wrinkle ridges; Europa: structure & lineation

FERNALD, ARTHUR T. **Jean**
Geologist
Br. of Special Projects
Denver **(303) 234-2391**
U. of New Hampshire-BS; M.I.T.; Harvard U.-PhD
Geomorphology, arctic & desert environments

FERREBEE, WAYNE M. **Patricia**
Geologist **(617) 548-1480**
95 Althea Road
N. Falmouth, Massachusetts 02556
Br. of Atlantic-Gulf of Mex. Geology
Woods Hole **(617) 548-8700**
West Virginia U.-BS, MS
Sedimentary petrology

FERRIANS, OSCAR J., JR. **Muriel**
Geologist **(415) 968-5339**
1230 Saint Joseph Avenue
Los Altos, California 94022
Br. of Alaskan Geology
Menlo Park **(415) 323-8111**
Washington State U.-BS, MS
Arctic environmental studies & earthquake hazards mapping
Quaternary geology; engineering geology & geocryology

FERRIER, SUSAN C.
Office Services Assistant
3325 Webley Court
Annandale, Virginia 22003
Office of Geochemistry & Geophysics
Reston **(703) 860-6585**
Northern Virginia Community Coll.

FICHTNER, WAYNE E. **Doris**
Administrative Officer **(301) 881-3737**
11020 Rosemont Drive
Rockville, Maryland 20852
Office of the Chief Geologist
Reston **(703) 860-6538**
American U.-BS

FICKLIN, WALTER H. **Ann**
Research Chemist **(303) 238-7264**
8019 W. 23rd Avenue
Lakewood, Colorado 80215
Br. of Exploration Research
Denver **(303) 234-6188**
Fort Lewis Coll.-BS; U. of California (Riverside)-MS
Research in water analysis
Analytical chemistry; water analysis

FIELD, MICHAEL E. **Kathy**
Geologist **(408) 733-2746**
1561 Samedra Street
Sunnyvale, California 94087
Br. of Pacific-Arctic Geology
Menlo Park **(415) 856-7057**
U. of Delaware-BS; Duke U.-MS; George Washington U.-PhD
Marine geology of the northern California continental margin & southern California
Marine geology; sedimentology

FILSON, JOHN R.
Geophysicist **(703) 860-2807**
Office of Earthquake Studies
Reston **(703) 860-6471**
Rice U.-BA; U. of California (Berkeley)-MA, PhD
Seismology

FINCH, WARREN I. **Mary**
Geologist **(303) 233-3372**
455 Dover Street
Lakewood, Colorado 80226
Br. of Uranium-Thorium Resources
Denver **(303) 234-5818**
S. Dakota Schl. of Mines-BS; U. of California-MS; Colorado Schl. of Mines
Uranium resource assessment group
Uranium & field geology

FINKELMAN, ROBERT B. **Judith**
Geologist **(703) 476-5722**
1936 Barton Hill Road
Reston, Virginia 22091
Br. of Analytical Laboratories
Reston **(703) 860-7543**
CCNY-BS; George Washington U.-MS; U. of Maryland-PhD
X-ray spectroscopy
Micromineralogy; coal geochemistry

FINN, CAROL A.
PST
Br. of Regional Geophysics
Denver **(303) 234-6959**
Wellesley Coll.-BA
Cascades geothermal assessment program
Volcanology; potential field theory (gravity)

FINNELL, TOMMY L. **Sue**
Geologist
717 Iowa Street
Golden, Colorado 80401
Br. of Central Mineral Resources
Denver **(303) 234-3293**
U. of Wyoming-BS
Petrology of volcanic rocks, New Mexico
Structural geology; mineral deposits

FISCHER, CHARLENE R.
Editorial Assistant
1312 Jefferson, #2
Redwood City, California 94062
Br. of Western Environmental Geology
Menlo Park **(415) 323-8111 x2203**
Idaho State U.

FISCHER, FREDERICK G.
Geophysicist **(415) 494-2374**
432 Margarita Avenue
Palo Alto, California 94306
Br. of Seismology
Menlo Park **(415) 323-8111 x2321**
U. of California (Berkeley)-BA
Garm (U.S.S.R.) earthquake source
Remote sensing; automatic data collection; processing

FISCHER, JEFFREY M.
GFA **(415) 493-2625**
946 Van Auken Circle
Palo Alto, California 94303
Br. of Pacific-Arctic Geology
Menlo Park **(415) 323-8111**
U. of Delaware-BS
Navarin Basin
Seismics; soil mechanics; computers

FISCHER, LYNN B. **Charles**
Chemist **(303) 399-6865**
637 Ash
Denver, Colorado 80220
Br. of Isotope Geology
Denver **(303) 234-3876**
U. of Colorado-BA
Analytical geochemistry; mass spectroscopy

FISHER, FREDERICK S. **Cynthia**
Geologist **(303) 674-7256**
4251 S. Aspen Lane
Evergreen, Colorado 80439
Br. of Central Mineral Resources
Denver **(303) 234-6294**
Wayne State U.-BS, MS; U. of Wyoming-PhD
Challis 2° quadrangle
Economic geology; geochemistry; igneous petrology

FISHER, MICHAEL A.
Geophysicist
Br. of Oil & Gas Resources
Menlo Park **(415) 856-7108**
U. of Utah-BS; Stanford U.-MS
Regional & petroleum geology of Cook Inlet; Kodiak shelf & northern Bering Sea
Geophysics, geology

FITCH, SHERRY L.
Secretary
11655 Charter Oak Court
Reston, Virginia 22090
Br. of Coal Resources
Reston **(703) 860-7734**
Frostburg State Coll.

FITTERMAN, DAVID V.
Geophysicist **(303) 444-0520**
858 9th Street
Boulder, Colorado 80302
Br. of Electromag. & Geomagnetism
Denver **(303) 234-5158**
Colorado Schl. of Mines-BS; MIT-PhD
Self-potential; geomagnetic variation; electrical earthquake phenomena

FITZHUGH, WILLIE M. **Muriel**
General Mechanic **(703) 775-7811**
Corbin, Virginia 22446
Br. of Electromag. & Geomagnetism
Corbin, Virginia **(703) 373-7601**

FLANAGAN, FRANCIS J. **Antoinette**
Chemist
6980 Oregon Avenue, N.W.
Washington, D.C. 20015
Br. of Analytical Laboratories
Reston **(703) 860-6688**
Catholic U.-BChE; George Washington U.
Rock and mineral standards
Rock & mineral standards; statistics; design of experiments

FLANIGAN, VINCENT J.
Geophysicist **(303) 838-4891**
757 Sleepy Hollow Drive
Boulder, Colorado 80421
Br. of Electromag. & Geomagnetism
Denver **(303) 234-5459**
U. of Washington-BS
Geophysical studies in mineral resource appraisal
EM methods

FLEISCHER, MICHAEL **Helen**
Chemist
3104 Chestnut Street, N.W.
Washington, D.C. 22015
Br. of Regional Geochemistry
Reston **(703) 860-7771**
Yale U.-BS, PhD
Data of geochemistry
Mineralogy, trace elements

FLEMING, HERSHELL L. **Helen**
Intnatl. Activities Officer
9600 Dangerfield Road
Clinton, Maryland 20735
Office of International Geology
Reston **(703) 860-6410**

FLEMING, ROBERT W. **Juv**
Geologist **(303) 279-8122**
2039 Crestvue Circle
Golden, Colorado 80401
Br. of Engineering Geology
Denver **(303) 234-5559**
Oklahoma U.-BS; Brown U.-MS; Stanford U.-PhD
Landslide processes
Engineering geology

FLEMING, STANLEY L., II **Glenda**
Physical Scientist
Br. of Analytical Laboratories
Reston **(703) 860-7147**
U. of North Carolina (Chapel Hill)-BA
Sample preparation
Abrasionless machining methods for ceramics (lasers, electron beam, ultrasonics)

FLETCHER, CHARLES H.
PST
Br. of Pacific-Arctic Geology
Menlo Park **(415) 856-7082**
Albion Coll.-BA
Coastal processes
Coastal morphology & sedimentation

FLETCHER, JANET D. **Boyd**
PST
Box 142
Dunn Loring, Virginia 22027
Br. of Analytical Laboratories
Reston **(703) 860-7655**
U. of Maryland; George Washington U.; Northern Va. Comm. Coll.
Spectrographic analysis of coal
Methods for coal analysis

FLORES, ROMEO M. **Marcia**
Geologist
12990 W. 21 Avenue
Golden, Colorado 80401
Br. of Coal Resources
Denver **(303) 234-3530**
U. of the Philippines-BS; U. of Tulsa-MS; Louisiana State U.-PhD
Tertiary, Cretaceous coal, Rocky Mts.
Sedimentology; stratigraphy; and sedimentary petrology

FLOT, TERRY R. **Donna**
PST **(303) 936-8609**
2396 S. Knox Court
Denver, Colorado 80219
Br. of Paleontology & Stratigraphy
Denver **(303) 234-5851**
Metro State Coll. & Colorado U.-BA
Palynology

FLUTY-HARVEY, LUANN
PST **(415) 752-9815**
#1 Edward
San Francisco, California 94118
Br. of Seismology
Menlo Park **(415) 323-8111**
U. of Florida-BS
Seismic patterns & their relation to central California faults

FOGLEMAN, KENT A.
Geophysicist **(415) 968-6312**
151 Calderon Avenue, #8
Mountain View, California 94041
Br. of Ground Motion & Faulting
Menlo Park **(415) 323-8111 x2579**
N. Carolina State U.-BS; U. of California (Berkeley)
Southern Alaska seismic studies
Seismicity of southern Alaska; location of earthquakes from regional nets.

FOLGER, DAVID W. Joan
Geologist
99 Nursery Road
Falmouth, Massachusetts 02540
Br. of Atlantic-Gulf of Mex. Geology
Woods Hole **(617) 548-8700**
Dartmouth-BA; Columbia U.-MA, PhD

FOORD, EUGENE E. Suzann
Geologist **(303) 278-8279**
14193 West 1st Drive
Golden, Colorado 80401
Br. of Central Mineral Resources
Denver **(303) 234-4742**
Stanford U.-PhD
Round Mountain-Manhattan, Nevada
Silicate mineralogy; single crystal x-ray diffraction; economic mineralogy

FOORD, SUZANN C. Eugene
PST
14193 West 1st Drive
Golden, Colorado 80401
Br. of Central Mineral Resources
Denver **(303) 234-2527**
Doane Coll.-BA
Pueblo 2° CUSMAP

FOOSE, MICHAEL P.
Geologist
Br. of Eastern Mineral Resources
Reston **(703) 860-7356**
Princeton U.-PhD
Geology of nickel & cobalt
Economic geology

FOOSE, PEGGY L. Michael
PST
Br. of Coal Resources
Reston **(703) 860-7734**
Colby Coll.-BA; George Mason U.

FOOTE, RICHARD Q. JoLynn
Geophysicist **(512) 852-0629**
4333 St. George
Corpus Christi, Texas 78413
Br. of Atlantic-Gulf of Mex. Geology
Corpus Christi **(512) 888-3294**
Texas A&M U.-BS
Stratigraphy & sedimentology of geopressured zones
Oil & gas resources; exploration geophysics

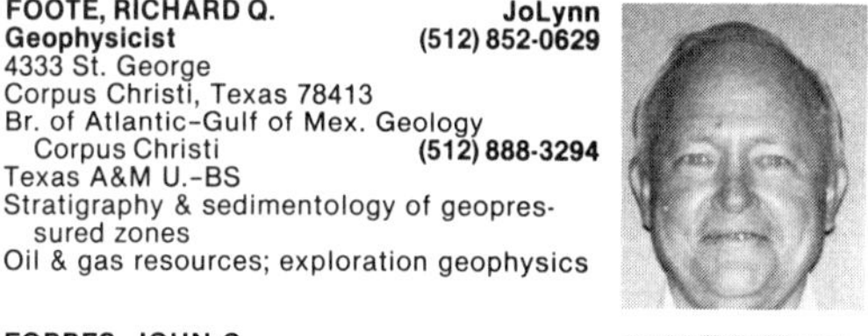

FORBES, JOHN C.
PST
Br. of Field Geochemistry & Petrology
Hawaiian Volcano Observ. **(808) 967-7485**

FORCE, ERIC R. Lucy
Geologist **(703) 435-1254**
471 Madison Street
Herndon, Virginia 22070
Br. of Eastern Mineral Resources
Reston **(703) 860-6913**
Occidental Coll.-BA; Lehigh U.-PhD
Titanium resources of U.S.
Provenance & dispersal of sediments

FORCE, LUCY M. Eric
Geologist **(703) 435-1254**
471 Madison Street
Herndon, Virginia 22070
Br. of Eastern Environmental Geology
Reston **(703) 860-6595**
Occidental Coll.-BA; Lehigh U.-MS, PhD
Charleston, S.C. project
Stratigraphy; sedimentology

FORD, ARTHUR B. Carole
Geologist **(415) 323-3652**
400 Ringwood Avenue
Menlo Park, California 94025
Br. of Alaskan Geology
Menlo Park **(415) 323-8111 x2275**
U. of Washington-BS, MS PhD
Dufek intrusion, Antarctica; Juneau, Alaska; Glacier Peak Wilderness Area, N. Cascades
Regional geology of Antarctica & Alaska; metamorphic & igneous petrology

FORRESTEL, PATRICIA L.
Scientific Illustrator
Br. of Atlantic-Gulf of Mex. Geology
Woods Hole **(617) 548-8700**
U. of Waterloo; U. of Colorado

FOSTER, HELEN L.
Geologist **(415) 323-7164**
270 O'Keefe Street, Apt. H
Palo Alto, California 94303
Br. of Alaskan Geology
Menlo Park **(415) 323-8111 x2331**
U. of Michigan-BS, MS, PhD
Circle, Alaska AMRAP (Yukon-Tanana Upland)
East Central Alaskan geology; geology of western Pacific islands

FOUCH, THOMAS D. Sally Ann
Geologist **(303) 986-7695**
1838 South Yank Place
Lakewood, Colorado 80228
Br. of Oil & Gas Resources
Lakewood **(303) 234-5745**
Portland State U.-BS; U. of Oregon-MS
Tertiary sedimentary basins of the western United States
Lacustrine & related sedimentation; unconventional hydrocarbon reservoir rocks

FOURNIER, ROBERT O.
Geologist
108 Paluma Road
Portola Valley, California 94025
Br. of Exper. Geochem. & Mineralogy
Menlo Park **(415) 323-8111 x2276**
Harvard-BA; U. of California (Berkeley)-PhD
Geochemistry of hydrothermal systems

FOX, D. DIANNE Donald
Secretary **(303) 422-5230**
5643 Garrison Street
Arvada, Colorado 80002
Br. of Oil & Gas Resources
Denver **(303) 234-5235**
U. of Northern Colorado-BS

FOX, JAMES E.
Geologist
Br. of Oil & Gas Resources
Denver **(303) 234-4750**
Gustavus Adolphus-BS; U. of South Dakota-MS; Virginia Polytechnic; U. of Wyoming-PhD
Sedimentology; stratigraphy

FOX, KENNETH F., JR. Shirley
Geologist
Br. of Western Environmental Geology
Menlo Park **(415) 323-8111 x2032**
U. of Idaho-BS; Montana Schl. of Mines-MS; Stanford-PhD
Okanogan 2° quadrangle
Tectonics; igneous & metamorphic petrology

FRANCZYK, KAREN J.
Geologist
1618 Violet Street
Golden, Colorado 80401
Br. of Uranium-Thorium Resources
Denver **(303) 234-5664**
U. of Illinois-BS
Upper Cretaceous sedimentology & uranium deposits-Black Mesa Basin, Arizona

FRAZEUR, W. SCOTT
Administrative Officer
Office of Earthquake Studies
Menlo Park **(415) 323-8111 x2566**
South Dakota State U.-BS
Administration
Management analysis

FREDERICKSEN, NORMAN O. **Elke**
Geologist **(301) 946-5214**
10911 Kenilworth Avenue
Garrett Park, Maryland 20766
Br. of Paleontology & Stratigraphy
Reston **(703) 860-7745**
Hamilton Coll.-AB; Penn. State U.-MS; U. of Wisconsin-PhD
Lower Tertiary palynology, Gulf & Atlantic Coastal Plains
Palynology; biostratigraphy; paleoecology

FREEBERG, JACQUELYN H.
Librarian
Library
Menlo Park **(415) 323-8111**
Tift Coll.-BA; U. of Denver-MLS

FREEMAN, VAL L.
Geologist **(303) 278-9294**
1409 Utah Street, Apt. B
Golden, Colorado 80401
Br. of Coal Resources
Denver **(303) 234-3578**
U. of California (Berkeley)-BS, MS
Western Colorado coal; stratigraphy

FREZON, SHERWOOD E.
Geologist **(303) 985-0831**
410 S. Queen Street
Lakewood, Colorado 80226
Br. of Oil & Gas Resources
Denver **(303) 234-3435**
U. of Michigan-BS, MS; U. of Colorado; Colorado Schl. of Mines
Domestic oil & gas resource appraisal
Oil & gas resource evaluation; stratig. paleo. rocks, N. Ark. & Okla.

FRIDRICH, DOUGLAS K. **Phyllis**
Administrative Officer **(703) 533-7768**
423 Lincoln Avenue
Falls Church, Virginia 22046
Office of Mineral Resources
Reston **(703) 860-6571**
Coll. of William & Mary-AB

FRIEDMAN, IRVING **Rita**
Geochemist
2620 Vivian Street
Lakewood, Colorado 80215
Br. of Isotope Geology
Denver **(303) 234-5531**
Montana State Coll.-BS; State Coll. of Washington-MS; U. of Chicago-PhD
Light stable isotope geochemistry
Volcanology; stable isotope geochemistry; climatology

FRIEDMAN, JULES D. **Kerstin**
Geologist **(303) 988-3719**
2675 S. Yukon Court
Lakewood, Colorado 80227
Br. of Petrophysics & Remote Sens.
Denver **(303) 234-3676**
Cornell-BA; Yale-MS, PhD
Remote sensing for geothermal investigations & radioactive waste disposal sites
Aerial infrared surveys in volcanology; geomorphology & economic geology

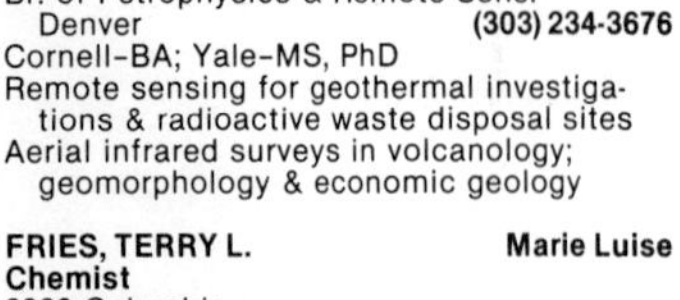

FRIES, TERRY L. **Marie Luise**
Chemist
2326 Columbia
Palo Alto, California 94306
Br. of Analytical Laboratories
Menlo Park **(415) 323-8111 x2949**
California State U. (Fresno)-BS

FRIESEN, WALTER B.
PST **(415) 366-0957**
P.O. Box 879
Menlo Park, California 94025
Br. of Pacific-Arctic Geology
Menlo Park **(415) 856-7128**
Sul Ross State U.-BA
Mineralogy; igneous petrology

FRISCHKNECHT, FRANK C. **Jachio**
Geophysicist
P.O. Box 136
Morrison, Colorado 80465
Br. of Electromag. & Geomagnetism
Denver **(303) 234-2588**
U. of Utah-BS, MS; U. of Colorado-PhD
Development of electromagnetic methods; minerals; geophysics

FRISKEN, JAMES G.
Geologist
Br. of Exploration Research
Denver **(303) 234-6148**
U. of Washington-MS
Wilderness evaluation
Geochemical exploration techniques; arid environments; porphyry copper

FRIZZELL, VIRGIL A., JR.
Geologist
Br. of Western Environmental Geology
Menlo Park **(415) 323-8111 x2987**
San Jose State U.-BS, BA; Stanford-MS, PhD
Wenatchee, Washington 2° sheet
Petrology; structure; fission track dating

FROELICH, ALBERT J. **Tina**
Geologist **(702) 759-9335**
10115 Squires Trail
Great Falls, Virginia 22066
Br. of Eastern-Environmental Geology
Reston **(703) 860-6421**
Ohio State U.-BSc, MSc
Environmental geology & hydrology of Culpepper Basin
Environmental geology; energy resources; foreign geology

FUGATE, JAMES K. **Gail Goo**
Mathematician **(512) 992-3854**
1914 Tara
Corpus Christi, Texas 78412
Br. of Atlantic-Gulf of Mex. Geology
Corpus Christi **(512) 888-3294**
U. of Hawaii-MA; Texas Christian U.-PhD
Geotechnical research
Simulation sedimentary processes; statistical applications to geology

FUIS, GARY S. **Stacey**
Geophysicist **(415) 856-1831**
745 La Para Avenue
Palo Alto, California 94306
Br. of Seismology
Menlo Park **(415) 323-8111 x2569**
Cornell U.-BA; Cal Tech-PhD
Southern Calif. coop. seismic network; Imperial Valley geothermal studies
Crustal structure; fault mechanics; earthquake prediction

FULLER, H. KIT **Wanda**
Geographer
2500 Juniper Avenue, Apt. C
Boulder, Colorado 80302
Office of Scientific Publications
Denver **(303) 234-3492**
Middlebury Coll.-BA
Map editor
Computerized Kern PGZ stereoplotting system; factors affecting land use

FULLERTON, DAVID S. **Elia**
Geologist **(303) 278-3549**
17216 W. 16th Avenue
Golden, Colorado 80401
Br. of Central Environmental Geology
Denver **(303) 234-4008**
Waynesburg Coll.-BS; Yale U.-MS; Princeton-AM, PhD
Quaternary stratigraphy; geomorphology

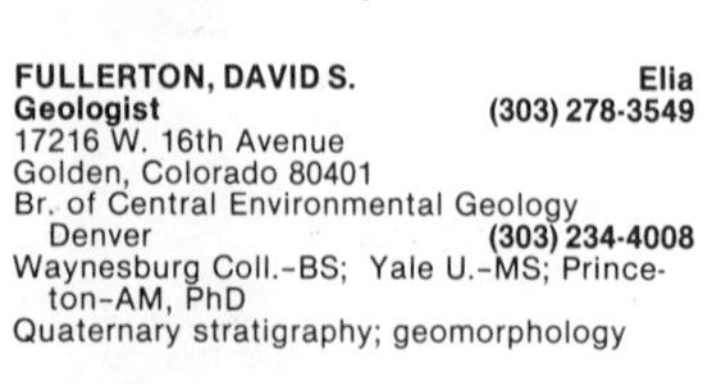

FULTON, PATRICIA A.
Mathematician
Office of Resource Analysis
Reston **(703) 860-6451**
U. of Pittsburgh-BS
Computer applications

FUTA, KIYOTO
Chemist
P.O. Box 15543
Lakewood, Colorado 80215
Br. of Isotope Geology
Denver **(303) 234-3876**
U. of Wyoming-BS, MS
Rb-Sr & Nd-Sm geochronology & trace element geochemistry

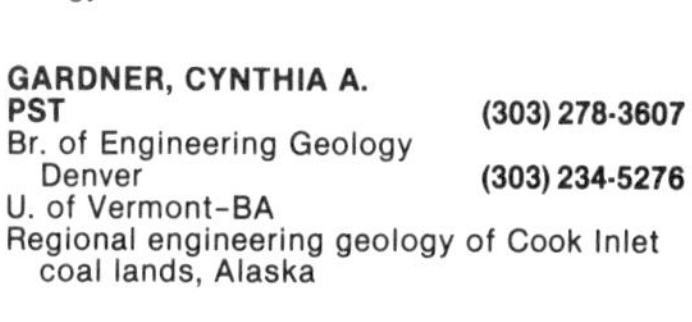

GABLE, DOLORES J. **James**
Geologist **(303) 424-2994**
7330 Iris Street
Arvada, Colorado
Br. of Central Environmental Geology
Denver **(303) 234-2825**
U. of Buffalo-BA; U. of Colorado
Central Wyoming Precambrian mapping
Precambrian geology; igneous & metamorphic petrology

GAIR, JACOB E. **Peggy Lou**
Geologist **(301) 949-7461**
Br. of Eastern Mineral Resources
Reston **(703) 860-6913**
U. of Rochester-BA; Johns Hopkins-PhD
Appalachian massive sulfides
Stratabound sulfide deposits; Precambrian iron deposits

GALANIS, S. PETER, JR.
Geologist **(415) 967-1343**
1794 San Luis Avenue
Mountain View, California 94043
Br. of Tectonophysics
Menlo Park **(415) 323-8111 x2840**
San Jose State U.-BA
Heat-flow studies
Petrology & structural geology

GALL, R. MICHAEL **Donna**
Computer Systems Admin. **(703) 860-9336**
2021 Golf Course Drive
Reston, Virginia 22091
Office of the Chief Geologist
Reston **(703) 860-6711**
George Washington U.-BCE; Georgia Inst. of Technology-MS

GALLANTHINE, STEVEN K.
PST
128 Promethean
Mountain View, California 94043
Br. of Tectonophysics
Menlo Park **(415) 323-8111 x2155**
California State U. (Hayward)-BS
Seismic refraction

GALLOWAY, JOHN PAUL
Archeaologist
Br. of Alaskan Geology
Menlo Park **(415) 323-8111 x2131**
San Francisco State U.-BA, MA
Western Arctic coastal plain Quaternary
Geoarchaeology; geomorphology

GANDY, GERALD D.
Engineering Technician
Br. of Analytical Laboratories
Reston **(703) 860-7442**
Instrument development

GARD, LEONARD M., JR.
Geologist **(303) 237-1533**
177 S. Upham Court
Lakewood, Colorado 80226
Br. of Special Projects
Denver **(303) 234-2261**
Colorado Coll.-BA; U. of Colorado-MS
Radioactive waste isolation, S.E. New Mexico
Engineering geology; Aleutian Islands geology

GARDNER, CYNTHIA A.
PST **(303) 278-3607**
Br. of Engineering Geology
Denver **(303) 234-5276**
U. of Vermont-BA
Regional engineering geology of Cook Inlet coal lands, Alaska

GARLOW, RICHARD A.
PST
626 Sheraton Drive
Sunnyvale, California 94087
Br. of Pacific-Arctic Geology
Menlo Park **(415) 323-8111 x2762**
San Jose State U.-BA
Navigation

GARRISON, LOUIS E. **Bette**
Geologist **(512) 851-0200**
101 Lake Shore
Corpus Christi, Texas 78413
Br. of Atlantic-Gulf of Mex. Geology
Corpus Christi **(512) 888-3294**
Texas Coll. Mines-BA; U. of California-MS; U. of Rhode Island-PhD
Mississippi Delta project
Continental shelf & slope sedimentation

GARTNER, ANNE E. **Jeffrey**
Geologist **(408) 296-4638**
3653 Williams Road
San Jose, California 95117
Br. of Exper. Geochem. & Mineralogy
Menlo Park **(415) 323-8111 x2063**
U. of South Florida-BA
Geologic simulation applied to radioactive waste disposal

GASSAWAY, JUDITH S.
Geologist **(602) 774-7436**
Route 4, Box 901
Flagstaff, Arizona 86001
Br. of Petrophysics & Remote Sens.
Flagstaff **(602) 774-7436**
San Diego State U.-BA, MS
Large mines of the world; data compilation
Sedimentation & tectonics; Basin and Range province

GAUTIER, DONALD L.
Geologist **(303) 278-1626**
501 9th Street
Golden, Colorado 80401
Br. of Oil & Gas Resources
Denver **(303) 234-4606**
U. of Colorado-BA, PhD
Western gas sands
Sedimentology

GAWARECKI, STEPHEN J. **Carolyn**
Geologist **(703) 532-5278**
7018 Vagabond Drive
Falls Church, Virginia 22042
Br. of Middle Eastern & Asian Geology
Reston **(703) 860-6555**
Rutgers-BS, MS; U. of Colorado-PhD
Remote sensing; photogeology; regional tectonics

GELINAS, JANET L.
Administrative Clerk
544 W. Falmouth
W. Falmouth, Massachusetts 02574
Br. of Atlantic-Gulf of Mex. Geology
Woods Hole **(617) 548-8700**
Bristol Coll.

GENT, CAROL A. **Sidney**
PST **(303) 985-7570**
332 South Moore Street
Lakewood, Colorado 80226
Br. of Analytical Laboratories
Denver **(303) 234-6401**
Alfred U.; U. of Colorado
Analytical services & research

GIBBONS, ANTHONY B. **Frances**
Geologist **(303) 238-1068**
12420 W. 35th Avenue
Wheat Ridge, Colorado 80033
Br. of Central Environmental Geology
Denver **(303) 234-5967**
Seattle U.-BA; U. of Washington-BS; U. of California
Hams Fork (Wyoming) environmental geology
Geologic mapping; surficial deposits

GIBBS, JAMES F. **JoAnn**
Geophysicist
6892 Chiala Lane
San Jose, California 95129
Br. of Ground Motion & Faulting
Menlo Park **(415) 323-2030**
U. of N. Dakota-BS; U. of Colorado-MS
Shear wave velocity measurements

GIBBS, JoANN **James**
Secretary
6892 Chiala Lane
San Jose, California 95129
Br. of Paleontology & Stratigraphy
Menlo Park **(415) 323-8111 x2261**
Minnesota Schl. of Business

GIBSON, THOMAS G. **Carol**
Geologist **(703) 430-7656**
744 Kentland Drive
Great Falls, Virginia
Br. of Paleontology & Stratigraphy
Washington, D.C. **(202) 343-3525**
U. of Wisconsin-BS, MS; Princeton-PhD
Paleontologic & stratigraphic synthesis of Paleogene & Neogene of eastern Gulf coast & Atlantic coast
Cenozoic forams & mollusks

GING, TOM G. **Anita**
Chemist **(303) 234-3342**
Br. of Oil & Gas Resources
Denver **(303) 234-3342**
New Mexico Highlands U.-MS
Organic geochemical laboratory
Analytical methods development; high performance liquid chromatography

GIRARD, OSWALD W., JR.
Geologist
Office of Energy Resources
Reston **(703) 860-6432**
U. of Florida
Petroleum geology

GLAISTER, CAROLE S.
Clerk-Typist
Office of the Chief Geologist
Reston **(703) 860-6544**

GLAMMEYER, ROY A. **Glenda**
Library Technician **(703) 822-5951**
Route 1, Box 225
Lovettsville, Virginia 22080
Library
Reston **(703) 860-6617**
American U.-BA; Washington Coll. of Law; Dept. of Agriculture Graduate Schl.

GLANZMAN, VIRGINIA M. **Dick**
Technical Publications Editor
Br. of Special Projects
Denver **(303) 234-2261**

GLICK, ERNEST E. **Carolyn**
Geologist **(303) 238-4812**
2263 Yellowstone Street
Golden, Colorado 80401
Br. of Central Environmental Geology
Denver **(303) 234-3353**
U. of Illinois; U. of Chicago; U. of Southern Cal.-BA
Structure of northeastern Arkansas
Structural geology; stratigraphy; petroleum geology

GODSON, RICHARD H. **Nan**
Supervisory Physical Scientist **(303) 988-8057**
12831 W. Florida Drive
Lakewood, Colorado 80228
Br. of Regional Geophysics
Denver **(303) 234-2623**
Notre Dame-BS
Computer programming; geophysical data processing; aeromagnetics

GOERLITZ, PATRICIA A. **Don**
Administrative Technician
Office of Geochemistry & Geophysics
Menlo Park **(415) 323-2626**

GOHN, GREGORY S.
Geologist
Br. of Eastern Environmental Geology
Reston **(703) 860-6421**
Juniata Coll.-BS; U. of Delaware-MS, PhD
Charleston project, geologic investigations
Atlantic Coastal Plain stratigraphy & tectonics

GOLDBERG, JERALD M.
Geologist **(202) 966-4589**
2844 Wisconsin Avenue, N.W.
Washington, D.C. 20007
Office of the Chief Geologist
Reston **(703) 860-6533**
Southern Methodist U.-BS, MS
Scientific management

GOLDHABER, MARTIN B.
Chemist **(303) 234-9665**
3470 Ward Road
Wheat Ridge, Colorado 80033
Br. of Uranium-Thorium Resources
Denver **(303) 234-2783**
UCLA-BS, PhD
Inorganic diagenesis of sedimentary uranium deposits
Sedimentary geochemistry; stable isotope geochemistry

GOLDSMITH, JUNE W. **Richard**
Geologist
105 E. Lenox Street
Chevy Chase, Maryland 20015
Office of Scientific Publications
Reston **(703) 860-6493**
Radcliffe-BA
Technical reports editor

GOLDSMITH, RICHARD **June**
Geologist **(301) 654-7926**
105 E. Lenox Street
Chevy Chase, Maryland 20015
Br. of Eastern Environmental Geology
Reston **(703) 860-6406**
U. of Maine-BA; U. of Washington-PhD
Charlotte 2° sheet
Metamorphic & igneous petrology; glacial geology

GOLIGHTLY, DANOLD W. **Marilyn**
Chemist **(703) 860-2062**
11960 Heathcote Court
Reston, Virginia 22091
Br. of Analytical Laboratories
Reston **(703) 860-7654**
Iowa State U.-PhD
Optical spectroscopy project
Analytical atomic spectroscopy

GONSALVES, JANET S. **John**
Administrative Technician **(303) 279-5093**
194 S. Holman Way
Golden, Colorado 80401
Office of Scientific Publications
Denver **(303) 234-3229**
Midland College

GOOD, ELIZABETH E. **Henry J. Noonan**
Geologist **(703) 860-5940**
11910 Winterthur Lane
Reston, Virginia 22091
Office of Scientific Publications
Reston **(703) 860-6494**
Smith Coll.-BA
Technical reports editor

GOODFELLOW, ROBERT W.
GFA
346-B 3rd Avenue
San Francisco, California 94118
Br. of Pacific-Arctic Geology
Menlo Park **(415) 856-7128**
Sonoma State U.-BS
Metallic deposits in ocean crust
Ore deposits; geothermal exploration; engineering geology

GOODMAN, MINNIE H. **Joseph**
Mathematician
12128 Basset Lane
Reston, Virginia 22091
Br. of Regional Geophysics
Reston **(703) 860-6507**
West Virginia State College-BS
Magnetic & gravity data

GOODWIN, GEORGE H., JR.
Librarian
Library
Reston **(703) 860-6619**
Syracuse U.-BA, MLS

GORDON, DAVID W.
Geophysicist
Br. of Global Seismology
Denver **(303) 234-4041**
Ohio State-MS, St. Louis U.
Earthquake locations in eastern U.S.

GORDON, MACKENZIE, JR. **Barbara**
Geologist **(202) 338-0176**
2905 Que Street, N.W.
Washington, D.C. 20007
Br. of Paleontology & Stratigraphy
Washington, D.C. **(202) 343-2045**
Stanford-AB
Biostratigraphic framework; coal basins of eastern U.S. & eastern Great Basin
Biostratigraphy; historical geology; paleontology (brachiopods & mollusks)

GOTTFRIED, DAVID
Geologist
5201 Augusta Street
Bethesda, Maryland 20016
Br. of Field Geochemistry & Petrology
Reston **(703) 860-7401**
Brooklyn Coll.-BS; New York U.; American U.-MS
Geochemistry of mafic igneous rocks
Geochemistry; petrology

GOUDARZI, GUS H. **Olga**
Geologist **(703) 471-4236**
658 Pemberton Court
Herndon, Virginia 22070
Office of Mineral Resources
Reston **(703) 860-6567**
Montana Coll. of Mineral Science & Technology-BS, MS
Wilderness coordinator
Economic geology

GOUGH, LARRY P. **Jane**
Botanist **(303) 423-0399**
5247 Bristol Street
Arvada, Colorado 80002
Br. of Regional Geochemistry
Denver **(303) 234-5241**
Carroll Coll.-BS; U. of Louisville-MS; U. of Colorado-PhD
Element availability-plants; geochem. Alaska
Biogeochemistry; plant ecology; application of statistical methods

GRAHAM, MARJORIE A. **David**
Secretary **(206) 364-8796**
11031 Second Avenue, N.W.
Seattle, Washington 98177
Br. of Pacific-Arctic Geology
Seattle **(206) 442-1995**
Seattle Community College

GRAMM, ALICE E. **Mel**
Administrative Technician
Office of Geochemistry & Geophysics
Denver **(303) 234-5461**

GRANGER, HARRY C. **Janet**
Geologist **(303) 935-8745**
2749 S. Patton Court
Denver, Colorado 80236
Br. of Uranium-Thorium Resources
Denver **(303) 935-8745**
U. of Oregon-BS; Queens U.
Uranium ore-forming processes
Uranium; ore deposition; geochemistry

GRANTZ, ARTHUR **Willene**
Geologist
Br. of Alaskan Geology
Menlo Park **(415) 323-8111**
Cornell U.-AB; Stanford-PhD
Geologic framework of Chukchi & Beaufort shelves, Alaska

GRAUCH, RICHARD I.
Geologist
955 S. Linda Lane
Evergreen, Colorado 80439
Br. of Uranium-Thorium Resources
Denver **(303) 234-5309**
Franklin & Marshall-BA; U. of Pennsylvania-PhD
Uranium in eastern U.S., Colorado
Economic geology; metamorphic petrology; northern Andean geology

GRAY, HELEN M.
Secretary
Office of International Geology
Reston **(703) 860-6551**

GRAY, KAREN J.
PST **(703) 860-0509**
11580 Woodhollow Court
Reston, Virginia 22091
Br. of Field Geochemistry & Petrology
Reston **(703) 860-7452**
SUNY (Albany)

GREEN, ARTHUR W., JR. Amy
Geophysicist
410 Woods Hole Road
Woods Hole, Massachusetts 02543
Br. of Atlantic-Gulf of Mex. Geology
Woods Hole (617) 548-6367
U. of Houston-BS, MS; Texas Christian U.-MS, PhD
Geophysics; geomagnetism

GREEN, MARSHA R. Ronald
Secretary
Br. of Alaskan Geology
Anchorage (907) 271-4150

GREEN, MORRIS W. Janis
Supervisory Geologist (303) 697-6177
8293 Surrey Drive
Morrison, Colorado 80465
Br. of Uranium-Thorium Resources
Denver (303) 234-3697
Texas Tech.-BS
Uranium in Jurassic rocks, San Juan basin, New Mexico
Stratigraphy; sedimentology; geologic mapping of uranium rocks

GREEN, RICHARD G. Barbara
Geophysicist (303) 776-4863
401 Grant Street
Longmount, Colorado 80501
Br. of Electromag. & Geomagnetism
Denver (303) 234-5505
U. of Vermont-BA
Magnetic observatory-field data processing & analysis
State isogonic charts

GREEN, WILLIAM N. Margaret
Geophysicist (505) 298-7273
509 Monte Alto Place, N.E.
Albuquerque, New Mexico 87123
Br. of Global Seismology
Albuquerque (505) 844-4637
St. Louis U.-BS

GREENE, GORDON W. Phyllis
Geophysicist (415) 494-2516
3790 Corina Way
Palo Alto, California 94303
Office of Earthquake Studies
Menlo Park (415) 323-8111 x2764
California State U. (Fresno)-AB; Stanford-MS
External research program
Earthquake hazard & loss

GREENE, ROBERT C. Edith
Geologist Jiddah 50119
USGS, American Embassy
APO New York, 09697
Br. of Latin Am. & African Geology
Jiddah, Saudi Arabia 691148, x447
Cornell U.-BA; U. of Tennessee-MS; Harvard-PhD
Quadrangle mapping
Stratigraphy, structure & petrology of Precambrian rocks

GREENE, ROXANN E. Larry
Clerk-Typist
1015 W. 23rd
Anchorage, Alaska 99503
Br. of Alaskan Geology
Anchorage (907) 271-4150

GREENHAUS, MICHAEL R.
Geophysicist
Box 3033
Eldorado Springs, Colorado 80025
Br. of Electromag. & Geomagnetism
Denver (303) 234-6591
Southampton Coll.-BS; Stanford-MS
Nuclear waste isolation
Paleomagnetism; gravity & magnetic interpretation; electrical methods

GREENLAND, L. PAUL Rima
Geochemist (808) 967-7668
Box 160
Volcano, Hawaii 96785
Br. of Field Geochem. & Petrology
Hawaiian Volcano Observ. (808) 967-7485
Australian National U.-PhD
Volcanic gases
Geochemistry; analytical chemistry

GRIFFIN, ELIZABETH A.
PST
Br. of Western Environmental Geology
Menlo Park (415) 323-8111
U. of California (Berkeley)-BA
Structural geology; igneous petrology

GRIFFITH, JEAN K.
PST
Br. of Coal Resources
Denver (303) 234-3530
U. of Colorado-BA
1:100,000 geologic map of Grants, New Mexico, coal folio program

GRIFFITTS, WALLACE R. Mary
Geologist (303) 442-3014
810 14th Street
Boulder, Colorado 80302
Br. of Exploration Research
Denver (303) 234-6162
U. of Michigan-BS, MS, PhD
Geochemical exploration; geology of beryllium deposits

GRIGGS, M. LOUISE LeRoy
Secretary (303) 422-7371
5199 Arbutus Street
Arvada, Colorado 80002
Office of the Chief Geologist
Denver (303) 234-2910
UCLA; Metropolitan State Coll.

GRIM, MURIEL S. Paul
Geologist (303) 442-5575
5164 Gallatin Place
Boulder, Colorado 80303
Br. of Regional Geophysics
Denver (303) 234-5631
Hunter Coll.-BA; Columbia-MA
Marine geology (continental margins)

GRIMES, DAVID J. Emily
Chemist
2610 Vivian Street
Lakewood, Colorado 80215
Br. of Exploration Research
Denver (303) 234-6152
U. of New Hampshire-BS; U. of Colorado
Mineral evaluation of CUSMAP & wilderness areas
Emission spectroscopy; exploration geochemistry; geobotany

GRISCOM, ANDREW
Geophysicist
1106 N. Lemon Avenue
Menlo Park, California 94025
Br. of Regional Geophysics
Menlo Park (415) 323-8111
Harvard-PhD
Interpretation of aeromagnetic data

GROLIER, MAURICE J.
Geologist (602) 774-6771
1909 N. Turquoise Drive
Flagstaff, Arizona 86001
Br. of Astrogeologic Studies
Flagstaff (602) 779-3311 x1455
Johns Hopkins-PhD
Desert geomorphology; wind processes
Desert geomorphology; climate change; field archeology

GROMMÉ, SHERMAN
Geologist
Br. of Petrophysics & Remote Sens.
Menlo Park **(415) 323-8111 x2357**
U. of California (Berkeley)-PhD
Microplate tectonics of western U.S.
Paleomagnetism; rock magnetism

GROSZ, ANDREW E. **Bonny**
Geologist
11946 Escalante Court
Reston, Virginia 22091
Br. of Eastern Mineral Resources
Reston **(703) 860-7356**
CUNY-BS
Southeastern Appalachian sediments
Mineral resource evaluation; radiometric exploration; Coastal Plain studies

GROW, JOHN A. **Alice**
Geophysicist **(617) 540-2968**
55 Meredith Drive
Falmouth, Massachusetts 02536
Br. of Atlantic-Gulf of Mex. Geology
Woods Hole **(617) 548-8700**
U. of Cincinnati-BS; Columbia U.-MS; U. of California (San Diego)-PhD
Resource assessment for Mid-Atlantic continental margin
Multichannel seismic reflection; gravity

GRUBB, FREDD V.
PST
22272 Cupertino Road
Cupertino, California 95014
Br. of Tectonophysics
Menlo Park **(415) 323-8111 x2840**
Sonoma State U.-BS

GRUNDY, WILBER D.
Geologist
Office of Resource Analysis
Denver **(303) 234-6281**
U. of Notre Dame-BS; U. of Arizona-MS
Mineral resource appraisal—CUSMAP
Geology of uranium & coal deposits; geostatistics

GRYBECK, DONALD J. **Ellen**
Geologist **(703) 476-4511**
11749 Indian Ridge Road
Reston, Virginia 22092
Office of Mineral Resources
Reston **(703) 928-6566**
U. of Alaska-BS; Colorado Schl. of Mines-DSc
Alaska programs
Economic geology; geology of Alaska; ore mineralogy

GUALTIERI, JAMES L.
Geologist
Br. of Coal Resources
Denver **(303) 234-5672**
U. of Washington-BS, MS
Coal geology of the Book Cliffs coal field
Mineral deposits; areal geology; sedimentary stratigraphy

GUDE, ARTHUR J., III **Bertha**
Geologist **(303) 237-7560**
845 Dudley Street
Lakewood, Colorado 80215
Br. of Sedimentary Mineral Resources
Denver **(303) 234-2991**
Colorado Schl. of Mines-MS
Zeolites
X-ray mineralogy & geology of zeolites

GUFFANTI, MARIANNE C.
PST
Br. of Field Geochemistry & Petrology
Menlo Park **(415) 323-8111 x2348**
Humboldt State U.-BA; San Jose State U.
Geothermal resources

GUILD, PHILIP W. **Terry**
Geologist **(301) 656-7829**
3609 Raymond Street
Chevy Chase, Maryland 20015
Office of Resource Analysis
Reston **(703) 860-6605**
Johns Hopkins-AB, PhD
Metallogenic maps
Mineral deposits

GULBRANDSEN, ROBERT A.
Geologist
Br. of Sedimentary Mineral Resources
Menlo Park **(415) 323-8111 x2313**
Stanford-PhD
Mineralogy & geochemistry of Phosphoria Fm.
Geochemistry; mineralogy; sedimentary petrology

GUNN, MARI L.
PST
Br. of Seismology
Menlo Park **(415) 323-8111**
U. of California (Santa Cruz)-BA

GUTMACHER, CHRISTINA E.
PST
Br. of Pacific-Arctic Geology
Menlo Park **(415) 856-7041**
U. of California (Santa Cruz)-BS
Deep sea fans

GWYN, MARY E. **Roger**
Sample Control Officer **(301) 552-3644**
6600 Cipriano Road
Lanham, Maryland 20801
Br. of Analytical Laboratories
Reston **(703) 860-6688**

HACKMAN, BETTIE S. **Robert**
Geologist **(703) 281-9728**
2714 Hunter Mill Road
Oakton, Virginia 22124
Br. of Coal Resources
Reston **(703) 860-6684**
U. of Cincinnati
Coal resources

HACKMAN, ROBERT J. **Bettie**
Geologist **(703) 281-9728**
2714 Hunter Mill Road
Oakton, Virginia 22124
Br. of Eastern Environmental Geology
Reston **(703) 860-6406**
Stanford U.; American U.-BS
Photogeology

HADLEY, DONALD G. **Clair**
Geologist **966-21-54435**
USGS American Embassy
APO New York, N.Y. 09697
Br. of Latin Am. & African Geology
Jiddah, Saudi Arabia **674188**
Eastern N.M. U.-BS; U. of Wisconsin-MS; Johns Hopkins-PhD
Sedimentology; stratigraphy; regional Precambrian geology

HAFFTY, JOSEPH **Joyce**
Chemist **(303) 238-1077**
45 South Carr Street
Lakewood, Colorado 80226
Br. of Analytical Laboratories
Denver **(303) 234-6406 (6407)**
U. of Mass.-BS; American U.-BA
Analysis of Pd, Pt, Rh, Ir, Ru, As, Bi, Se, Sb, and Te
Optical emission and atomic absorption spectroscopy analysis

HAGSTRUM, JONATHAN T.
Geophysicist **(303) 278-0867**
918 9th Street
Golden, Colorado 80401
Br. of Petrophysics & Remote Sens.
Denver **(303) 234-5387**
Cornell U.-BA; U. of Michigan-MS
Paleomagnetism; rock magnetism; tectonics

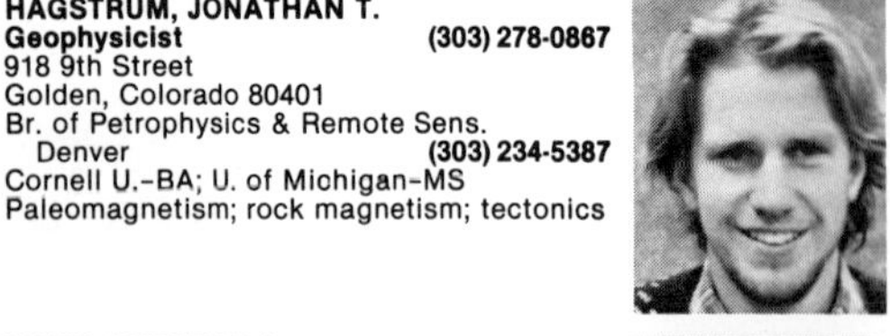

HAHN, DEBORAH A.
Geological Technician
Br. of Oil & Gas Resources
Flagstaff **(602) 779-3311 x1464**
Glendale Community Coll.-AA; Northern Arizona U.-BS
Aeromagnetics in northern Alaska
Field mapping; geophysical & remote sensing research; drafting

HAIL, WILLIAM J. **Dorothy**
Geologist
3083 S. Ivan Way
Denver, Colorado 80277
Br. of Sedimentary Mineral Resources
Denver **(303) 234-2787**
Wooster Coll.-BA; Washington U.-MA
Central Roan Cliffs, Colorado
Tertiary & Cretaceous stratigraphy

HAIT, MORTIMER H., JR. **Brenda**
Geologist
Pine, Colorado 80470
Br. of Engineering Geology
Denver **(303) 234-3624**
Lafayette Coll.-AB; Penn. State U.-PhD
Reactor site investigations
Cenozoic structure; stratigraphy

HALEY, BOYD R. **Ruth**
Geologist **(501) 225-4362**
7 Warwick Road
Little Rock, Arkansas 72205
Br. of Oil & Gas Resources
Little Rock **(501) 371-1616**
Princeton U.-BS
Tight gas sands of Pennsylvanian age in Arkansas
Lithology; structure; fossil fuels

HALL, PHILIP C.
PST
Br. of Seismology
Menlo Park **(415) 323-8111 x2632**
San Jose State U.-BA
CALNET

HALL, RAYMOND E.
Geologist **(617) 548-8313**
80 Carl Landi Circle
Waquoit, Massachusetts 02536
Br. of Atlantic-Gulf of Mex. Geology
Woods Hole **(617) 548-8700**
St. Joseph's Coll.-BS; Brown U.-MS
Water mass & sea level fluctuations, western Atlantic
Water mass & sea level fluctuations; plantik foraminifers

HALL, ROBERT B. **Anna**
Geologist
Br. of Central Mineral Resources
Denver **(303) 234-5171**
U. of Arkansas-BS; Northwestern U.-MS
Economic geology; mineral resource investigations

HALL, VIRGINIA M. **Mark**
Travel Clerk
Br. of Astrogeologic Studies
Flagstaff **(602) 779-3311 x1382**

HALL, WAYNE E. **Dorothy**
Geologist
1028 Wilmington Way
Redwood City, California 94062
Br. of Field Geochemistry & Petrology
Menlo Park **(415) 323-8111 x2501**
U. of Washington-BS; Harvard-MA, PhD
Genesis of base metal deposits
Genesis of ore deposits; structure; mineral resource evaluation

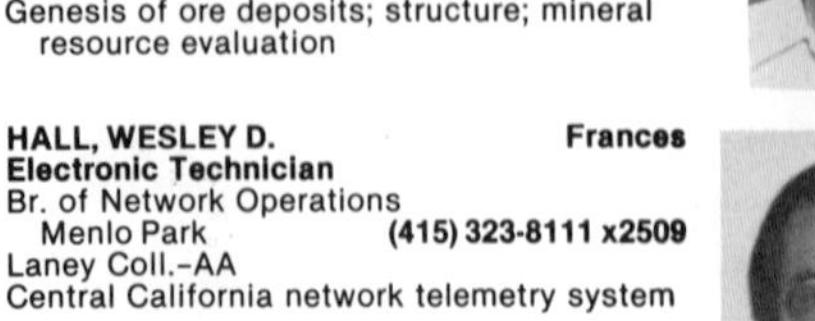

HALL, WESLEY D. **Frances**
Electronic Technician
Br. of Network Operations
Menlo Park **(415) 323-8111 x2509**
Laney Coll.-AA
Central California network telemetry system

HALLEY, ROBERT B.
Geologist
Br. of Oil & Gas Resources
Denver **(303) 234-3624**
Oberlin Coll.-AB; Brown U.-MS; SUNY (Stony Brook)-PhD
Carbonate hydrocarbon reservoirs
Carbonate sediments & diagenesis

HAMBLIN, IRENE F.
Editorial Assistant **(303) 233-6787**
2886 Gray Street
Denver, Colorado 80214
Br. of Special Projects
Denver **(303) 234-2391**
Fort Collins Commercial Coll.

HAMILTON, JOHN C. **Evelyn**
Chemist **(303) 934-4878**
3405 S. Raleigh Street
Denver, Colorado 80236
Br. of Analytical Laboratories
Denver **(303) 234-6406**
Nebraska State Teachers Coll.-BS; U. of Colorado
Spectrographic services & research

HAMILTON, ROBERT M. **Mary Hudson**
Geophysicist **(703) 860-9370**
2020 Mock Orange Court
Reston, Virginia 22091
Br. of Earthquake Tectonics & Risk
Reston **(703) 860-6529**
U. of California (Berkeley)-PhD; Colorado Schl. of Mines-GE
Eastern U.S. seismotectonics
Seismicity; seismic reflection profiling; seismotectonics

HAMILTON, THOMAS D. **Lisa**
Geologist **(415) 494-6457**
777 Christine Drive
Palo Alto, California 94303
Br. of Alaskan Geology
Menlo Park **(415) 323-8111 x2156**
U. of Idaho-BS; U. of Wisconsin-MS; U. of Washington-PhD
Surficial geology of the central Brooks Range
Quaternary geology; Arctic & alpine geomorphology; environmental geology

HAMILTON, WARREN B. **Alicita**
Geologist **(303) 233-6091**
425 Garland
Lakewood, Colorado 80226
Br. of Regional Geophysics
Denver **(303) 234-3247**
UCLA-BA, PhD; U. of Southern California-MS
Tectonics

HAMMOND, DAVID J.
Geologist
6144 Pierson Street
Arvada, Colorado 80004
Br. of Uranium-Thorium Resources
Denver **(303) 234-5611**
Hartwick Coll.-BA
Uranium geochemical sampling; x-ray fluorescence

HAMPSON, JOHN C., JR.
Geologist
P.O. Box 307
Falmouth, Massachusetts 02540
Br. of Atlantic-Gulf of Mex. Geology
Woods Hole **(617) 548-8700**
Princeton U.-BA; U. of Rhode Island
Environmental assessment; Middle Atlantic U.S. continental slope
Seismic interpretation and mapping

HANLEY, JOHN H. **Mary Lou**
Geologist **(303) 232-1029**
1995 Nelson Street
Lakewood, Colorado 80215
Br. of Paleontology & Stratigraphy
Denver **(303) 234-5861**
Iowa State U.; U. of Wyoming-BA, PhD
Taxonomy, biostratigraphy, evolution, & paleoecology of Mesozoic & Cenozoic non marine Mollusca; stratigraphy, sedimentology of nonmarine sediments

HANLEY, J. THOMAS
Geologist
7228 Timber Lane
Falls Church, Virginia 22046
Office of Resource Analysis
Reston **(703) 860-6451**
Virginia Tech-BS; Syracuse U.-MS
Statistical analysis of mineral assessment data
Economic geology, geomathematics; sedimentary petrology

HANNA, WILLIAM F. **Vivian**
Geophysicist **(303) 278-8572**
14351 W. 2nd Place
Golden, Colorado 80401
Br. of Regional Geophysics
Denver **(303) 234-2623**
Indiana U.-BS, MA, PhD
Mineral resources, earthquake hazards, Montana, California
Magnetism; gravity; mineralogy

HANSEN, DAN E. **Carolyn**
Geologist **(303) 988-3495**
Br. of Coal Resources
Denver **(303) 234-3536**
U. of North Dakota-BS, MS
Coal resources, Hanna Basin
Stratigraphy; structure; environmental geology

HANSEN, ELLEN E. **Wallace**
Cartographic Technician
Office of Scientific Publications
Denver **(303) 234-3624**

HANSEN, WALLACE R. **Ellen**
Geologist **(303) 233-4250**
70 South Everett Street
Lakewood, Colorado 80226
Br. of Central Environmental Geology
Denver **(303) 234-3495**
U. of Utah-BS
Wasatch-Uinta tectonics; Colorado Front Range urban corridor
Areal, engineering, & environmental geology

HANSHAW, PENELOPE M.
Geologist
Office of the Chief Geologist
Reston **(703) 860-7429**
Wellesley-BA
Triassic & Quaternary geology of New England

HARDEN, JENNIFER W.
Soil Scientist
Br. of Western Environmental Geology
Menlo Park **(415) 323-8111 x2318**
U. of California (Berkeley)-MS
Soil correlation & dating
Soil genesis; soil morphology; soil chemistry

HARDIE, JOHN K.
Geologist **(303) 922-2276**
Br. of Coal Resources
Denver **(303) 234-8533**
Coal geology

HARDING, SAMUEL T.
Geophysicist
Br. of Earthquake Tectonics & Risk
Denver **(303) 234-5090**
U. of North Carolina-BS
Numerical methods
Numerical methods

HARMON, FORREST L.
Engineering Technician
472 Dogwood Road
Winchester, Virginia 22601
Br. of Analytical Laboratories
Reston **(703) 860-7442**

HARMS, THELMA F.
Chemist
919 Third Street
Golden, Colorado 80401
Br. of Regional Geochemistry
Denver **(303) 234-3453**
Southwestern State Coll.-BS
Trace element analysis in vegetation
Methods of trace element analysis

HARMSEN, STEPHEN C.
Mathematician **(303) 279-4471**
1130 10th
Golden, Colorado 80401
Br. of Ground Motion & Faulting
Denver **(303) 234-3624**
U. of Washington-BA; Central Washington U.-MS
Modeling sedimentary basins
Finite difference wave modeling

HARNER, JOY L.
GFA **(415) 325-0231**
16 Waverly Court, #1
Menlo Park, California 94025
Br. of Western Mineral Resources
Menlo Park **(415) 323-8111 x2251**
Chaffey Community Coll.-AA; U. of California (Santa Barbara)-BA
CRIB, RARE II wilderness project
Walker Lake 2° sheet

HARP, EDWIN L.
Geologist
Br. of Engineering Geology
Menlo Park **(415) 856-7124**
Montana State U.-MS; U. of Utah-PhD
Earthquake-induced landslides
Structural geology; engineering geology; soil mechanics

HARPER, BILLIE J. **Leonard**
Budget Clerk **(602) 526-4291**
5250 N. Highway 89, #2
Flagstaff, Arizona 86001
Br. of Astrogeologic Studies
Flagstaff **(602) 779-3311 x1362**

HARPER, KENNETH R. **Marion E.**
Engineering Technician **(415) 257-7749**
21905 Hyannisport Drive
Cupertino, California 95014
Br. of Tectonophysics
Menlo Park **(415) 323-8111 x2586**

HARRACH, GEORGE H. **Magdalina**
Geologist
294 Short Place
Louisville, Colorado 80027
Br. of Regional Geochemistry
Denver **(303) 234-6566**
U. of Colorado-BA
Computer Science

HARRIS, ANITA G. **Leonard**
Geologist **(301) 776-6807**
6007 Windham Road
Laurel, Maryland 20810
Br. of Paleontology & Stratigraphy
Washington, D.C. **(202) 343-3495**
Brooklyn Coll.-BS; Ohio State U.-PhD
Conodont biostratigraphy of the Appalachian orogen (lower & middle Paleozoic)
Conodont biostratigraphy; organic maturations; Ordovician-Devonian stratigraphy

HARRIS, DAVID M.
Physical Scientist **(703) 476-6997**
11909 Winterthur Lane, Apt. 108
Reston, Virginia 22091
Br. of Exper. Geochem. & Mineralogy
Reston **(703) 860-6661**
Whitman Coll.; U. of Chicago-BA
Volcanology; igneous petrology

HARRIS, JOSEPH L.
Chemist **(202) 726-8203**
1110 Buchanan Street, N.W.
Washington, D.C. 20011
Br. of Analytical Laboratories
Reston **(703) 860-7655**
Virginia State Coll.-BS
Spectrographic services & research

HARRIS, LEONARD D. **Anita**
Geologist **(301) 776-6807**
6007 Windham Road
Laurel, Maryland 20810
Br. of Oil & Gas Resources
Reston **(703) 860-6634**
U. of Missouri-BA
Structure of the Appalachian Orogen; Devonian black shales, Appalachian Basin
Structural geology; carbonate sedimentology; regional basin analysis

HARRIS, LINDA D. **Jim**
Administrative Clerk
Office of Environmental Geology
Reston **(703) 860-6416**

HARRISON, D. GEORGE
PST
4118 Capri Drive
Corpus Christi, Texas 78411
Br. of Atlantic-Gulf of Mex. Geology
Corpus Christi **(512) 888-3241**
Texas A&M-BS, MS
Trace metals, biophysics

HARRISON, GLENDA K.
Clerk-Typist
Br. of Central Mineral Resources
Denver **(303) 234-3836**

HARRISON, JACK E. **Joan**
Geologist **(303) 237-7094**
Lakewood, Colorado 80226
Br. of Central Environmental Geology
Denver **(303) 234-3090**
DePauw U.-AB; U. of Illinois-PhD
Belt Basin
Belt Supergroup, stratabound ores in Belt

HARSH, PHILIP W.
Geophysicist
P.O. Box 3151
Stanford, California 94305
Br. of Tectonophysics
Menlo Park **(415) 323-8111 x2714**
Stanford U.-BS
Parkfield prediction experiment
Fault displacement, fault zone deformation

HART, PATRICK E.
GFA
Br. of Pacific-Arctic Geology
Menlo Park **(415) 323-8111 x2695**
Stanford U.-BS, MS
Beaufort-Chukchi Seas resource assessment
Gas hydrates

HARTLEY, DEBRA L.
Administrative Clerk **(303) 420-1813**
8504 Chase Drive
Arvada, Colorado 80003
Office of Geochemistry & Geophysics
Denver **(303) 234-2515**

HARTZ, ROGER WILLIAM
Geologist
135 Jenne Street
Santa Cruz, California
Br. of Alaskan Geology
Menlo Park **(415) 323-8111 x2962**
Ball State U.-BS
Offshore permafrost studies, Beaufort Sea, Alaska
Coastal morphology; coastal processes; Quaternary geology

HARWOOD, DAVID S.
Geologist **(415) 493-6120**
3324 Vernon Terrace
Palo Alto, California 94303
Br. of Western Environmental Geology
Menlo Park **(415) 323-8111 x2959**
Dartmouth Coll.-AB; Harvard-PhD
Tectonic history of the Northern Sierran foothills
Structure & petrology of metamorphic rocks; regional geologic synthesis

HASBROUCK, WILFRED P. **Barbara**
Geophysicist **(303) 424-9108**
6060 Wright Street
Arvada, Colorado 80004
Br. of Petrophysics & Remote Sens.
Denver **(303) 234-5018**
Colorado Schl. of Mines-BS, PhD
Coal geophysics research
Coal geophysics

HASCHKE, LAURA R.
PST
Br. of Coal Resources
Denver **(303) 234-3578**
Western State Coll.; Metro. State Coll.-BS
Geology of Fence Lake area, New Mexico; Broadus, Montana 1:100,000 map
Stratigraphy; depostional environments; coal geology

HASLER, J. WILLIAM **Marjorie**
Geologist **(303) 279-2326**
1815 Pinal Road
Golden, Colorado 80401
Br. of Central Mineral Resources
Denver **(303) 234-2451**
BYU-BA; U. of Utah-MA; U. of Wisconsin
Mineral exploration, economics of mineral deposits
Mineral deposits, resource appraisal

HASSEMER, JERRY H.
PST **(303) 232-7994**
10605 W. 9th Place
Lakewood, Colorado 80215
Br. of Regional Geophysics
Denver **(303) 234-2710**
Northern Arizona U.
Gravity of Butte 1x2° quad.
Gravity; karst hydrology

HATCH, NORMAN L., JR. **Sally**
Geologist **(301) 530-4991**
8514 Irvington Avenue
Bethesda, Maryland 20034
Br. of Eastern Environmental Geology
Reston **(703) 860-6421**
Harvard U.-AB, MA, PhD
Lewiston (Me-NH-Vt) 2° sheet
Northern Appalachians stratigraphy, structure, tectonic history

HATCHER, PATRICK G. **Susan**
Chemist
Br. of Coal Resources
Reston **(703) 860-7158**
North Carolina State U.-BS; U. of Miami-MS; U. of Maryland
Geochemistry of fossil fuels & related substances
Organic geochemistry of humic substances, coal & kerogen

HATFIELD, D. BROOKE
PST
5367 Carr Street, #201
Arvada, Colorado 80002
Br. of Analytical Laboratories
Denver **(303) 234-6401**
Soils availability

HATHAWAY, JOHN C. **Ilene**
Geologist **(617) 548-6321**
16 Emmons Road
Falmouth, Massachusetts 02540
Br. of Atlantic-Gulf of Mex. Geology
Woods Hole **(617) 548-8700**
M.I.T., Colgate U.-AB; U. of Illinois-MS
Stratigraphy-Atlantic OCS
Marine sediments; clay mineralogy

HAUBERT, ADOLPH W.
PST
Br. of Analytical Laboratories
Denver **(303) 234-3624**

HAVACH, GEORGE A. **Ellen**
Technical Pub. Editor **(415) 285-9727**
550-27th Street, #202
San Francisco, California 94131
Office of Scientific Publications
Menlo Park **(415) 323-8111 x2302**
Carnegie-Mellon U.-BS; U. of California (Berkeley)-MS
Scientific literature

HAVENER, ALICE F. **Shannon**
Library Technician
Library
Reston **(703) 860-6673**

HAWKINS, BERNARD W. **Claire**
Cartographic Technician
12238 West Ohio Drive
Lakewood, Colorado 80228
Office of Scientific Publications
Denver **(303) 234-3648**

HAWKINS, FRED F.
PST **(303) 447-2737**
1475 Folsom Street, #2012
Boulder, Colorado 80302
Br. of Central Environmental Geology
Denver **(303) 234-5371**
U. of Connecticut-BS; U. of Colorado-MS
Quaternary geology; glacial geomorphology; glaciology

HAXEL, GORDON B. **Terrie**
Geologist **(415) 792-5022**
5214 Ramsgate
Newark, California 94560
Br. of Western Mineral Resources
Menlo Park **(415) 323-8111 x2169**
U. of Ill.-BS, MS; U. of Cal. (Santa Barb.)-PhD
Geology & mineral resources, Papago Indian Reserv. and Ajo 2° quad. s. Arizona
Regional geology, tectonics, & geologic history of the Southwest

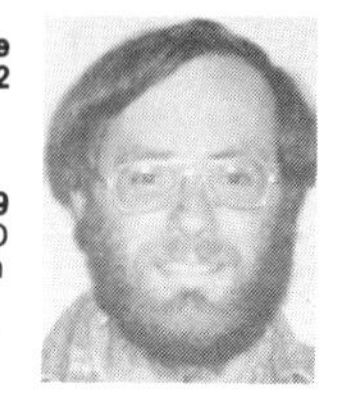

HAYES, JEANNETTE M. **Bob**
Library Technician
261A Edwards Ferry Road, Apt. 3
Leesburg, Virginia 22075
Library
Reston **(703) 860-6616**

HAYS, WILLIAM H. **Jane**
Geologist
Br. of Engineering Geology
Denver **(303) 234-4697**
Stanford U.-BS; Yale U.-MS, PhD
Geologic & hazard evaluation & mapping
Structural geology; stratigraphy

HAZEL, JOSEPH E. **Marilyn**
Geologist **(703) 451-4063**
5915 Veranda Drive
Springfield, Virginia 22152
Br. of Paleontology & Stratigraphy
Reston **(703) 860-7745**
U. of Missouri-BA, MA; Louisiana St. U.-PhD
Ostracode zonation of the Cretaceous & Tertiary, Atlantic & Gulf Coastal Province
Biostratigraphy; systematic paleontology (Ostracoda); computer techniques

HAZLEWOOD, ROBERT M. **Judy**
Geophysicist **(408) 245-4169**
1255 Rousseau Drive
Sunnyvale, California 94087
Br. of Ground Motion & Faulting
Menlo Park **(415) 323-8111**
U. of Texas-BA
Shear wave velocity measurements
Seismic refraction-reflection; gravity

HEALEY, DON L. **Betty**
Geologist **(303) 935-8946**
2761 S. Utica Street
Denver, Colorado 80236
Br. of Special Projects
Denver **(303) 234-2146**
Utah State U.-BS
Gravity surveys, Nevada test site
Gravity anomalies & subsurface structure

HEARD, IRVIN, JR.
Physicist
Br. of Isotope Geology
Reston **(703) 860-7662**
Southern U.-BS; Howard U.-MS
IR study of radiation damage & iron in zircon; electrolysis of coal

HEARN, B. CARTER **Mary**
Geologist
4709 Argyle Avenue
Garrett Park, Maryland 20766
Br. of Field Geochemistry & Petrology
Reston **(703) 860-7408**
Wesleyan U.-BA; Johns Hopkins U.-PhD
Clear Lake geothermal area, CA; kimberlites
Kimberlites; volcanology; alkalic rocks

HEARN, PAUL P.
Geologist
Br. of Analytical Laboratories
Reston **(703) 860-7543**
Duke U.-BA; George Washington U.-MS
Sediment & water chemistry; X-ray fluorescent spectroscopy

HEDGE, CARL E.
Geologist
Br. of Isotope Geology
Denver **(303) 234-3876**
Colorado Schl. of Mines-DSc
Geochronology; isotope geology

HEDRICKS, LOUISE S. **Joseph**
PST
Br. of Central Mineral Resources
Denver **(303) 234-2679**
Colorado Schl. of Mines; U. of Utah-BS; McGill University; Oregon State Coll.
Geologic Division Photolab & Ore Microscopy Lab
Ultra-violet luminescence; mineral microscopy; geoscience teaching

HEGGEM, FLORA A.
Administrative Officer **(303) 935-6152**
340 Osceola Street
Denver, Colorado 80219
Br. of Central Environmental Geology
Denver **(303) 234-5643**
Business College

HEIDEL, ROBERT H.
Physicist **(303) 232-6493**
180 S. Hoyt Street
Lakewood, Colorado 80226
Br. of Central Mineral Resources
Denver **(303) 234-2963**
Mankato State U.-BE; U. of Minnesota-MA; Iowa State U.-MS
Denver Electron Microprobe Facility
Electron probe & X-ray analyzer

HEIN, JAMES R.
Geologist **(408) 427-0208**
Br. of Pacific-Arctic Geology
Menlo Park **(415) 856-7127**
Oregon State U.-BSc; U. of California (Santa Cruz)-PhD
Circum-Pacific siliceous deposits; clay mineral authigenesis in the deep sea
Marine sedimentology; low temperature geochemistry-diagenesis; argillaceous & fine-grained siliceous deposits

HELLER, JOAN S.
PST
2551 Harlan Street
Edgewater, Colorado 80214
Br. of Petrophysics & Remote Sens.
Denver **(303) 234-6592**
Wesleyan U.-BA

HELLER, PAUL L.
Geologist
Br. of Western Environmental Geology
Menlo Park **(415) 323-8111 x2223**
Western Washington U.-MS
Neotectonic synthesis of the U.S.
Quaternary geology; sedimentology; stratigraphy

HELLEY, EDWARD J. **Kay**
Geologist **(415) 797-2813**
35181 Donegal Court
Newark, California 94560
Br. of Western Environmental Geology
Menlo Park **(415) 323-8111 x2462**
Occidental Coll.-BA; U. of Cal. (Berkeley)
Quaternary geology, Sacramento valley & northern Sierran foothills
Quaternary geology; fluvial process & sediment transport; dendrochronology

HELTON, SUZANNE M.
PSA **(415) 328-5942**
430 Blake Street
Menlo Park, California 94025
Br. of Ground Motion & Faulting
Menlo Park **(415) 323-8111 x2579**
U. of Michigan-BA
Southern Alaska seismic studies

HELZ, ROSALIND T. **George**
Geologist
4120 Everett Street
Kensington, Maryland 20795
Br. of Exper. Geochem. & Mineralogy
Reston **(703) 860-6668**
Stanford U.-BS; Penn State U.-MS, PhD
Experimental petrology of basalt
Igneous petrology; experimental petrology; volcanology

HEMLEY, J. JULIAN **Virginia**
Geologist **(703) 620-2284**
11701 Foxvale Court
Oakton, Virginia 22124
Office of Geochemistry & Geophysics
Reston **(703) 860-6531**
Texas Mines & Metal.-BS; Northwestern U.-MS; U. of California (Berkeley)-PhD
Ore deposits; geochemical processes

HEMPENIUS, MARGARET W. **Rudolph**
Secretary **(617) 540-1398**
36 Grace Court
East Falmouth, Massachusetts 02536
Br. of Atlantic-Gulf of Mex. Geology
Woods Hole **(617) 548-8700**

HENDRICKS, JOHN D.
Geophysicist
Br. of Oil & Gas Resources
Denver **(602) 779-3311 x1464**
Northeast Okla. State U.-BS; Northern Arizona U.-MS
Remote detection of hydrocarbons
Gravity; aeromagnetic surveys

HENRY, CHARLES P.
PST
Br. of Electromag. & Geomagnetism
Reston **(703) 860-7233**
U. of Chicago-BA; Michigan State U.-MS
Geophysical properties of serpentine & applications for mineral prospecting

HENRY, MITCH E. **Patricia**
Geologist
6045 North Smokerise Drive
Flagstaff, Arizona 86001
Br. of Oil & Gas Resources
Flagstaff **(602) 779-3311 x1368**
Midwestern U.-BS; Texas A&M-MS
Marine petroleum prospecting

HENRY, THOMAS W. **Donna**
Geologist **(703) 591-2974**
3224 Atlanta Street
Fairfax, Virginia 22030
Br. of Coal Resources
Reston **(703) 860-7734**
U. of Oklahoma-BSc, MSc, PhD
Pennsylvanian system stratotype & Appalachian Basin studies
Coal geology; biostratigraphy; paleontology

HERAN, WILLIAM D.
PST
4618 W. 33rd Avenue
Denver, Colorado
Br. of Electromag. & Geomagnetism
Denver **(303) 234-5397**
Colorado State U.-BS
Massive sulfides; airborne EM.

HERD, DARRELL G.
Geologist
36367 Fremont Boulevard
Fremont, California
Br. of Western Environmental Geology
Menlo Park **(415) 323-8111 x2951**
Indiana U.-AB; U. of Washington-MS, PhD
Neotectonics of the San Fran. Bay Region; earthquake tectonics of Ariz., Sonora, & Chihuahua
Neotectonics; Quaternary geology; pedology

HEREFORD, RICHARD
Geologist
Br. of Central Environmental Geology
Flagstaff **(602) 261-1455**
Northern Arizona U.-BS, MS
Pleistocene stratigraphy of the Little Colorado River valley
Stratigraphy & sedimentology

HEROPOULOS, CHRIS **Helen**
Chemist **(408) 248-1045**
2328 Glendenning Avenue
Santa Clara, California 95050
Br. of Analytical Laboratories
Menlo Park **(415) 323-8111 x2949**
Niagara U.¢BS; San Jose State
Spectrographic analysis on silicate rocks & sulfide minerals
Methods development for determing element concentrations

HERRING, JAMES R.
Geologist, Geochemist **(303) 232-2669**
875 Flower Street
Lakewood, Colorado 80215
Br. of Regional Geochemistry
Denver **(303) 234-5248**
UCSD-Scripps Inst. Oceanography-PhD
Geochemistry of naturally burning coal
Geochemistry

HERRIOT, JAMES W.
Computer Specialist
1302 Channing Avenue
Palo Alto, California 94301
Br. of Tectonophysics
Menlo Park **(415) 323-8111 x2932**
Stanford U.
Geolab/Unix

HERSHISER, ROBERT W. **Sharon**
Museum Specialist **(415) 369-6103**
227 Rutherford Avenue
Redwood City, California 94061
Br. of Paleontology & Stratigraphy
Menlo Park **(415) 323-8111 x2540**
Oregon State U.-BS

HERZOG, DONALD C. **Mary**
Geophysicist
Br. of Electromag. & Geomagnetism
Denver **(303) 234-5499**
U. of South Florida-BA; UCLA-MS

HERZON, PAIGE LEIGH
PST
Br. of Alaskan Geology
Menlo Park **(415) 323-8111 x2471**
U. of California (Santa Cruz)-BS
Mt. Hayes Quadrangle (AMRAP)

HESS, GORDON R. **Patricia**
Geologist **(408) 733-4681**
20567 Sunnymount Avenue
Sunnyvale, California 94087
Br. of Pacific-Arctic Geology
Menlo Park **(415) 856-7046**
U. of Minnesota-BS, MS; Stanford U.
Deep sea fans
Sedimentary petrology; abyssal sedimentation

HEYL, ALLEN V. **Maxine**
Geologist **(303) 674-5829**
Paint Brush Drive, P.O. Box 1052
Evergreen, Colorado 80439
Br. of Central Mineral Resources
Denver **(303) 234-3423**
Penn State U.-BS; Princeton U.-PhD
Quadrangle mapping, Black Range, New Mexico
Economic geology; structure; geochemistry

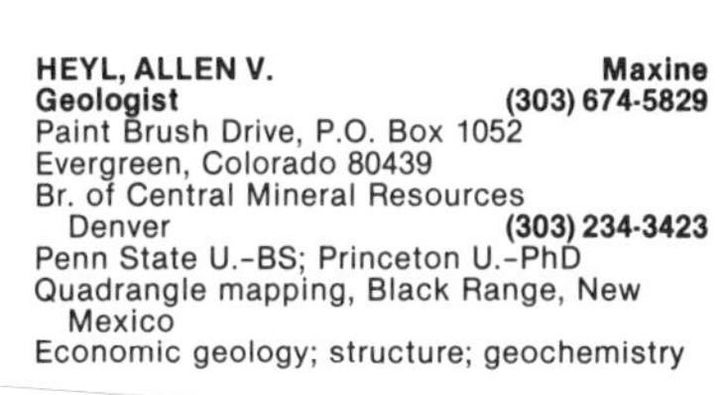

HICKLING, NELSON L.
Geologist
Rt #4, Harrisville Road
Mt. Airy, Maryland 21771
Br. of Coal Resources
Reston **(703) 860-6652**
Bowling Green State U.-BS; American U.-MS
Coal resources of Wind River Indian Reservation-Wyoming
Mineralogy; petrography; coal resources

HIETANEN-MAKELA, ANNA M.
Geologist
1134 Palo Alto Avenue
Palo Alto, California 94301
Br. of Western Environmental Geology
Menlo Park **(415) 323-8111 x2382**
U. of Helsinki, Finland-PhD
Northern Sierra Nevada
Petrology; structure; mineralogy

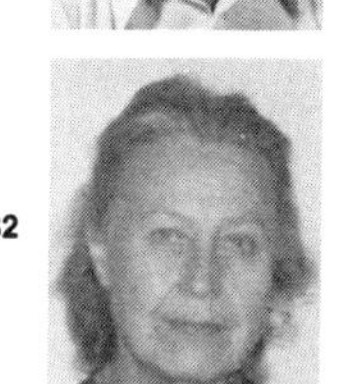

HIGGINS, MICHAEL W.
Geologist **(404) 498-0706**
19-A2 Chatfield Drive
Stone Mountain, Georgia 30083
Br. of Eastern Environmental Geology
Atlanta, Georgia **(404) 656-3214**
Emory U.-BS, MS; U. of California (Santa Barbara)-PhD
Atlanta 2° sheet
Structure; metamorphic stratigraphy; geochemistry

HILDEBRAND, RICKY T. **Diane**
PST **(303) 238-6843**
11775 W. Security Avenue
Golden, Colorado 80401
Br. of Coal Resources
Denver **(303) 234-5280**
U. of Colorado-BA
Geology and geochemistry of coal

HILDENBRAND, THOMAS G. **Mary Jo**
Geophysicist
13966 W. Warren Drive
Lakewood, Colorado 80228
Br. of Regional Geophysics
Denver **(303) 234-5464**
U. of California (Berkeley)-BS, MS, PhD
Mississippi embayment geophysics
Gravity and magnetic studies

HILDRETH, EDWARD W.
Geologist
49 Pearce Mitchell Place
Stanford, California 94305
Br. of Field Geochemistry & Petrology
Menlo Park **(415) 323-2643**
Harvard-BA; U. of California (Berkeley)-PhD
Petrology & geochem. of Yellowstone, Katmai
Behavior of elements & isotopes; evolution of magmas & magmatic ores

HILL, DAVID P. **Ann**
Geophysicist
Br. of Seismology
Menlo Park **(415) 323-8111 x2891**
San Jose State U.-BS; Colo. Schl. of Mines-MS; Cal Tech-PhD
Earth structure; wave propagation; earthquakes

HILL, GARY W. **Nancy J.**
Oceanographer **(408) 374-0259**
1673 Adrien Drive
Campbell, California 95008
Br. of Pacific-Arctic Geology
Menlo Park **(415) 856-7080**
U. of Corpus Christi-BS; Texas A&I U.-MS; U. of California (Santa Cruz)-PhD
Biogenic sedimentary processes
Geobiology; sedimentology

HILL, PATRICIA L. **Gilmore**
PST **(303) 237-3810**
1420 Kingsbury Court
Golden, Colorado 80401
Br. of Regional Geophysics
Denver **(303) 234-6143**
U. of Wisconsin-BS
Aeromagnetics
Aeromagnetics and gravity

HILLHOUSE, JOHN W.
Geophysicist **(415) 366-3559**
435 Seventh Avenue
Menlo Park, California 94025
Br. of Petrophysics & Remote Sens.
Menlo Park **(415) 323-8111**
U.of Cal. (Berkeley)-AB; Stanford U.-MS, PhD
Paleomagnetic methods applied to dating & plate tectonics
Paleomagnetism; plate tectonics

HILLIER, BARBARA C. **Donald**
Geologist
13537 West Exposition Drive
Lakewood, Colorado 80228
Office of Scientific Publications
Denver **(303) 234-3281**
U. of Michigan-BS

HILLS, F. ALLAN **Sonia**
Geologist **(303) 423-3783**
3850 Hoyt Street
Wheat Ridge, Colorado 80033
Br. of Uranium-Thorium Resources
Denver **(303) 234-5653**
U. of North Carolina-BS; Yale U.-PhD
Uranium potential Precambrian rocks, regional variations of uranium in granites
Precambrian regional geology & tectonics; geochronology

HINKLE, MARGARET E. **Roger**
Chemist
Arvada, Colorado 80005
Br. of Exploration Research
Denver **(303) 234-6174**
Wayne State U.-BS; U. of Michigan-MS
Volatile elements & compounds in geochemical exploration
Geochemical exploration

HINKLEY, TODD K.
Geologist **(303) 674-4966**
5243 Gertrude Road
Evergreen, Colorado 80439
Br. of Regional Geochemistry
Denver **(303) 234-5628**
Cal Tech-PhD
Masses & trajectories of atmospheric dusts
Element cycling in geologic systems; geochemistry of rocks, soils & waters; igneous & metamorphic petrology

HINRICHS, E. NEAL **Gertrude**
Geologist **(303) 233-0790**
1005 Field Street
Lakewood, Colorado 80215
Br. of Engineering Geology
Denver **(303) 234-3894**
Oberlin Coll.-BA; Cornell U.-MS
Engineering geologic mapping, Sheridan-Buffalo area, Wyoming
Holistic economic geology

HIROZAWA, CAROL A.
PST
14790 Manvela Road
Los Altos Hills, California 94022
Br. of Pacific-Arctic Geology
Menlo Park **(415) 856-7043**
U. of California (Santa Cruz)-BS
Northern California; Oregon

HIRSCHER, ELSIE H.
Secretary **(415) 967-7364**
793 Gantry Way
Mt. View, California 94040
Br. of Seismology
Menlo Park **(415) 323-8111 x2892**
Foothill College

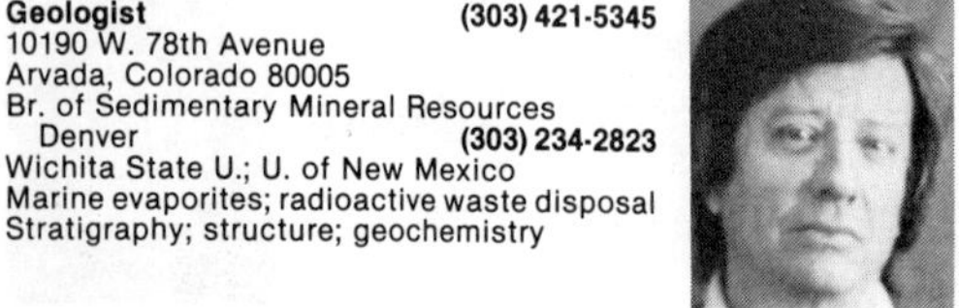

HITE, ROBERT J.
Geologist **(303) 421-5345**
10190 W. 78th Avenue
Arvada, Colorado 80005
Br. of Sedimentary Mineral Resources
Denver **(303) 234-2823**
Wichita State U.; U. of New Mexico
Marine evaporites; radioactive waste disposal
Stratigraphy; structure; geochemistry

HOARE, JOSEPH M. **Janet**
Geologist **(415) 851-7050**
220 Highland Terrace
Woodside, California 94062
Br. of Alaskan Geology
Menlo Park **(415) 323-8111 x2372**
Augustana Coll.-BA; U. of Chicago; Columbia
Tikchik Lakes, S.W. Alaska
Geology of S.W. Alaska; Bering Sea Cenozoic basalts

HOBBS, ROBERT G. **Wanda**
Mining Engineer
Br. of Coal Resources
Denver **(303) 234-3519**
U. of Utah-BS, MS
Coal exploratory drilling, drill support group
Coal geology & mining engineering; geophysical logging (coal)

HOBBS, S. WARREN **Louise**
Geologist **(303) 674-3843**
29663 Paintbrush Drive
Evergreen, Colorado 80439
Br. of Central Environmental Geology
Denver **(303) 234-4369**
U. of Washington-BS; Yale U.-PhD
Challis CUSMAP
Geologic mapping; mineral deposits; structural geology

HOBLITT, RICHARD P. **Marian**
Geologist
3661 Chase Court
Boulder, Colorado 80302
Br. of Engineering Geology
Denver **(303) 234-3546**
U. of Washington-BS; U. of Colorado-MS, PhD
Volcanic hazards
Rock magnetism & paleomagnetism; volcanology

HODGEN, LAURIE D.
Geologist
Office of Scientific Publications
Menlo Park **(415) 323-8111 x2301**
U. of the Pacific-BA

HODGES, CARROLL ANN
Geologist
1367 Canada Road
Woodside, California 94062
Br. of Astrogeologic Studies
Menlo Park **(415) 323-8111 x2361**
U. of Texas-BA; U. of Wisconsin-MS; Stanford U.-PhD
Geologic history of Moon & Mars
Planetary geology; geomorphology

HODGSON, HELEN E. **James M. Soule**
Technical Pub. Editor **(303) 756-5641**
3401 S. Cherry Street
Denver, Colorado 80222
Office of Scientific Publications
Denver **(303) 234-2866**
U. of Michigan-BA; U. of Denver-MA, PhD
Techniques for teaching technical writing

HOFFMAN, JOHN P. **Kitty**
Geophysicist
Br. of Global Seismology
Albuquerque **(505) 844-4637**
U. of Wisconsin-BS; U. of Utah-MS
Global digital seismograph network
Processing digital seismic data; seismograph instrumentation

HOFFMAN, SUSAN T. **Charlie**
Clerk-Typist **(303) 674-6221**
2821 Olympia Lane
Evergreen, Colorado 80439
Office of the Chief Geologist
Denver **(303) 234-4790**
Fresno State U.-BA

HOGGAN, ROGER D. Karen
Geologist (208) 356-5636
587 Gemini Drive
Rexburg, Idaho 83440
Br. of Central Environmental Geology
Rexburg, Idaho (208) 356-1519
Weber State Coll.-BS; Washington State U.; Brigham Young U.-MS, PhD
Mapping-Pocatello 2°
Geologic field mapping; stratigraphy

HOGGATT, WENDY C.
PST
Br. of Western Environmental Geology
Menlo Park (415) 323-8111 x2223
U. of California (Berkeley)-BA
Peninsular ranges tectonics
K-Ar & fission track dating; field mapping

HOLCOMB, L. GARY Emily
Electronic Engineer (505) 296-2702
3723 Espejo NE
Albuquerque, New Mexico 87111
Br. of Global Seismology
Albuquerque (505) 844-4637
U. of Nebraska-BS, MS, PhD
Long period seismic background digital analysis

HOLCOMB, ROBIN T. Annette
PST (415) 327-7129
82-B Escandido Village
Stanford, California 94305
Br. of Field Geochemistry & Petrology
Menlo Park (415) 323-8111
Cornell-BA; U. of Ariz.-MS; Stanford-PhD
Holocene volcanism of Kilauea Volcano, Hawaii
Volcanic geomorphology; paleomagnetism; lunar geology

HOLMES, CHARLES W. Barbara
Geologist (512) 387-6160
Route 1, Box 75A
Robstown, Texas 78380
Br. of Atlantic-Gulf of Mex. Geology
Corpus Christi (512) 888-3294
St. Joseph's Coll.; Florida State U.-PhD
Carbonate environment-West Florida shelf
Sedimentation; geochemistry

HOLMSTROM, CARL A. Joyce
Engineering Technician (408) 338-3592
P.O. Box 454
Boulder Creek, California 95006
Br. of Isotope Geology
Menlo Park (415) 323-8111 x2035
U. of California
Isotope geology; instrument building
Tool-making; designing

HOLT, HENRY E. Mary
Geologist
2036 Crescent Drive
Flagstaff, Arizona 86001
Br. of Astrogeologic Studies
Flagstaff (602) 779-3311 x1485
U. of Idaho-BS; U. of Colorado-PhD
Planetology; stratigraphy

HOLZER, THOMAS L. Mary
Geologist (415) 321-7346
151 Walter Hays Drive
Palo Alto, California 94303
Br. of Engineering Geology
Menlo Park (415) 323-8111 x2909
Princeton U.-BSE; Stanford U.-MS, PhD
Fissuring-subsidence research
Engineering geology; hydrogeology

HOLZLE, ALVIN F. Mary
Supervising Geologist (301) 589-3106
642 Bennington Drive
Silver Spring, Maryland 20910
Office of International Geology
Reston (703) 860-6551
U. of Buffalo-BA
Remote sensing; photointerpretation

HONMA, KENNETH T.
PST (808) 967-7215
P.O. Box 24
Volcano, Hawaii 96765
Br. of Field Geochem. & Petrology
Hawaiian Volcano Observ. (808) 967-7528

HOOSE, SEENA N.
Geologist (408) 252-5811
10394 Bret Avenue
Cupertino, California 95014
Br. of Engineering Geology
Menlo Park (415) 856-7116
U. of California (Santa Barbara)-BA
Liquefaction potential mapping
Liquefaction

HOOVER, DAVID L. Carol
Geologist (303) 238-7277
1900 Yukon
Lakewood, Colorado 80215
Br. of Special Projects
Denver (303) 234-3624
Colorado Schl. of Mines-BS
Seismicity, tectonics & volcanism-Nevada nuclear waste studies
Tectonic geomorphology; alluvial & volcanic stratigraphy

HOOVER, DONALD B. Lucille
Geophysicist
Br. of Electromag. & Geomagnetism
Denver (303) 234-2950
Case Institute-BS; U. of Michigan-MS; Colorado Schl. of Mines-DSc
Electrical techniques applied to geothermal exploration
Audio-magnetotellurics; self potential

HOOVER, LINN Joan
Geologist (301) 652-2016
6902 Oakridge Avenue
Chevy Chase, Maryland 20015
Office of Energy Resources
Reston (703) 860-6432
U. of North Carolina-AB; U. of Michigan-MA; U. of California (Berkeley)-PhD
IVGS, IGCP & UNESCO affairs
International scientific activities; research administration

HOPKINS, DAVID M. Rachel
Geologist (415) 341-5932
1712 Yorktown Road
San Mateo, California 94402
Br. of Alaskan Geology
Menlo Park (415) 323-8111 x2659
U. of New Hampshire-BS; Harvard-MS, PhD
Quaternary geology of northern Alaska
Quaternary geology & paleoecology of Beringia; Late Cenozoic history & paleo-oceanography of Arctic Ocean

HOPKINS, ROY T. Geraldine
Physical Scientist
1830 S. Valentine Street
Lakewood, Colorado 80228
Br. of Exploration Research
Denver (303) 234-4440
U. of Houston-BS
Butte 2° sheet & Iron River 2° sheet
Emission spectroscopy

HOPPER, MARGARET G.
Geophysicist
7865H Barbara Ann Drive
Arvada, Colorado 80004
Br. of Earthquake Tectonics & Risk
Denver (303) 234-2820
U. of North Carolina (Greensboro)-BA; Virginia Poly. Inst. & State U.-MS
Intensity studies-central U.S.
Puget Sound seismicity

HORAN, CAROL L.
Librarian
Library
Reston (703) 860-6671
SUNY (Oneonta)-BA; SUNY (Albany)-MLS

HORTON, J. WRIGHT **Beverly**
Geologist **(703) 437-1991**
1410 Sadlers Wells Drive
Herndon, Virginia 22070
Br. of Eastern Environmental Geology
Reston **(703) 860-6421**
Furman U.-BS; U. of North Carolina-MS, PhD
Kings Mountain Belt & adjacent areas
Structural geology; hard rock petrology; regional geology

HOSS, KAREN L.
Administrative Officer
Office of Scientific Publications
Menlo Park **(415) 323-8111 x2563**

HOSTERMAN, JOHN W. **Margaret**
Geologist
9510 Rockport Road
Vienna, Virginia 22180
Br. of Eastern Mineral Resources
Reston **(703) 860-6913**
Pennsylvania State U.-BS, MS
Clay mineralogy

HOUSER, BRENDA B.
Geologist
Br. of Central Mineral Resources
Denver **(303) 234-6810**
U. of Florida-BS, MS; Virginia Polytech Inst.-PhD
Sedimentology; geomorphology; sedimentary petrology

HOUSER, FRED N.
Geologist
Br. of Engineering Geology
Denver **(303) 234-4697**
Michigan State U.-BS; U. of Michigan; U. of Arizona
Information exchange for dams and other critical installations
Areal geology-volcanic terranes; causes and mechanisms of subsidence

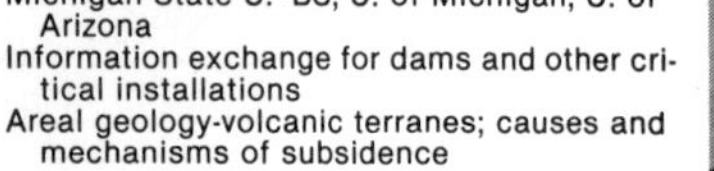

HOWARD, KEITH A. **Jean**
Geologist **(415) 591-4680**
700 Alameda de las Pulgas
Belmont, California 94002
Br. of Western Environmental Geology
Menlo Park **(415) 323-8111 x2282**
U. of California (Berkeley)-BS, MS; Yale U.-PhD
Needles 2° sheet
Tectonics; volcanology; astrogeology

HOWD, PETER A.
PST **(415) 365-2280**
636 Southdale Way
Woodside, California 94062
Br. of Pacific-Arctic Geology
Menlo Park **(415) 856-7072**
Williams Coll.-BA
Wave-current-profile interactions on beaches
Edge waves, infragravity waves & effects on beaches; morphology high energy beaches

HOWELL, DAVID G. **Susan**
Geologist
3825 Grove Avenue
Palo Alto, California
Br. of Oil & Gas Resources
Menlo Park **(415) 856-7011**
Colgate U.-BA; U. of California (Santa Barbara)-MA, PhD
Continental margin tectonics & sedimentation

HOWELL, TREVOR H.
PST
Box 22194
GMF, Guam 96921
Br. of Global Seismology
Guam Observatory **(671) 355-5259**
Box 8001, Agana, Guam 96910
Sacramento State College-AA

HUBER, DONALD F.
Geologist
Office of Resource Analysis
Menlo Park **(415) 323-8111 x2906**
California State U. (Hayward)-BS
CRIB
Mineral resource assessment

HUBER, N. KING **Nan**
Geologist **(415) 968-5475**
220 Diablo Avenue
Mountain View, California 94043
Br. of Western Environmental Geology
Menlo Park **(415) 323-8111**
Franklin & Marshall Coll.-BS; Northwestern U.-MS, PhD
Pinecrest & Dardanelles Cone quads., Sierra Nevada
Sierra NV geology; Lake Super. Precambrian

HUBERT, ARTHUR E.
Chemist **(303) 422-4139**
6146 Queen Street
Arvada, Colorado 80004
Br. of Exploration Research
Denver **(303) 234-6168**
Iowa State U.-BS
Research on trace methods, x-ray analysis

HUBERT, MARILYN L.
Geologist **(703) 243-8132**
3000 Spout Run Parkway (D-601)
Arlington, Virginia 22201
Office of Scientific Publications
Reston **(703) 860-6511**
Colby Coll.-AB; U. of Maryland
Lexicon of stratigraphic names
Stratigraphic nomenclature

HUDSON, J. HAROLD
Biologist
8325 SW 68 Street
Miami, Florida 33143
Br. of Oil & Gas Resources
Fisher Island Station **(305) 672-1784**
U. of Miami-BA
Environmental effects in areas of carbonate deposition
Growth banding in corals

HUEBNER, J. STEPHEN **Bunny**
Geologist
Br. of Exper. Geochem. & Mineralogy
Reston **(703) 860-6693**
Princeton U.-BA; Johns Hopkins U.-PhD
Geological thermometry
Experimental petrology & mineralogy

HUEBNER, MARK
PST
Br. of Isotope Geology
Denver **(303) 234-3624**
San Jose State U.

HUESTIS, GARY M. **Carla**
PST **(303) 989-5934**
454 S. Wright, #210
Lakewood, Colorado 80228
Br. of Isotope Geology
Denver **(303) 234-4201**
U. of Montana-BA
Uranium-thorium isotope geochemistry of soils

HUFF, WILLIAM E.
Electronic Technician
Br. of Petrophysics & Remote Sens.
Reston **(703) 860-7233**
Landsat image processing

HUFFMAN, A. CURTIS — **Susan**
Geologist — **(303) 233-2852**
2378 Van Gordon Street
Lakewood, Colorado 80215
Br. of Uranium-Thorium Resources
Denver — **(303) 234-5004**
Virginia Polytech. Inst.-BS; George Washington U.-MS, PhD
San Juan Basin
Areal mapping; stratigraphy; sedimentology

HUGHES, ERIC M. — **Lorraine**
PST — **(303) 642-7215**
11975 Vonnie Claire Road
Golden, Colorado 80401
Br. of Petrophysics & Remote Sens.
Denver — **(303) 234-5490**
U.S. Naval Academy-BS; Colorado Schl. of Mines
Uranium geophysics
Recovery of marine minerals

HUIE, CARL
PST — **(415) 325-5203**
Br. of Alaskan Geology
Menlo Park — **(415) 323-8111 x2275**
U. of Montana-BS, MS
Dufek intrusion, Antarctica; Riddle Peaks Intrusion, Glacier Peak Wilderness
Petrology of layered intrusions & ophiolites

HUMPHREYS, D. DARLENE — **Jay**
Admin. Opr. Clerk — **(703) 368-3369**
10821 Daisy Court
Manassas, Virginia 22110
Office of Scientific Publications
Reston — **(703) 860-6784**

HUMPHREYS, RICHARD D.
PST
2303 Ramona Street
Palo Alto, California
Br. of Sedimentary Mineral Resources
Menlo Park — **(415) 323-8111 x2306**
San Jose State U.-BS
Minerals; exploration; geochemistry

HUNT, GRAHAM R. — **Jeanette**
Chemist — **(303) 278-2941**
3255 N. Braun Court
Golden, Colorado 80401
Br. of Petrophysics & Remote Sens.
Denver — **(303) 234-5386**
Sydney U., Australia-BS, MS, PhD
Mineral & rock spectroscopy
Spectroscopy & remote sensing

HUNT, SUSAN J. — **Douglas**
Geologist — **(415) 854-2356**
191 Hillside Avenue
Menlo Park, California 94025
Br. of Alaskan Geology
Menlo Park — **(415) 323-8111 x2477**
U. of Arizona-BS
Petersburg Project—SE Alaska
Geologic mapping; structural geology; petrology

HUNTER, JUDY F.
Computer Technician — **(703) 670-9742**
15117 Campbell Lane
Woodbridge, Virginia 22193
Br. of Coal Resources
Reston — **(703) 860-7734**

HUNTER, PATRICIA A.
Administrative Officer — **(703) 241-0542**
6571 Snowbell Lane
Falls Church, Virginia 22042
Br. of Paleontology & Stratigraphy
Reston — **(703) 860-7745**
U. of Maine; George Mason U.

HUNTER, RALPH E.
Geologist
445 Encinal
Menlo Park, California 94025
Br. of Pacific-Arctic Geology
Menlo Park — **(415) 856-7078**
Johns Hopkins U.-PhD
Coastal sedimentary processes
Sedimentology

HUNTER, ROGER N. — **Wanda**
Geophysicist — **(303) 494-3933**
5419 Omaha Place
Boulder, Colorado 80303
Br. of Global Seismology
Denver — **(303) 234-4041**
U. of Wichita-BA
Prediction monitoring & evaluation program
Computer programming; automated event detection

HUSK, ROBERT H, JR. — **Patricia Gomez**
PST — **(408) 988-1348**
Br. of Tectonophysics
Menlo Park — **(415) 323-8111 x2154**
San Jose State U.-BS, MS
Soil-gas radon water level fluctuation

HUTCHINSON, DEBORAH R.
Geologist
18 Maker Lane
Falmouth, Massachusetts 02540
Br. of Atlantic-Gulf of Mex. Geology
Woods Hole — **(617) 548-8700**
Middlebury Coll.-BA; U. of Toronto-MS
Atlantic continental margin; gravity; Great Lakes

HUTCHISON, LA VERNNE W.
Secretary
Br. of Pacific-Arctic Geology
Menlo Park — **(415) 856-7014**

HUTT, CHARLES R. — **Hiroko**
Geophysicist — **(505) 298-9359**
13728 Pruitt, NE
Albuquerque, New Mexico 87112
Br. of Global Seismology
Albuquerque — **(505) 844-4637**
Montana Tech-BS; U. of N. Mexico
Global seismograph network operations
Seismic instrumentation & seismic data processing

IANNACITO, SHERRI E.
Editorial Assistant
Office of Scientific Publications
Denver — **(303) 234-2445**
U. of Colorado (Denver)-BS
Word processing & telecommunications

IMLAY, RALPH W.
Geologist — **(301) 946-9317**
3913 Jeffry Street
Silver Springs, Maryland
Br. of Paleontology & Stratigraphy
Washington, D.C. — **(202) 343-3448**
U. of Montana-BS; U of Michigan-PhD
Upper Jurassic mollusks, western interior & southeast U.S.
Jurassic & Cretaceous biostratigraphy; correlations; mollusks

IMLAY, WILMA G.
Librarian
1801 W. 82nd Place
Denver, Colorado 80221
Library
Denver — **(303) 234-4133**

INGE, JAY L. **Mimi**
Cartographer **(602) 774-6148**
2746 W. Darleen Drive
Flagstaff, Arizona 86001
Br. of Astrogeologic Studies
Flagstaff **(602) 779-3311 x1537**
UCLA-BA
Planetary airbrush cartography
Airbrush techniques; photo interpretation; photo processes

IRWIN, WILLIAM P. **Norma**
Geologist **(415) 322-3873**
480 Lemon Street
Menlo Park, California 94025
Br. of Earthquake Tectonics & Risk
Menlo Park **(415) 323-8111 x2065**
New Mexico Tech-BS; Cal Tech-MS
Tectonics of northern California
Tectonics; structural geology; mineral deposits

ISAACS, CAROLINE M.
Geologist
Br. of Oil & Gas Resources
Menlo Park **(415) 856-7010**
U. of California (Berkeley)-BA; Stanford-PhD
Petroleum geology of siliceous rocks
Sedimentary petrology; siliceous rocks; field petrology

IWATSUBO, EUGENE Y.
PST
Br. of Tectonophysics
Menlo Park **(415) 323-8111 x2705**
Humboldt State U.-BA
Creep operations; instrumental strain
Instrumentation of fault creep

IYER, H. MAHADEVA
Geophysicist
Br. of Seismology
Menlo Park **(415) 323-8111 x2685**
Kerala U. (India)-BS, MS; Imperial Coll. (London)-PhD
Seismic studies in geothermal areas
Seismology; geothermal studies; volcanology

IZETT, GLEN ARTHUR **Gretchen**
Geologist
11122 West 27th Place
Lakewood, Colorado 80215
Br. of Central Environmental Geology
Denver **(303) 234-2835**
U. of Colorado-BA; U. of Alaska; Colorado Schl. of Mines
Tephrochronology of central region

JACHENS, ROBERT C.
Geophysicist
10488 Bonny Drive
Cupertino, California 95014
Br. of Regional Geophysics
Menlo Park **(415) 323-8111 x2168**
San Jose State U.-BS; Columbia-PhD
Crustal deformation research; mineral resource assessment

JACKSON, CHARLOTTE E. **Robert**
Secretary **(303) 466-3654**
4641 West 109th Avenue
Westminster, Colorado 80030
Br. of Engineering Geology
Denver **(303) 234-3721**

JACKSON, DALLAS B. **Beverly**
Geophysicist
P.O. Box 599
Volcano Hawaii
Br. of Field Geochem. & Petrology
Hawaiian Volcano Observ. **(808) 967-7328**
U. of Colorado-BA
Geoelectrical studies at Kilauea Volcano, Hawaii
Electrical geophysics; geothermal studies

JACKSON, LARRY L.
Chemist
Br. of Analytical Laboratories
Denver **(303) 234-2521**
Kansas State U.-BS; Colorado State U.-PhD
Analytical chemistry

JACKSON, THOMAS C.
Computer Specialist **(415) 494-6973**
3498 Janice Way
Palo Alto, California 94303
Br. of Network Operations
Menlo Park **(415) 323-8111 x2974**
Dartmouth-BA; Stanford-MS
Process control computer programming

JACOBS, WILLIS S.
Geophysicist **(671) 362-4279**
Box 8001, MOU #3
Agana, Guam 96910
Br. of Global Seismology
Box 8001, Guam Observatory **(671) 355-5259**
Agana, Guam 96910
Kansas State U.-BS

JAKSHA, LAWRENCE A. **Mary**
Geophysicist
P.O. Box 448
Tijeras, New Mexico 87059
Br. of Global Seismology
Albuquerque **(505) 844-4637**
U. of Northern Colorado-BA
Seismic studies in New Mexico
Crustal structure; seismicity

JAMES, HAROLD L. **Ruth**
Geologist **(206) 385-0878**
1617 Washington Street
Port Townsend, Washington 98368
Br. of Western Mineral Resources
Menlo Park **(415) 323-2214**

JAMES, ODETTE B.
Geologist **(703) 476-9738**
10715 Midsummer Drive
Reston, Virginia 22091
Br. of Exper. Geochem. & Mineralogy
Reston **(703) 860-6641**
Stanford-BA; PhD
Lunar highland breccias
Petrology of igneous rocks; lunar breccias; ultramafic rocks

JENKINS, EVAN C. **Ann**
Geologist **(303) 722-4785**
301 S. Williams Street
Denver, Colorado 80209
Br. of Special Projects
Denver **(303) 234-2391**
U. of Colorado-BA; U. of Texas-MA
Geology of the Nevada test site
Volcanic stratigraphy & petrology, Basin & Range structural geology

JENKINS, LILLIE B.
Chemist **(202) 882-7573**
6101 Sixteenth Street, N.W.
Washington, D.C. 20011
Br. of Analytical Laboratories
Reston **(703) 860-6975**
Fisk U.-BS
Classical analysis

JENKINS, ROSEMARY
Secretary
Office of Geochemistry & Geophysics
Reston **(703) 860-6582**
Radford Coll.; Northern Virginia Community College

JENSEN, E. GRAY **Sandy**
Geophysicist
Br. of Network Operations
Menlo Park **(415) 323-8111**
U. of California (Santa Barbara)-BS, MS
Geophysical instrument development

JENSEN, RICHARD E.
Electronics Engineer
Br. of Network Operations
Menlo Park **(415) 323-8111 x2587**
U. of California (Davis)-BS
Central California 30 Hz seismic telemetry network
Telemetry processing electronics

JIMENEZ, IRENE S.
Editorial Assistant **(415) 493-2382**
765-2 San Antonio Avenue
Palo Alto, California 94303
Br. of Engineering Geology
Menlo Park **(415) 323-8111**

JOHANNESEN, DANN C. **JoAnn Menard**
PST **(408) 268-7177**
6529 Camden Avenue
San Jose, California 95120
Br. of Western Mineral Resources
Menlo Park **(415) 323-8111 x2668**
San Jose State U.-BA
Walker Lake 2° sheet
Field mapping; petrography

JOHN, BARBARA E.
PST **(415) 327-5999**
359 Embarcadero Road
Palo Alto, California 94301
Br. of Western Environmental Geology
Menlo Park **(415) 323-8111 x2282**
U. of California (Berkeley)-BA; U. of Oregon

JOHNSON, BRUCE R. **Kate**
Geologist
3153 South Court
Palo Alto, California 94306
Br. of Alaskan Geology
Menlo Park **(415) 323-8111 x2769**
U. of Southern Colorado-BS; U. of Montana-PhD
Wilderness mineral resources
Mineral resources; petrology; geochemistry

JOHNSON, DARLINE E. **Bill**
Secretary **(602) 525-1750**
223 E. Comanche
Flagstaff, Arizona 86001
Br. of Central Environmental Geology
Flagstaff **(602) 779-3311 x1539**

JOHNSON, GORDON R. **Mary**
Geophysicist **(303) 233-5837**
12101 View Point Drive
Golden, Colorado 80401
Br. of Petrophysics & Remote Sens.
Denver **(303) 234-5389**
Colorado State U.-BS
Petrophysics

JOHNSON, KATHLEEN M. **Bruce**
Geologist **(415) 493-2239**
3153 South Court
Palo Alto, California 94306
Br. of Alaskan Geology
Menlo Park **(415) 323-8111 x2485**
Smith Coll.-BA; Syracuse U.-MS
Quaternary geology for Valdez AMRAP project
Igneous petrology

JOHNSON, MAUREEN G.
Geologist
1640 Crestview Drive
Los Altos, California 94022
Office of the Chief Geologist
Menlo Park **(415) 323-8111 x2215**
CCNY-BS; Columbia; U. of California (Berkeley)-MA
Deputy Regional Geologist
Mineral resource evaluation; computerized resource studies; petrology of metavolcanics

JOHNSON, MICHELINA J. **Harvey**
Secretary **(703) 430-9241**
800 West Redwood Road
Sterling, Virginia 22170
Br. of Engineering Geology
Reston **(703) 860-7404**
Northern Virginia Community Coll.

JOHNSON, ROBERT G.
Chemist
Br. of Analytical Laboratories
Reston **(703) 860-7543**
U. of Maryland-MS
X-ray fluorescence
Energy dispersive XRF; Mossbauer spectroscopy

JOHNSON, RONALD C.
Geologist **(303) 988-1666**
423 South Kline Street
Lakewood, Colorado 80226
Br. of Sedimentary Mineral Resources
Denver **(303) 234-5118**
SUNY (Buffalo)-BA, MA
Strat. of the Green River formation, Piceance Creek Basin
Stratigraphy; sedimentary environments; basin analysis

JOHNSON, WILLIAM D., JR **Margaret**
Geologist **(303) 279-7870**
486 Devinney Court
Golden, Colorado, 80401
Br. of Central Environmental Geology
Denver **(303) 234-4227**
U. of North Carolina-BS
Paducah 2° map
Areal geology; midcontinent stratigraphy

JOHNSTON, MALCOLM J.
Geophysicist **(415) 366-9622**
527 Encina Avenue
Menlo Park, California 94025
Br. of Tectonophysics
Menlo Park **(415) 323-8111 x2132**
U. of Queensland (Australia)-BS, PhD
Tilt, strain & magnetic field measurements
Active crustal dynamics; mechanics of faults & volcanoes

JONES, ALAN C. **Olga**
Geophysicist **(415) 325-5471**
458 Central Avenue
Menlo Park, California 94025
Br. of Tectonophysics
Menlo Park **(415) 323-8111 x2532**
U. of California (Berkeley)-BS
Strain studies
Earth strain; photography

JONES, CHARLES L. **Vera**
Geologist **(303) 423-3430**
3665 Holland Court
Wheat Ridge, Colorado 80033
Br. of Sedimentary Mineral Resources
Denver **(303) 234-2146**
U. of North Carolina-BS; Northwestern U.-MS
Radioactive waste investigations—Delaware Basin, New Mexico
Evaporites—petrology; stratigraphy; structure

JONES, D.R.
Cartographer
Br. of Pacific-Arctic Geology
Menlo Park **(415) 856-7102**
San Jose State U.-BA, MA
Oblique diagrams; base maps

JONES, DAVID M.
PST
311 Leland Avenue
Menlo Park, California 94025
Br. of Pacific-Arctic Geology
Menlo Park **(415) 856-7097**
Stanford U.-BS; Amherst Coll.

JONES, DIANE N. **Clifford**
Geologist **(703) 938-1582**
9638 Masterworks Drive
Vienna, Virginia 22180
Office of Scientific Publications
Reston **(703) 860-6492**
Ohio Wesleyan U.-BA; Purdue-MS
Engineering geology

JONES, LIBBY L. **Jim**
Administrative Technician
1220 Norman Drive
Santa Clara, California 96051
Br. of Western Environmental Geology
Menlo Park **(415) 323-8111 x2001**

JONES, VICKY L.
Management Technician
Br. of Coal Resources
Denver **(303) 234-3578**
Texas Tech U.-BSED

JONES, WILLIAM J.
Geophysicist
1405 Northgate Square, #22B
Reston, Virginia 22090
Br. of Regional Geophysics
Reston **(703) 860-7233**
West Virginia State Coll.-BS; U. of Northern Colorado-MS

JONES-CECIL, MERIDEE
Geologist
Office of Earthquake Studies
Denver **(303) 234-5087**
Oberlin-BA; U. of Michigan-MS
Seismotectonics of eastern United States
Structural geology; pattern recognition; paleomagnetism

JORDAN, JAMES N. **Peggy**
Geophysicist **(303) 494-8274**
4065 Apache Road
Boulder, Colorado 80303
Br. of Global Seismology
Denver **(303) 234-4041**
Princeton; Dartmouth; U. of North Carolina; American U.
Travel time, magnitude & seismicity studies

JORDAN, RAYMOND
Physical Scientist **(602) 774-6334**
3831 N. Paradise
Flagstaff, Arizona 86001
Br. of Astrogeologic Studies
Flagstaff **(602) 779-3311 x1515**
Siena Coll.-BS
Mars topographic mapping
Unconventional photogrammetry; custom mapping

JOYNER, WILLIAM B. **Mary Lou**
Geophysicist **(415) 343-8479**
472 Virginia Avenue
San Mateo, California 94402
Br. of Ground Motion & Faulting
Menlo Park **(415) 323-8111 x2754**
Harvard-BA, MA, PhD
Ground motion prediction
Engineering seismology; exploration seismology, gravity & magnetics

JUSSEN, VIRGINIA M.
Geologist **(703) 532-0797**
210 East Fairfax Street, Apt. 401
Falls Church, Virginia 22046
Office of Scientific Publications
Reston **(703) 860-6511**
Bryn Mawr-BA
Geologic Names Committee, lexicon

KACHADOORIAN, REUBEN
Geologist
Br. of Alaskan Geology
Menlo Park **(415) 323-8111**
Cal. Tech-BS; Stanford U.-MS
Engineering geology; NPRA; Alaska gas pipeline
Engineering geology; environmental geology

KADERABEK, RONALD M. **Carol**
Electronics Technician **(408) 255-0953**
6223 Rainbow Drive
San Jose, California 95129
Br. of Network Operations
Menlo Park **(415) 323-8111 x2443**

KADISH, RUTH L.
Administrative Technician
Br. of Astrogeologic Studies
Menlo Park **(415) 323-8111 x2361**
Wright Coll.-AA

KANE, JEAN S. **Robert**
Chemist **(703) 938-0264**
2501 Babcock Road
Vienna, Virginia 22180
Br. of Analytical Laboratories
Reston **(703) 860-6939**
Keuka Coll.-BA; Mount Holyoke-MA
Atomic absorption spectroscopy; graphite furnace atomization mechanisms

KANE, MARTIN F. **Jacqueline**
Geophysicist **(303) 986-5421**
1347 S. Routt Way
Lakewood, Colorado 80226
Br. of Regional Geophysics
Denver **(303) 234-2623**
St. Francis Xavier U.-BS; St. Louis U.-PhD
Tectonics
General geophysics; gravity; geodynamics

KANGAS, REINO **Micki**
Geophysicist **(303) 279-7734**
158 S. Holman Way
Golden, Colorado 80401
Br. of Global Seismology
Denver **(303) 234-3994**
Montana Schl. of Mines-BS
Global seismology

KANIZAY, STEPHEN P.
Geologist
Br. of Engineering Geology
Denver **(303) 234-2798**
Miami U. (Ohio)-BA, MS; Colorado Schl. of Mines-DSc
Regional geotechnical study, Powder River Basin
Engineering geology

KARAKOV, YELENA M.
Librarian
Library
Reston **(703) 860-6613**
Queens Coll. of the City of New York-MLS

KARKLINS, OLGERTS L. **Vija**
Geologist
Br. of Paleontology & Stratigraphy
Washington, D.C. **(202) 343-3741**
Columbia-BS; U. of Minnesota-MS, PhD
Middle & Late Ordovician bryozoan biostratigraphy of eastern North America
Ordovician biostratigraphy; Paleozoic bryozoan biostratigraphy & systematics

KARL, SUSAN M.
Geologist
c/o Geology Dept., Stanford
Stanford, California 94305
Br. of Alaskan Geology
Menlo Park **(415) 323-8111 x2477**
Middlebury Coll.-BA
Micropaleontology; sedimentology; geochemistry

KARLSON, KATHRYN H.
PST
Br. of Paleontology & Stratigraphy
Washington, D.C. **(202) 343-2524**
National Art Schl.

KARLSTROM, THOR N.V. **Florence**
Geologist **(602) 774-6530**
628 N. Bertrand Street
Flagstaff, Arizona 86001
Br. of Central Environmental Geology
Flagstaff **(602) 779-3311**
Augustana Coll.-BA; U. of Chicago-PhD
Holocene paleoclimate
Paleoclimate; Quaternary geology; surficial geology

KASMEN, NANCY G.
Secretary **(303) 753-0149**
5875 East Iliff, #214-D
Denver, Colorado 80222
Br. of Sedimentary Mineral Resour.
Denver **(303) 234-3785**

KAUFMANN, HAROLD E.
Geophysicist
Br. of Regional Geophysics
Denver **(303) 234-4938**
U. of Illinois-BS; Brigham Young U.-MS; Penn State
Dillon 2° sheet CUSMAP
Magnetics & gravity projects

KAWAKITA, GARY M.
PST
114 N. Idaho Street
San Mateo, California 94401
Br. of Analytical Laboratories
Menlo Park **(415) 323-8111 x2948**
U. of California (Berkeley)-BA

KAYS, M. ALLAN **Dorothy**
Geologist **(503) 686-1851**
2260 E. 15th Street
Eugene, Oregon 97403
Br. of Field Geochemistry & Petrology
Menlo Park **(415) 323-8111**
Southern Illinois U.-BS; Washington U.-MS, PhD
Geology of Marble Mts. Wilderness metamorphic facies, north-central Klamath Mts.
Field geochemistry & petrology

KEATEN, BARBARA A.
Chemist
Br. of Analytical Laboratories
Denver **(303) 234-4201**
Illinois State U.; U. of California (Davis)-BA
Radiochemistry, Denver

KEEFER, DAVID K. **Karen**
Geologist **(415) 961-8148**
766 Gantry Way
Mountain View, California 94040
Br. of Engineering Geology
Menlo Park **(415) 856-7115**
Stanford-BS, MS, PhD; U. of Illinois-MS
Landslides & liquefaction in historic earthquakes
Engineering geology; surficial processes & deposits; soil & rock mechanics

KEEFER, ELEANOR K. **William**
Geologist **(303) 988-5346**
773 S. Youngfield Court
Lakewood, Colorado 80228
Office of Resource Analysis
Denver **(303) 234-**
U. of Wyoming-BA
Computerization of geologic data

KEEFER, MARCIA MERGNER
Geologist
Br. of Earthquake Tectonics & Risk
Menlo Park **(415) 323-8111 x2935**
Carleton Coll.-BA; Colorado Schl. of Mines-MS
Seismogenic source zones of the southeast United States
Neo-tectonics

KEEFER, WILLIAM R. **Eleanor**
Geologist **(303) 988-5346**
773 S. Youngfield Court
Lakewood, Colorado 80228
Office of the Chief Geologist
Denver **(303) 234-3625**
U. of Wyoming-BA, MA, PhD
Stratigraphy, structure, & geologic history of Rocky Mountain sedimentary basins

KEHN, THOMAS M. **Eleanor**
Geologist
2617 Mountain Laurel Place
Reston, Virginia 22091
Br. of Coal Resources
Reston **(703) 860-7467**
U. of Arkansas-BS
Western Kentucky coal field
Coal geology & stratigraphy

KEIGHIN, C. WILLIAM **Margaret**
Geologist **(303) 986-2694**
1666 S. Holland Court
Lakewood, Colorado 80226
Br. of Sedimentary Mineral Resources
Denver **(303) 234-5119**
Oberlin Coll.-BA; U. of Colorado-MS, PhD
Low-permeability, gas-bearing sands
Sedimentary petrology; diagenesis in sedimentary rocks

KEITH, JOHN R.
Geochemist
902 N. Watford Street
Sterling, Virginia 22170
Office of Geochemistry & Geophysics
Reston **(703) 860-6582**
Georgetown Coll.-BA; U. of Kentucky-MS
Geochemistry & health; vegetation mapping

KEITH, TERRY E.C. **William**
Geologist
Br. of Field Geochemistry & Petrology
Menlo Park **(415) 323-8111 x2562**
U. of Arizona-BS; U. of Oregon-MS
Hydrothermal alteration in the Cascades
Hydrothermal alteration mineralogy

KEITH, WILLIAM J. **Terry**
Geologist
149 Bay Road
Menlo Park, California 94025
Br. of Western Mineral Resources
Menlo Park **(415) 323-8111 x2665**
U. of Arizona-BS; San Jose State U.-MS
Mineral resource evaluation
Tertiary volcanic rocks; volcanology

KELLER, MARGARET A.
Geologist
870 Bruce Way
Palo Alto, California 94303
Br. of Oil & Gas Resources
Menlo Park **(415) 856-7067**
Occidental Coll.-BA; California State (LA)

KELLER, VINCENT G.
PST
Br. of Tectonophysics
Menlo Park **(415) 323-8111 x2694**
Coll. of San Mateo-AA; San Jose State U.-BS
Portable magnetometer surveys along major' faults in California

KELLEY, EDMUND **Shirley**
Electronics Technician **(602) 774-1939**
1705 E. Linda Vista
Flagstaff, Arizona 86001
Br. of Petrophysics & Remote Sens.
Flagstaff **(602) 774-7436**
Phoenix Coll.
Paleomagnetics
Geophysical instrumentation; electronics R&D & communications

KELLEY, M. LEA
Chemist **(301) 977-7891**
19626 Enterprise Way
Gaithersburg, Maryland 20760
Br. of Isotope Geology
Reston **(703) 860-6113**
Rosemont Coll.-BA
Carbon 14 dating
Radiochemistry; Hawaii geochronology

KELLOGG, KARL S. **Nancy**
Geologist
918 E. Willamette Avenue
Colorado Springs, Colorado 80903
Br. of Latin Am. & African Geology
Jidda, Saudi Arabia
U. of Cal.(Berkeley)-BA; U. of Colorado-PhD
Geology & paleomagnetism of western Saudi Arabia
Paleomagnetism; Saudi Arabian geology; Antarctic geology

KELSEY, JAMES **Jesse**
PST
202-Ascot Place, N.E.
Washington, D.C. 20002
Br. of Analytical Laboratories
Reston **(703) 860-7473**

KEMPEMA, EDWARD W.
PST
Br. of Pacific-Arctic Geology
Menlo Park **(415) 856-7005**
Humboldt State U.-BA
Beaufort Sea geohazards
Arctic nearshore bathimetry

KENNEDY, GEORGE L. **Rose Marie Sanchez**
Geologist
Br. of Ground Motion & Faulting
Menlo Park **(415) 323-8111**
San Diego State U.-BS; U. of California (Davis)-MS, PhD
Paleontology in coastal tectonics project, Western USA
Tertiary & Quaternary paleontology & zoogeography; amino acid geochronology; molluscan systematics

KENNEDY, MARK W.
Electronics Technician **(415) 366-1710**
431-A Lincoln Avenue
Redwood City, California 94061
Br. of Isotope Geology
Menlo Park **(415) 323-8111 x2354**
Stanford-BS
Instrumentation; microprocessor applications

KENNELLY, JOHN P., JR. **Eileen**
Engineering Technician
21700 Via Regina
Saratoga, California 95070
Br. of Tectonophysics
Menlo Park **(415) 323-8111 x2386**
Geothermal studies

KENT, BION H. **Lois**
Geologist **(303) 985-8864**
13511 W. Alaska Place
Lakewood, Colorado 80228
Br. of Coal Resources
Denver **(303) 234-3536**
Cornell-BS; Stanford-MS
Recluse 1° quadrangle, Wyoming
Coal geology; photogeology

KENT, KATHLEEN M.
Geophysicist
P.O. Box 4
Pocasset, Massachusetts 02559
Br. of Atlantic-Gulf of Mex. Geology
Woods Hole **(617) 548-8700**
Northeastern U.-BS; U. of Delaware-MS
Marine geology; geophysics

KEPFERLE, ROY C. **Rhua**
Geologist **(513) 242-0671**
915 Egan Hills Drive
Cincinnati, Ohio 45229
Br. of Oil & Gas Resources
Cincinnati **(513) 684-3541**
U. of Colorado-BA; S. Dakota Schl. of Mines-MS; U. of Cincinnati-PhD
Eastern gas shale
Stratigraphy; sedimentology; basin analysis

KERR, PATRICIA T.
Coal Data Assistant **(703) 430-3369**
1049 Temple Court
Sterling, Virginia 22170
Br. of Coal Resources
Reston **(703) 860-7734**

KERRY, LEONARD E. **Helen**
Geophysicist **(509) 447-4677**
Route 4, Box 475
Newport, Washington 99156
Br. of Global Seismology
Newport, Washington **(509) 447-3195**
Louisiana State U.-BS
Seismology; geomagnetism; tsunami warning system

KESECKER, PATRICIA A.
Secretary
Office of Resource Analysis
Reston **(703) 860-6446**
Northern Virginia Community Coll.

KETCH, JANE K.
Library Technician
Library
Reston **(703) 860-6613**

KETNER, KEITH B.
Geologist
10705 W. 73rd Place
Arvada, Colorado 80005
Br. of Sedimentary Mineral Resources
Denver **(303) 234-3365**
U. of Wisconsin-BA, MA, PhD
General geology

KETTELL, NAN L.
Secretary
3615 Rose Lane
Annandale, Virginia 22003
Office of the Chief Geologist
Reston **(703) 860-6531**

KHAN, ABDUL S.
Geologist **(303) 279-9350**
1200 Golden Circle
Golden, Colorado 80401
Br. of Oil & Gas Resources
Denver **(303) 234-3435**
U. of Panjab (Pakistan)-BS, MS; U. of London-PhD, DIC
Resource appraisal, eastern coastal states & Atlantic offshore
Petroleum geology; sedimentation

KIBLER, JOHN D. **Joyce**
PST
16359 W. 10th Avenue, #V-6
Golden, Colorado 80401
Br. of Special Projects
Denver **(303) 234-2528**

KIBLER, JOYCE E. **John**
PST
16359 W. 10th Avenue, #V-6
Golden, Colorado 80401
Br. of Special Projects
Denver **(303) 234-2371**
Utica Coll.-BS
Computer analysis of geophysical data

KIEFFER, HUGH H. **Susan**
Geophysicist
Route 4, Box 710
Flagstaff, Arizona 86001
Br. of Astrogeologic Studies
Flagstaff **(602) 779-3311 x1357**
Cal Tech-PhD
Thermal behavior
NIMS; thermal infrared; comet

KIEFFER, SUSAN WERNER **Hugh**
Geologist
Route 4, Box 710
Flagstaff, Arizona 86001
Br. of Exper. Geochem. & Mineralogy
Flagstaff **(602) 261-1455**
Allegheny Coll.-BS; Cal Tech-MS, PhD
Multiphase fluid flow in geothermal & volcanic systems
Mineral thermodynamics; shock wave & meteorite impact physics

KIILSGAARD, THOR H. **Martha**
Geologist **(509) 448-1985**
S. 4604 Napa
Spokane, Washington 99203
Office of Mineral Resources
Spokane **(509) 456-4677**
U. of Idaho-BS; U. of California-MS
Challis, CUSMAP
Mineral deposits; mineral resource appraisal; mapping

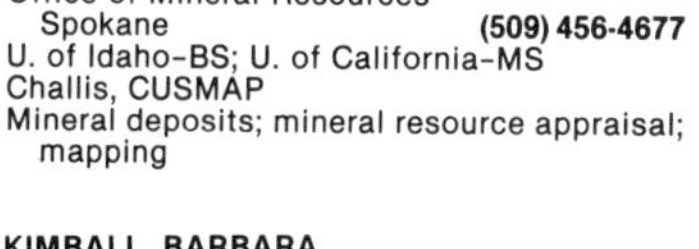

KIMBALL, BARBARA
Operations Support Technician (typing)
608 Figueroa N.E.
Albuquerque, New Mexico 87123
Br. of Global Seismology
Albuquerque **(505) 844-4637**

KINDINGER, JACK L. **Mary**
Oceanographer **(512) 991-5528**
6017 Idylwood
Corpus Christi, Texas 78412
Br. of Atlantic-Gulf of Mex. Geology
Corpus Christi **(512) 888-3294**
Corpus Christi State U.-MS
Intraslope basins, Gulf of Mexico
Biogeology

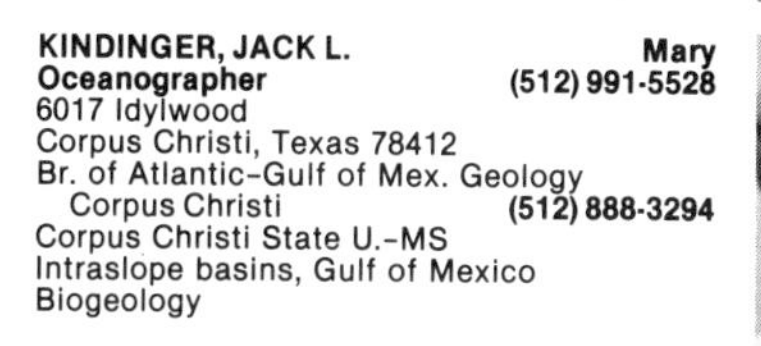

KINDINGER, MARY E
Biological Technician **(512) 991-5528**
6017 Idylwood
Corpus Christi, Texas 78412
Br. of Atlantic-Gulf of Mex. Geology
Corpus Christi **(512) 888-3294**
Texas A&I U.-BS; Corpus Christi State U.
Taxonomy; marine ecology

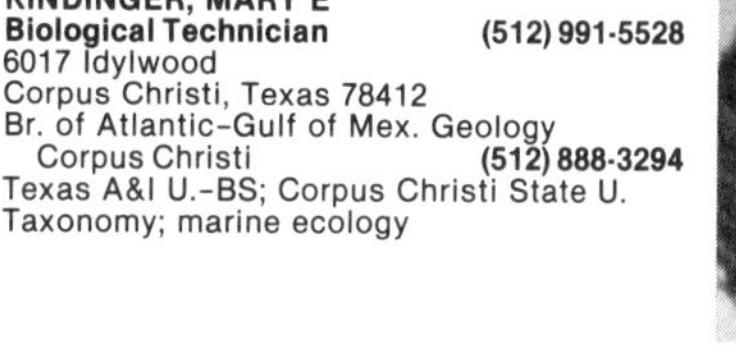

KING, BI-SHIA **Chi-Yu**
Chemist **(415) 948-4438**
381 Hawthorne Avenue
Los Altos, California 94022
Br. of Analytical Laboratories
Menlo Park **(415) 323-8111**
Taiwan Chung-Hsin U.-BS; Cornell-MS
X-ray spectroscopy
XRF analysis

KING, CHI-YU **Bi-Shia**
Geophysicist **(415) 948-4438**
381 Hawthorne Avenue
Los Altos, California 94022
Br. of Tectonophysics
Menlo Park **(415) 323-8111 x2706**
National Taiwan U.-BS; Duke-MS; Cornell-PhD
Earthquake prediction
Seismology; tectonics

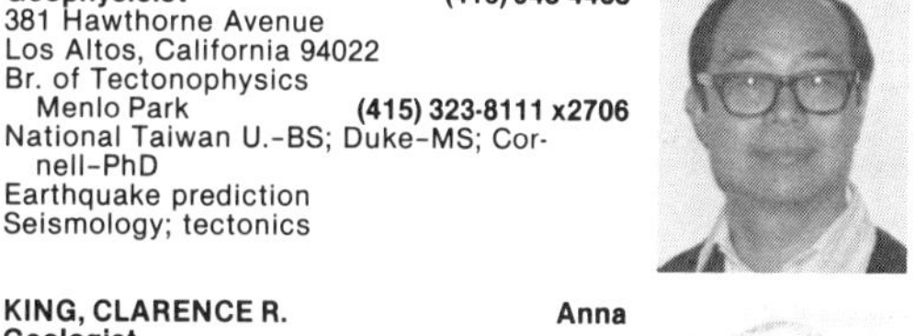

KING, CLARENCE R. **Anna**
Geologist
Office of the Director
Reston **(703) 860-6531**
Yale-PhB; Sheffield Scientific Schl. (Yale)
Geology & tectonics, Sierra Nevada, Great Basin, Rocky Mtns.; diamonds
Scientist; administrator; consultant; literateur; art collector; raconteur; gourmet

KING, ELIZABETH R.
Geophysicst **(301) 588-9604**
8403 Hartford Avenue
Silver Spring, Maryland 20910
Br. of Regional Geophysics
Reston **(703) 860-7233**
Smith-BA
Midcontinent geophysical studies
Interpretation of aeromagnetic & other geophysical data

KING, HARLEY D. **Carol**
Geologist
Br. of Exploration Research
Denver **(303) 234-6186**
John Muir Coll.; Pasadena City Coll.; Utah State U.-BS, MS
Alaska Mineral Resource Assessment Program
Geochemical & biogeochemical exploration techniques

KING, HELEN M.
Library Technician
920 S. Dale Court
Denver, Colorado 80219
Library
Denver **(303) 234-4133**

KING, NANCY E.
Mathematician **(415) 321-2419**
932 Crane Street
Menlo Park, California 94025
Br. of Tectonophysics
Menlo Park **(415) 323-8111 x2961**
U. of California (San Diego)-BA
Crustal strain
Adjustment theory; dislocation modeling

KING, PHILIP B. **Helen**
Geologist **(415) 948-0603**
670 Covington Road
Los Altos, California 94022
Br. of Western Environmental Geology
Menlo Park **(415) 323-8111 x2375**
U. of Iowa-BA, MS; Yale-PhD
General geology; regional tectonics

KING, ROBERT U. **Oleta**
Geologist **(303) 424-7551**
4551 Vance Street
Wheat Ridge, Colorado 80033
Br. of Central Mineral Resources
Denver **(303) 234-3455**
MIT-BS
Molybdenum and rhenium resources; mineral exploration

KINGSTON, MARGUERITE J.
Geochemist **(301) 657-3215**
4500 South Chelsea Lane
Bethesda, Maryland 20015
Br. of Petrophysics & Remote Sens.
Reston **(703) 860-7407**
Dunbarton Coll.-BA; George Washington U.-MS
Space shuttle multispectral radiometry
Remote sensing; geochemistry

KINNEY, DOUGLAS M. **Jeddie**
Geologist **(301) 652-4772**
5221 Baltimore Avenue
Bethesda, Maryland 20016
Office of Scientific Publications
Reston **(703) 860-6575**
Occidental Coll.-BA; Yale-PhD
Publication of colored geological maps

KIPFINGER, ROY P., JR. **Leila**
Electronics Technician **(303) 344-5025**
423 S. Lima Circle
Aurora, Colorado 80012
Br. of Petrophysics & Remote Sens.
Denver **(303) 234-2590**
Metropolitan State Coll.-BS
Operation, maintenance & modification of airborne infrared scanner
Mobile high density recording

KIRBY, JOHN R.
PST **(617) 563-5391**
Box 382
N. Falmouth, Massachusetts 02556
Br. of Atlantic-Gulf of Mex. Geology
Woods Hole **(617) 548-8700 x192**

KIRK, ALLAN R. **Cheryl**
Geologist **(303) 234-5003**
Route #1, Box 567
Conifer, Colorado 80433
Br. of Uranium & Thorium Resources
Denver **(303) 234-5003**
U. of New Hampshire-BA; SUNY (Buffalo)-MA
North Church Rock uranium, New Mexico
Geologic mapping; sedimentology; geochemistry

KIRKEMO, HAROLD **Peggy**
Geologist **(301) 530-2962**
5807 Folkstone Road
Bethesda, Maryland 20034
Br. of Eastern Mineral Resources
Reston **(703) 860-6681**
U. of Washington-BS
Office of Minerals Exploration
Mineral exploration

KIRSCHBAUM, MARK A.
Geologist
9700 W. 51st Place, D-102
Arvada, Colorado 80002
Br. of Coal Resources
Denver **(303) 234-3578**
U. of Miami-BS

KIRSCHENBAUM, HERBERT **Susan**
Chemist **(301) 299-7491**
11828 Enid Drive
Potomac, Maryland 20854
Br. of Analytical Laboratories
Reston **(703) 860-6144**
CCNY-BS
Conventional rock analysis
Gas chromatography; computer programming; particle X-ray diffraction

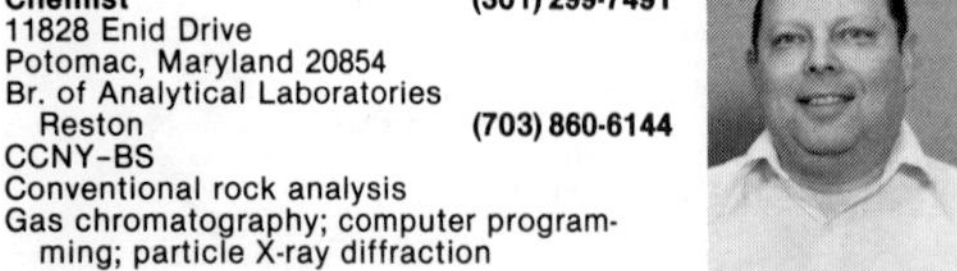

KIRSCHMER, BEVERLY A. **Jack**
Computer Technician **(303) 979-2233**
7162 S. Johnson Street
Littleton, Colorado 80123
Office of the Chief Geologist
Denver **(303) 234-6488**
Denver Community Coll.

KISTLER, RONALD W.
Geologist
Br. of Isotope Geology
Menlo Park **(415) 323-8111 x2294**
Johns Hopkins-BA; U. of California (Berkeley)-PhD
K-Ar & Rb-Sr geochronology; isotope tracers, tectonics

KLEIN, DOUGLAS P. **Sueko**
Geophysicist **(303) 986-0683**
931 S. Briarwood Drive
Lakewood, Colorado 80226
Br. of Electromag. & Geomagnetism
Denver **(303) 234-5763**
Iowa State U.-BS; U. of Hawaii-MS, PhD
Mineral resource studies, southern Arizona
Natural source electromagnetics; gravity; magnetics

KLEIN, FRED W.
Geophysicist
Br. of Field Geochem. & Petrology
Hawaiian Volcano Observ. **(808) 967-7485**
Columbia-PhD
Hawaiian seismic studies
Seismology; tectonics; computer application

KLEINHAMPL, FRANK J.
Geologist
Br. of Western Mineral Resources
Menlo Park **(415) 323-8111 x2650**

KLEINKOPF, DEAN **Kitty**
Geophysicist **(303) 986-9688**
1022 S. Beech Drive
Lakewood, Colorado 80228
Br. of Electromag. & Geomagnetism
Denver **(303) 234-2695**
U. of Missouri (Rolla)-BS; Columbia-PhD
Geophysics of northern Rocky Mountains
Magnetics; gravity; minerals geophysics

KLICK, DONALD W. **Barbara**
Physical Scientist **(703) 573-2632**
8624 Redwood Drive
Vienna, Virginia 22180
Office of Geochemistry & Geophysics
Reston **(703) 860-6581**
U.S. Naval Academy-BS; U. of Illinois-MS
Geothermal research program
Research management & program coordination

KLITGORD, KIM D. **Wivine**
Geophysicist **(617) 540-3723**
11 Mullen Way
Falmouth, Massachusetts 02540
Br. of Atlantic-Gulf of Mex. Geology
Woods Hole **(617) 548-8700**
Penn State-BS; Scripps Inst. of Oceanography-PhD
Magnetic studies of the Atlantic continental margin U.S.
Marine geophysics; continental margins

KLOCK, PAUL R. **Katherine**
Chemist
7953 Wentworth Place
Newark, California 94560
Br. of Analytical Laboratories
Menlo Park **(415) 323-8111**
U. of San Francisco-BS
Analytical services & research
Instrumentation for geochemical analysis

KNEBEL, HARLEY J.
Geologist
44 Cachalot Lane
Falmouth, Massachusetts 02540
Br. of Atlantic-Gulf of Mex. Geology
Woods Hole **(617) 548-8700**
U. of Iowa-BA; U. of Washington-MS, PhD
Environmental assessment of continental margin, east coast U.S.
Marine geology; sedimentation; nearshore & shelf sediments

KNEPPER, DANIEL H., JR. **Nancy**
Geologist **(303) 279-9420**
2592 Braun Court
Golden, Colorado 80401
Br. of Petrophysics & Remote Sens.
Denver **(303) 234-5460**
Bowling Green State U.-BS; U. of Kansas-MS; Colorado Schl. of Mines-PhD
Remote sensing; tectonics

KNIGHT, ROY J.
Chemist
Br. of Analytical Laboratories
Denver **(303) 234-4201**
U. of Colorado-BA; Colorado Schl. of Mines
Radiochemistry, Denver
Analytical geochemistry

KNIPE, DIANE E.
GFA
248 Seaside Street
Santa Cruz, California 95060
Br. of Pacific-Arctic Geology
Menlo Park **(415) 323-8111 x7039**
U. of California (Santa Cruz)-BA

KOCH, RICHARD D.
Geologist
Br. of Alaskan Geology
Menlo Park **(415) 323-8111 x2134**
San Jose State U.-BS, MS
Reconnaissance mapping of Coast Range, southeast Alaska

KOEPPEN, ROBERT P.
Geologist **(703) 471-5065**
P.O. Box 2876
Reston, Virginia 22090
Br. of Field Geochemistry & Petrology
Reston **(703) 860-7452**
U. of California (Santa Cruz)-BS, MS
Geology of Long Valley-Mono Basin
Volcanology; igneous petrology

KOESTERER, CHARLES L. **Sheryl**
Electronics Technician **(213) 339-5937**
1121 Viceroy Street
Covina, California 91722
Br. of Seismology
Pasadena **(213) 795-6811 x2916**
San Bernardino Valley Coll.; Citrus Coll.
Southern Cal. seismic network

KOESTERER, MARY ELLEN
PST
Br. of Central Mineral Resources
Denver **(303) 234-6359**
Metropolitan State Coll.-BS
Selway-Bitterroot Wilderness Area

KOHLER, WILLIAM M.
Geophysicist **(408) 246-4991**
1700 Civic Center Drive, #510
Santa Clara, California 95050
Br. of Seismology
Menlo Park **(415) 323-8111 x2237**
U. of California (Berkeley)-MS
Active seismology

KOJIMA, GEORGE **Lily**
PST **(808) 968-6932**
P.O. Box 376
Mountain View, Hawaii 96711
Br. of Field Geochem. & Petrology
Hawaiian Volcano Observ. **(808) 967-7328**
U. of Hawaii-BA
Maintenance of seismic instrumentation & telemetry

KOLLMANN, AURIEL C.
GFA
1121 Byrant, #4
Palo Alto, California 94301
Br. of Pacific-Arctic Geology
Menlo Park **(415) 856-7020**
San Jose State U.; Foothill Coll.-AA
Eastern Gulf of Alaska

KONNERT, JUDITH A. **John**
Chemist
2408 Sweetbay Lane
Reston, Virginia 22091
Br. of Exper. Geochem. & Mineralogy
Reston **(703) 860-6668**
Coll. of Wooster-BA
Crystallography

KOOK, ANN E.
Scientific Illustrator
Office of Scientific Publications
Menlo Park **(415) 323-8111 x2303**
West Valley Coll.-AA; San Jose State U.

KOONTZ, ALLEN H.
Engineering Technician
10104 Chapala Court, N.E.
Albuquerque, New Mexico 87111
Br. of Global Seismology
Albuquerque **(505) 844-4637**

KOPF, RUDOLPH W. **Doris**
Geologist **(415) 323-2391**
1233 Harrison Avenue
Redwood City, California 94062
Office of Scientific Publications
Menlo Park **(415) 323-8111 x2391**
U. of Buffalo-BA, MA
Geologic Names Committee, Western Region
Hydrotectonics, including development and flow of fault, injection, & extrusion slurry

KORK, JOHN O. **Carole**
Mathematician **(303) 279-4722**
300 Plateau Parkway
Golden, Colorado 80401
Office of Resource Analysis
Denver **(303) 234-5874**
Colorado State U.-BS, MS; U. of Tennessee-PhD
National uranium resource evaluation
Geostatistics; computer graphics; large data bases

KOSANKE, ROBERT M. **Avalonne**
Geologist **(303) 238-2998**
12085 Applewood Knolls Drive
Lakewood, Colorado 80215
Br. of Paleontology & Stratigraphy
Denver **(303) 234-3624**
Coe Coll.-BA; U. of Cincinnati-MA; U. of Illinois-PhD
Late Paleozoic biostratigraphy of major coal basins, eastern U.S.
Paleozoic palynology; stratigraphy

KOSKI, RANDOLPH A. **Nancy**
Geologist
3307 Vernon Terrace
Palo Alto, California 94303
Br. of Pacific-Arctic Geology
Menlo Park **(415) 856-7130**
U. of Minnesota-BA; Stanford-MS, PhD
Metallic deposits in oceanic crust
Volcanogenic sulfide deposits; ophiolites & related mineral deposits; metallogenesis along plate boundaries

KOTEFF, CARL Dianne
Geologist
Office of the Chief Geologist
Reston (703) 860-6544
Ohio State-BS
Glacial geology; geomorphology

KOVACH, JACK Frances
Geologist
Dept. of Geology, Muskingum College
New Concord, Ohio 43762
Br. of Isotope Geology
Denver (303) 234-5531
Waynesburg Coll.-BSc; Ohio State-MS, PhD
Isotopic composition of strontium in sea water
Strontium isotope geology; geochemistry; biogeochemistry

KOVER, ALLAN N. Debby
Geologist (301) 593-2714
10904 Lombardy Road
Silver Spring, Maryland 20901
Office of Geochemistry & Geophysics
Reston (703) 860-6581
CCNY-BS; Hamilton Coll.; George Washington U.
Geothermal research program
Remote sensing; radar geology; special maps

KOYANAGI, ROBERT Y. Judith
Geophysicist (808) 935-8950
419 Huali Place
Hilo, Hawaii 96720
Br. of Field Geochem. & Petrology
Hawaiian Volcano Observ. (808) 967-7328
Michigan Coll. of Mining & Tech-BS
Seismic monitoring of Hawaiian volcanoes
Volcanic seismicity

KRASNOW, MARTA R. David
Chemist
Br. of Coal Resources
Reston (703) 860-7551
U. of Havana-BS
Organic geochemistry of fossil fuels

KRAUSKOPF, KONRAD B. Kathryn
Geologist (415) 854-4506
806 La Mesa Drive
Menlo Park, California 94025
Br. of Field Geochem. & Petrology
Menlo Park (415) 497-3325
U. of Wisconsin-BA; U. of California-PhD; Stanford-PhD
Mapping Mariposa 15' quad
Geochemistry of ore deposits; igneous petrology

KRIVOY, HAROLD L. Glade
Geophysicist (703) 860-0733
11562 Shadbush Court
Reston, Virginia 22091
Br. of Oil & Gas Resources
Reston (703) 860-6634
U. of Utah-BS
Foreign energy resource appraisal
Offshore geophysical exploration; earthquakes & volcanology

KRIZMAN, ROBERT W.
Electronics Technician (303) 237-2172
1388 Garrison Street, E-101
Lakewood, Colorado 80215
Br. of Electromag. & Geomagnetism
Denver (303) 234-2589
Magnetotellurics

KRODEL, LIDA L.
Editorial Assistant (415) 327-9048
Br. of Western Environmental Geology
Menlo Park (415) 323-8111 x2475

KROHN, KATHLEEN KOZEY Dennis
Geologist
Br. of Coal Resources
Reston (703) 860-7580
Bryn Mawr-BA
National coal resources data system
Coal geology; sedimentary geology; regional planning

KROHN, M. DENNIS Kathy
Geologist
Br. of Petrophysics & Remote Sens.
Reston (703) 860-6994
Johns Hopkins-BA; Penn State-MS
Shuttle imaging radar
Remote sensing; structural geology; geobotany

KROPSCHOT, SUSAN J.
Technical Editor
Office of Scientific Publications
Denver (303) 234-3551
U. of Colorado-BA; U. of Denver-MA

KRUSHENSKY, RICHARD D. Arvene
Geologist
8106 Timber Valley Court
Dunn Loring, Virginia 22027
Office of Environmental Geology
Reston (703) 860-6411
Wayne State U.-BS, MS; Ohio State-PhD
Stratigraphy; structural geology of volcanic terrains; volcanology

KUBERRY, RICHARD W. Barbara
Geophysicist (303) 278-3797
95 Flora Way
Golden, Colorado 80401
Br. of Electromag. & Geomagnetism
Denver (303) 234-5064
LaSalle Coll.-BA; George Washington U.-MS
Geomagnetic observatory automation
Geophysical instrumentation; geomagnetism & observatory operations

KUCKS, ROBERT P.
PST (303) 443-7637
948 North Street
Boulder, Colorado 80302
Br. of Regional Geophysics
Denver (303) 234-4816
Wittenberg U.-BA

KUNTZ, MEL A.
Geologist
29126 Histead Drive
Evergreen, Colorado 80439
Br. of Central Environmental Geology
Denver (303) 234-3927
Carleton Coll.-BA; Northwestern U.-MS; Stanford-PhD
Snake River Plain basalts
Petrology; mineralogy; geochemistry

KVENVOLDEN, KEITH A. Mary Ann
Geologist (415) 328-0414
2433 Emerson Street
Palo Alto, California 94301
Br. of Pacific-Arctic Geology
Menlo Park (415) 856-7150
Colorado Schl. of Mines-GE; Stanford-MS, PhD
Marine organic geochemistry
Geochemical prospecting; geochemical geochronology; geohazards

KWAK, LORETTA M.
Chemist (303) 674-4337
P.O. Box 37
Idaho Springs, Colorado 80452
Br. of Isotope Geology
Denver (303) 234-3876
St. Xavier U.-BA
Zircon analysis for Pb-U-Th
Analytical chemistry & mass spectroscopy

KYSER, T. KURTIS
Geochemist **(303) 278-8376**
274 Holman Way, #4E
Golden, Colorado 80401
Br. of Isotope Geology
Denver **(303) 234-5531**
U. of California (San Diego)-BS; U. of California (Berkeley)-PhD
Origin of basic lavas & mantle xenoliths
Isotopes; igneous petrology; mantle geochemistry

LACHENBRUCH, ARTHUR H. **Edith**
Geophysicist
Br. of Tectonophysics
Menlo Park **(415) 323-8111 x2272**
Johns Hopkins U.-BA; Harvard-MA, PhD
Geothermal studies
Tectonophysics; terrestrial heat flow

LAHR, JOHN C. **Janice**
Geophysicist **(415) 328-8721**
128 Elliott Drive
Menlo Park, California 94025
Br. of Ground Motion & Faulting
Menlo Park **(415) 323-8111 x2510**
Rensselaer-BS; Columbia U.-MS, PhD
Seismic potential & tectonic framework of southern Alaska
Seismology; tectonics & seismic data processing

LAJOIE, KENNETH R. **Andrea**
Geologist **(415) 322-9791**
275 Oakhurst Place
Menlo Park, California 94025
Br. of Ground Motion & Faulting
Menlo Park **(415) 323-8111**
U. of California (Berkeley)-BA, PhD
Coastal tectonics, western United States
Quaternary geology, neotectonics

LAMB, BETH M.
PST **(415) 325-0231**
16 Waverly Court Apt. #1
Menlo Park, California 94025
Br. of Alaskan Geology
Menlo Park **(415) 323-8111 x2131**
U. of California (Santa Barbara); Universitet I Bergen (Norway)
Surficial geology, central Brooks Range
Quaternary geology; sedimentology; paleoecology

LAMOTHE, PAUL J.
Chemist **(408) 737-0532**
897 Russet Drive
Sunnyvale, California 94087
Br. of Analytical Laboratories
Menlo Park **(415) 323-8111 x2945**
U. of San Francisco-BS; Marquette U.-PhD
Emission spectroscopy
Optical emission spectroscopy; computer interfacing

LANDER, DIANE L. **Bruce**
PST **(415) 965-2166**
465 Victory Avenue
Mountain View, California 94043
Br. of Pacific-Arctic Geology
Menlo Park **(415) 856-7051**
U. of California (Berkeley)-BA
Oregon—Washington continental shelf

LANDIS, EDWIN R. **Pat**
Geologist
Br. of Coal Resources
Denver **(303) 234-3579**
Missouri Schl. of Mines-BS; U. of Colorado
Stratigraphy & resource studies of fuels and other commodities in sedimentary & volcanic rocks

LANDIS, GARY P.
Geologist
Br. of Central Mineral Resources
Denver **(303) 234-3830**
Occidental Coll.-BA; U. of Minnesota-PhD
Stable isotope studies of hydrothermal systems
Stable isotope & ore deposit geochemistry; thermodynamics

LANE, BRENT L.
PST
2183 Creighton Drive
Golden, Colorado 80401
Br. of Oil & Gas Resources
Denver **(303) 234-3435**
U. of Iowa; Metro State College
Oil & gas resource appraisal methods
Geological analysis for oil & gas resource appraisal

LANE, MARILYN K. **Thomas**
Administrative Assistant **(703) 860-8282**
2160 Quincy Adams
Herndon, Virginia 22070
Office of Energy Resources
Reston **(703) 860-6434**

LANGBEIN, JOHN O.
Geophysicist
Br. of Tectonophysics
Menlo Park **(415) 323-8111**
U. of Washington-PhD
Crustal deformation

LANGENHEIM, VIRGINIA A.
Geologist
32 Kent Place
Menlo Park, California 94025
Office of Scientific Publications
Menlo Park **(415) 323-8111 x2391**
U. of California (Berkeley)-BA, MA
Geologic Names Committee
Stratigraphy; Pennsylvanian & Permian corals

LANGER, CHARLEY J. **Susan**
Geophysicist
2510 Cragmoor Road
Boulder, Colorado 80303
Br. of Earthquake Tectonics & Risk
Denver **(303) 234-5091**
U. of Utah-BS; Virginia Poly. Inst.-MS
Post-earthquake investigations
Aftershock studies; near-field observations & seismic source studies

LANPHERE, MARVIN A. **Joyce**
Geologist
1036 Oakland Avenue
Menlo Park, California 94025
Br. of Isotope Geology
Menlo Park **(415) 467-2649**
Montana Schl. of Mines-BS, MS; Cal. Tech.-PhD
Geochronology; isotope geochemistry

LANTZ, ROBERT J. **Joyce**
Geologist **(206) 632-2161**
3820 Meridian Avenue N.
Seattle, Washington, 98103
Office of Energy Resources
Seattle **(206) 442-1995**

LARSEN, FREDERICK D.
Geologist **(802) 485-7715**
9 Slate Avenue
Northfield, Vermont 05663
Br. of Eastern Environmental Geology
Northfield, Vermont **(802) 485-5011 x242**
Middlebury Coll.-BA; Boston U.-MA; U. of Massachusetts-PhD
Massachusetts cooperative
Glacial geology; geomorphology

LARSEN, MATTHEW C.
PST
Br. of Pacific-Arctic Geology
Menlo Park **(415) 323-2214**
Antioch Coll.-BS
Northern Bering Sea

LARSON, JOHN C.
Electronics Technician
77 Canterbury Circle
Hyannis, Massachusetts 02601
Br. of Atlantic-Gulf of Mex. Geology
Woods Hole **(617) 548-8700**

LARSON, RICHARD R. **Mildred**
Physicist **(301) 572-7993**
3144 Castleleigh Road
Silver Spring, Maryland 20904
Br. of Analytical Laboratories
Reston **(703) 860-7543**
Oregon State U.-BS, MS
X-ray spectroscopy
Scanning electron microscopy; X-ray spectroscopy; electronics

LAW, BEN E.
Geologist
Br. of Oil & Gas Resources
Denver **(303) 234-5876**
San Diego State U.-BS, MS
Western gas sands—greater Green River Basin
Stratigraphy & sedimentology

LAYMAN, LAWRENCE R.
Chemist **(303) 232-7704**
2440 Otis Court
Edgewater, Colorado 80214
Br. of Analytical Laboratories
Denver **(303) 234-6405**
Occidental Coll.-BA; Indiana U.-PhD
Laboratory automation
Atomic spectrophotometry—sources & systems

LEACH, DAVID L. **Kay**
Geologist **(303) 985-5681**
13221 W. Montana Avenue
Lakewood, Colorado 80228
Br. of Exploration Research
Denver **(303) 234-6556**
Virginia Poly. Inst.-BS; U. of Missouri-PhD
Geochemistry; economic geology

LEARNED, ROBERT E. **Abby**
Geologist **(303) 279-4684**
614 Wyoming Street
Golden, Colorado 80401
Br. of Exploration Research
Denver **(303) 234-6156**
Occidental Coll.-BA; UCLA-MA; U. of California (Riverside)-PhD
Exploration geochem., tropical regions
Geochemical exploration; metallic mineral deposits

LEAVY, BRIAN D.
Geologist
Br. of Eastern Environmental Geology
Reston **(703) 860-6421**
U. of Rhode Island-BS, MS
Culpeper Basin environmental geology & geohydrology
Igneous petrology; geochem.; tectonics

Le COMPTE, JAMES R.
Geologist **(415) 326-5918**
400 Ravenswood Avenue, #8
Menlo Park, California 94025
Br. of Alaskan Geology
Menlo Park **(415) 323-8111 x2025**
San Jose State U.-BA, MS
Alaskan mineral resource assessment program (AMRAP)—Landsat imagery interp.
Remote sensing; areal geology; sedimentary geology (carbonates)

LEE, CLEARTHUR
Electronics Technician
Br. of Network Operations
Menlo Park **(415) 323-8111**
Alabama State Coll.-BA
Computer science & digital design

LEE, DONALD E. **Joyce**
Geologist **(303) 986-8345**
12979 W. Ohio Avenue
Lakewood, Colorado 80228
Br. of Regional Geochemistry
Denver **(303) 234-3286**
Carleton Coll.-BA; U. of Minnesota-MS; Stanford-PhD
Basin-Range granites

LEE, FITZHUGH, T. **Peggy**
Geologist **(303) 237-6100**
1710 Alkire Court
Golden, Colorado 80401
Br. of Engineering Geology
Denver **(303) 234-5185**
U. of Virginia-BA; Virginia Polytech U.-MS
Underground openings for oil & radioactive waste storage
In situ rock stresses; engineering geology of underground excavations

LEE, HOMA J. **Kathleen**
Civil Engineer
341 Redwood Avenue
Redwood City, California 94062
Br. of Engineering Geology
Menlo Park **(415) 856-7107**
MIT-BS, MSCE; Scripps Inst. of Oceanography
Marine geotechnology

LEE, JACK W. **Kathryn**
Electronics Technician **(415) 968-1511**
11605 Farndon Avenue
Los Altos, California 94022
Br. of Pacific-Arctic Geology
Menlo Park **(415) 856-7093**
California State Polytech. U.
Marine geology research electronics
Design; development; interface

LEE, KWANG Y. **Phyllis**
Geologist **(703) 524-3428**
715 North George Mason Drive
Arlington, Virginia 22203
Br. of Eastern Environmental Geology
Reston **(703) 860-6595**
National Peking U.-BS; Ohio State-MS, PhD
Sedimentology, stratigraphy, petrology & economic geology

LEE, WILLIAM H. K. **Shirley**
Geophysicist
Br. of Seismology
Menlo Park **(415) 323-8111 x2630**
U. of Alberta-BS; UCLA-PhD
Microearthquake data analysis
Seismology

LeFEBER, JOSEPHINE A.
Clerk-Stenographer
Br. of Oil & Gas Resources
Denver **(303) 234-2328**
U. of New Mexico

LEINZ, REINHARD W. **Sandra**
Chemist **(303) 423-9020**
7360 W. 74th Place
Arvada, Colorado 80003
Br. of Exploration Research
Golden **(303) 234-6152**
Regis Coll.-BS; Colorado Schl. of Mines; Colorado U.
Geochemical exploration, western U.S.; analytical chemistry applied to geochemical exploration

LEISTER, JOHN W., JR. **Lois**
Engineering Technician **(301) 735-7708**
4112 Ryon Road
Upper Marlboro, Maryland 20870
Br. of Analytical Laboratories
Reston **(703) 860-7441**
Design & development, lab, equipment

LeLANGE, JOHN E.
PST
Br. of Isotope Geology
Menlo Park (415) 323-8111 x2857
USC-BS

LEMASTER, MARY E. Billy
Coal Data Assistant (703) 437-3097
1115 S. Williamsburg Court
Sterling, Virginia 22170
Br. of Coal Resources
Reston (703) 860-7306

LEO, GERHARD W. Elaine
Geologist (703) 938-3229
Vienna, Virginia 22180
Br. of Eastern Environmental Geology
Reston (703) 860-6503
U. of Iowa; Stanford-BS, PhD
Geochemistry & petrology of Oliverian domes
Petrology; geochemistry; field geology

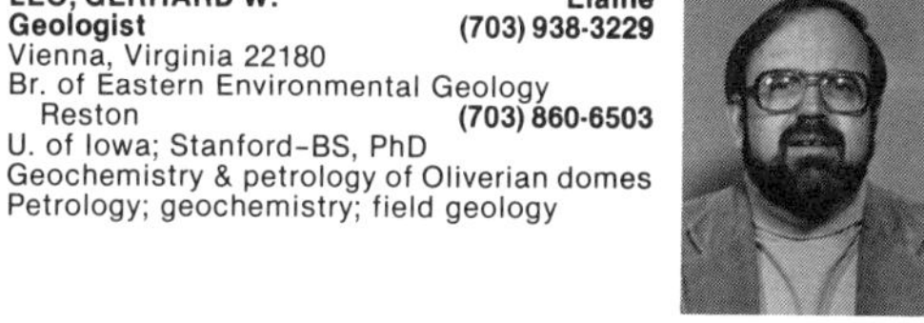

LEONARD, BENJAMIN F., III Eleanor
Geologist (303) 279-1373
2907 Sunset Drive
Golden, Colorado 80401
Br. of Central Mineral Resources
Denver (303) 234-4207
Hamilton Coll.-BS; Princeton-MA, PhD
Big Creek—Yellow Pine, Idaho
Mineral deposits; ore minerals; regional geology, central Idaho

LEONARD, SUSAN E. Michael
Administrative Assistant (202) 543-6165
445 4th Street, NE
Washington, D.C. 20002
Br. of Paleontology & Stratigraphy
Washington, D.C. (202) 343-4348
Drew U.-BA

LEONE, LAUREL E. Steve Wegener
PST (415) 854-5887
P.O. Box 1678
Palo Alto, California 94302
Br. of Seismology
Menlo Park (415) 323-8111 x2043
Brown U.; California State (Hayward)
Fault zone properties
Computer programming

LEONG, KAM W. Siu
Geochemist (415) 365-5854
2707 Jefferson Avenue
Redwood City, California 94062
Br. of Pacific-Arctic Geology
Menlo Park (415) 856-7154
U. of Hawaii-BS
Ocean floor mineral resources
Marine geochemistry; analytical chemistry

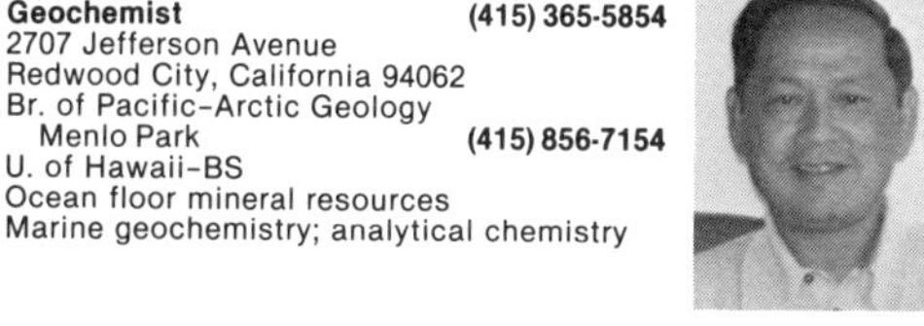

LERCH, HARRY E., III Marcia
Chemist (304) 725-4177
Rt. 1, Box T175
Charlestown, West Virginia 25414
Br. of Atlantic-Gulf of Mex. Geology
Reston (703) 860-6965
Sheperd College-BS
Marine organic geochemistry; resource & environmental assessment
Marine geochemistry; instrumental analysis

LESURE, FRANK G. Nancy
Geologist (703) 356-7102
6210 Loch Raven Drive
McLean, Virginia 22101
Br. of Eastern Mineral Resources
Reston (703) 860-6913
Virginia Polytech. Inst.-BS; Yale-MS, PhD
Resource assessments, eastern wilderness areas
Mineral resources, eastern United States

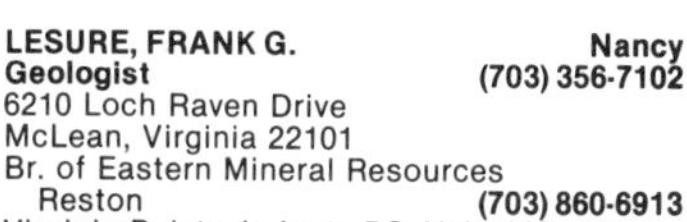

LETTIS, WILLIAM R. Heidi
Geologist (415) 482-5037
4051 Maybelle Avenue
Oakland, California 94619
Br. of Western Environmental Geology
Menlo Park (415) 323-8111 x2098
Humboldt State U.-BS; U. of California (Berkeley)-MA
San Joaquin Valley
Tectonics; Quaternary geology; mineral resource exploration

LEVENTHAL, JOEL S.
Research Chemist (303) 423-4255
7685 Lewis Street
Arvada, Colorado 80005
Br. of Uranium-Thorium Resources
Denver (303) 234-3285
California State U. (Los Angeles)-BS; U. of Arizona-MS, PhD
Organic geochemistry & uranium
Trace elements in sediments; carbon & sulfur cycles; pyrolysis gas chromatography

LEWIS, DIANE M.
Librarian
2200 Castlerock Square, 11-C
Reston, Virginia 22091
Library
Reston (703) 860-6613
Seton Hill Coll.-BA; Catholic U.-MLS

LEWIS, ELIZABETH M. William
Editorial Assistant (303) 279-1548
1953 Goldenvue Drive
Golden, Colorado 80401
Br. of Petrophysics & Remote Sens.
Denver (303) 234-5177
Colorado State U.-BS

LICHTE, FREDERICK E. Julianne
Chemist
Br. of Analytical Laboratories
Denver (303) 234-2521
Wartburg Coll.-BA; Colorado State U.-PhD
Atomic spectroscopy

LICHTMAN, GRANT S.
PST (415) 854-1094
1692 Altschul Avenue
Menlo Park, California 94025
Br. of Pacific-Arctic Geology
Menlo Park (415) 856-7041
U. of California (San Diego); Stanford-BS, MS
Mid-ocean ridge exploration

LIDKE, DAVID J.
Geologist (303) 477-0564
4618 W. 33rd Avenue
Denver, Colorado 80212
Br. of Central Environmental Geology
Denver (303) 234-5371
Colorado State U.-BS
Tectonics; structural geology; sedimentation

LIDZ, BARBARA H.
Geologist
5776 S.W. 27th Street
Miami, Florida 33155
Br. of Oil & Gas Resources
Miami Beach (305) 672-1784
Smith College; Columbia U.; USC; U. of Miami-BS
Biostratigraphy (Neogene); logging well cuttings; carbonate diagenesis & coral banding studies

LIGON, DENNIS T., JR.
PST
Br. of Atlantic-Gulf of Mex. Geology
Reston (703) 860-6965
Marine organic geochemistry (resource & environmental assessment)
Hydrocarbon geochemistry

LIKE, LINDA L.
Administrative Officer
Office of the Chief Geologist
Menlo Park **(415) 467-2214**
U. of California (Santa Barbara); San Jose State U.-BA

LILLARD, NEAL M.
Motor Boat Operator **(512) 992-8194**
5110 Janssen
Corpus Christi, Texas 78411
Br. of Atlantic-Gulf of Mex. Geology
Corpus Christi **(512) 888-3294**
Texas A&I-BS

LINDSAY, JAMES R. **Sharon**
Chemist **(703) 364-2433**
Cool Springs Farm
Delaplane, Virginia 22025
Br. of Analytical Laboratories
Reston **(703) 860-7543**
Rutgers-BA; U. of Maryland-MS
X-ray spectroscopy project
X-ray spectroscopy; electron spectroscopy; trace element analytical methods

LINDSEY, DAVID A. **Barbara**
Geologist
Br. of Central Mineral Resources
Denver **(303) 234-2721**
U. of Nebraska-BS; Johns Hopkins-PhD
Metal resources in redbed sequences, Colorado & New Mexico
Mineral deposits; volcanic rocks; clastic sedimentary rocks

LINDVALL, ROBERT M. **Phoebe**
Geologist **(303) 986-8272**
2812 South Depew Street
Denver, Colorado 80227
Br. of Engineering Geology
Denver **(303) 234-3819**
Augustana Coll.-BA; U. of Iowa-MS
Energy lands program, Powder River basin
Engineering geology; urban geology

LINE, MILDRED W.
Administrative Technician **(303) 455-0706**
4895 Julian Street
Denver, Colorado 80221
Office of Geochemistry & Geophysics
Denver **(303) 234-5461**
Colorado State U.; Community Coll. of Denver-AA; Metropolitan State Coll.

LINENBERGER, WILLIAM
PST
14 Colorado Boulevard
Denver, Colorado 80206
Br. of Oil & Gas Resources
Denver **(303) 234-5105**
Metropolitan State College-BS

LINTON, JIMMIE R. **Nell**
Cartographic Technician **(303) 237-6851**
825 Oak Street, Apt. 3
Lakewood, Colorado 80215
Br. of Regional Geophysics
Denver **(303) 234-4816**
Arkansas Polytechnic U.
Computer processing of data

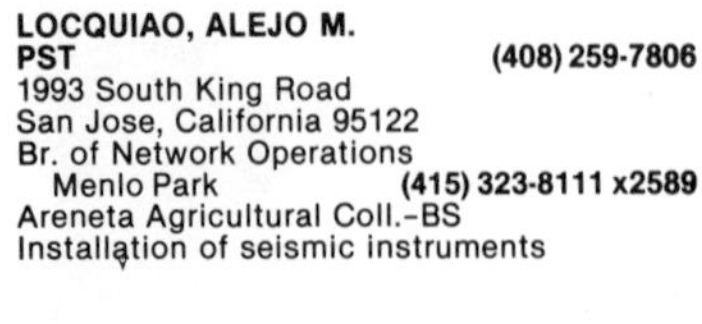

LIPIN, BRUCE R. **Cookie**
Geologist **(703) 860-0091**
2308 Archdale Road
Reston, Virginia 22091
Br. of Eastern Mineral Resources
Reston **(703) 860-7356**
City Coll. of New York-BS; Penn State-PhD
Geology of chromium
Economic geology; petrology; phase equilibria

LIPMAN, PETER W. **Beverly**
Geologist **(303) 526-1517**
26596 Columbine Glen
Golden, Colorado 80401
Br. of Central Environmental Geology
Denver **(303) 234-2901**
Yale-BS; Stanford-MS, PhD
Roots of calderas
Volcanology; petrology; structural geology

LISOWSKI, MICHAEL
Geophysicist **(415) 321-1635**
540 Lowell Avenue
Palo Alto, California 94301
Br. of Tectonophysics
Menlo Park **(415) 323-8111 x2715**
U. of California (Berkeley)-BA
Crustal strain
Geodetic surveys & analysis

LISTER, JEAN H. **James**
Geologist **(303) 232-4652**
Red Cloud Way
Conifer, Colorado 80433
Br. of Oil & Gas Resources
Denver **(303) 234-5235**
U. of Montana-BS, BA
Sedgwick Basin (oil & gas appraisals)
Oil & gas resource appraisal, logging

LISZEWSKI, EDWARD H. **Frances**
Librarian **(301) 926-6936**
9400 Union Place
Gaithersburg, Maryland 20760
Library
Reston **(703) 860-6618**
Loyola Coll.-BS; Syracuse U.-MLS

LIU, HSI-PING
Geophysicist
Br. of Tectonophysics
Menlo Park **(415) 323-8111 x2731**
Cal. Tech.-PhD
Anelastic properties of rocks; numerical modeling of crustal phenomena; field investigation of earthquake precursory phenomena

LOCKNER, DAVID A.
Geophysicist
Br. of Tectonophysics
Menlo Park **(415) 323-8111 x2597**
U. of Rochester
Rock mechanics

LOCKWOOD, JOHN P. **Martha**
Geologist **(808) 967-7357**
P.O. Box 69
Volcano, Hawaii 96785
Br. of Field Geochem. & Petrology
Hawaiian Volcano Observ. **(808) 967-7328**
U. of California (Riverside)-BA; Princeton-PhD
Eruptive history of Mauna Loa volcano, Hawaii
Volcanology; field geology; igneous petrology

LOCQUIAO, ALEJO M.
PST **(408) 259-7806**
1993 South King Road
San Jose, California 95122
Br. of Network Operations
Menlo Park **(415) 323-8111 x2589**
Areneta Agricultural Coll.-BS
Installation of seismic instruments

LOFERSKI, PATRICIA J.
Geologist **(703) 860-8854**
11817 Breton Court, 12-C
Reston, Virginia 22091
Br. of Eastern Mineral Resources
Reston **(703) 860-7356**
Boston College-BS
Geology of chromium
Mineralogy; petrology; economic geology

LOMBARD, CHARLES W.
Warehouseman **(617) 888-6841**
28 Beach Road
East Sandwich, Massachusetts 02537
Br. of Atlantic-Gulf of Mex. Geology
Woods Hole **(617) 548-8700**

LONDON, ELIZABETH B.H.
Geologist
Br. of Eastern Environmental Geology
Reston **(703) 860-6503**
Tufts-BS; Arizona State U.-MA
Surficial geologic map of Connecticut
Surficial deposits

LONEY, ROBERT A.
Geologist
12112 Foothill Lane
Los Altos Hills, California 94022
Br. of Alaskan Geology
Menlo Park **(415) 323-2384**
U. of Washington-BS, MS; U. of California (Berkeley)-PhD
Petrology of LaPerouse layered intrusion
Structural petrology; igneous & metamorphic petrology

LONG, ALLAN T.
Geophysicist
Br. of Oil & Gas Resources
Menlo Park **(415) 323-8111 x2069**
Humboldt State U.-BS; U. of Colorado-MS, MSEE
Marine geology multichannel seismic data
Exploration geophysics; digital signal processing

LONG, CARL L. **Ruth Marie**
Geophysicist **(303) 985-2736**
12090 W. Alabama Place
Lakewood, Colorado 80228
Br. of Electromag. & Geomagnetism
Denver **(303) 234-3341**
U. of Colorado-BA
Wilderness area studies
Electromagnetics applied to geothermal & minerals exploration

LONG, KEITH, R.
PST **(408) 241-2691**
3245 Catalina Avenue
Santa Clara, California 95051
Office of Resource Analysis
Menlo Park **(415) 323-8111 x2906**
U. of California (Santa Cruz)-BA, BS
CUSMAP, Walker Lake Quadrangle
Porphyry-type ore deposits; metallogenesis

LONG, SHIRLEY L. **Gerard**
Program Assistant **(301) 681-6252**
1809 Myrtle Road
Silver Spring, Maryland 20902
Office of International Geology
Reston **(703) 860-6418**

LOPEZ, DAVID A. **Carolyn**
Geologist
Br. of Central Environmental Geology
Denver **(303) 234-5260**
U. of Colorado-BA; U. of New Mexico-MS; Colorado Schl. of Mines-PhD
Dillon 2° (Montana-Idaho)
Silicic volcanism; stratigraphy & sedimentation

LOVE, ALONZA H. **Eva**
Chemist **(303) 771-2570**
3872 S. Quebec
Denver, Colorado 80237
Br. of Oil & Gas Resources
Denver **(303) 234-2399**
Langston U.-BS; Colorado Schl. of Mines-MS
Kerogen studies; organic petrology

LOVE, ELEANOR M. **James**
Budget & Accounting Asst. **(703) 777-7096**
358 Fort Evans Drive
Leesburg, Virginia 22075
Office of Scientific Publications
Reston **(703) 860-6536**

LOVE, J. DAVID **Jane**
Geologist **(307) 745-4436**
309 South 11th Street
Laramie, Wyoming 82070
Br. of Central Environmental Geology
Laramie, Wyoming **(307) 745-4495**
U. of Wyoming-BA, MA, LLD (Hon); Yale-PhD
Geologic map of Wyoming & geology of Grand Teton National Park, Wyoming
Stratig.; Rocky Mt. structure; commodities (oil & gas, vanadium, gold, uranium)

LOW, DORIS L.
PST **(617) 693-3441**
Vineyard Haven, Massachusetts 02568
Br. of Paleontology & Stratigraphy
Woods Hole **(617) 548-8700 x138**
American U.
Stratigraphy of the Atlantic coastal margin
Micropaleontology (Cenozoic—Mesozoic smaller Foraminifera)

LUCCHITTA, BAERBEL K. **Ivo**
Geologist **(602) 774-0136**
1111 Navajo Drive
Flagstaff, Arizona 86001
Br. of Astrogeologic Studies
Flagstaff **(602) 779-3311 x1356**
Kent State U.-BS; Penn State U.-MS, PhD
Planetary geology
Lunar & planetary geology; geomorphology & structural geology

LUCCHITTA, IVO **Baerbel**
Geologist
1111 Navajo Drive
Flagstaff, Arizona 86001
Br. of Central Environmental Geology
Flagstaff **(602) 779-3311 x1383**
Cal. Tech-BS; Penn State-PhD'
Western Arizona tectonics
Tectonics; Cenozoic deposits; geomorphology

LUCE, ROBERT W. **Patricia**
Geologist **(703) 860-3795**
11500 Purple Beech Drive
Reston, Virginia 22091
Br. of Eastern Mineral Resources
Reston **(703) 860-7356**
Dartmouth-BA; U. of Illinois-MS; Stanford-PhD
Economic geology; alteration geochemistry

LUDINGTON, STEPHEN D.
Geologist **(303) 377-5758**
1111 Adams Street
Denver, Colorado 80206
Br. of Central Mineral Resources
Denver **(303) 234-2773**
Stanford-BS; U. of Colorado-MS, PhD
Mineral resource appraisal of Questa, N.M. volcanic field
Ore deposits; igneous petrology

LUEDKE, ROBERT G. **Elaine**
Geologist **(301) 656-2680**
4522 Cheltenham Drive
Bethesda, Maryland 22014
Br. of Field Geochemistry & Petrology
Reston **(703) 860-7165**
U. of Colorado-BA, MS
Regional volcanic geology
Areal geology; mineral deposits

LUEPKE, GRETCHEN
Geologist
350 Sharon Park Drive, #T-12
Menlo Park, California 94025
Br. of Pacific-Arctic Geology
Menlo Park **(415) 856-7001**
U. of Arizona-BS, MS
Coastal sedimentary processes
Sedimentology; heavy minerals

LUETSCHER, JOHN D.
Electronic Technician
Br. of Isotope Geology
Menlo Park **(415) 323-2214**

LUFT, STANLEY J. **Anita**
Geologist **(303) 986-3098**
870 S. Miller Court
Lakewood, Colorado 80226
Br. of Central Environmental Geology
Denver **(303) 234-5966**
Syracuse-BA; Penn State-MS
Uranium potential, Browns Park Formation, Colorado & Utah
Stratigraphy; volcanology & economic geology

LUGN, RICHARD V. **Barbara**
PST **(415) 964-3176**
2450 Benjamin Drive
Mountain View, California 94040
Br. of Astrogeologic Studies
Menlo Park **(415) 323-8111 x2361**
U. of Nebraska-BS, MS
Stereoplotters
Planetary data

LUPE, ROBERT D. **Carolyn**
Geologist **(303) 722-2088**
1219 S. Race Street
Denver, Colorado 80210
Br. of Uranium-Thorium Resources
Denver **(303) 234-5038**
Pomona College; U. of Washington
Triassic rocks, Colorado Plateau; uranium resource assessment
Sedimentology & stratigraphy; geology & public policy; resource assessment

LUTTER, WILLIAM J.
PST
Br. of Tectonophysics
Menlo Park **(415) 323-2214**
U. of California (Berkeley)-BS

LUTTRELL, GWENDOLYN W. **John**
Geologist **(301) 933-7686**
9408 Byeforde Road
Kensington, Maryland 20795
Office of Scientific Publications
Reston **(703) 928-6511**
Wellesley-BA; U. of New Mexico
Lexicon of Geologic Names
Metallic ore deposits of southeastern U.S.

LYONS, PAUL C. **Arlene**
Geologist **(703) 620-3166**
2335 Millennium Lane
Reston, Virginia 22091
Br. of Coal Resources
Reston **(703) 860-6684**
Boston U.-AA, BA, MA, PhD
Maryland coal resources
Coal geology; coal stratigraphy; paleobotany

LYTTLE, PETER T.
Geologist
Middleburg, Virginia 22117
Br. of Eastern Environmental Geology
Reston **(703) 860-6421**
Syracuse-BS; Harvard-MA, PhD
Newark 2° Quadrangle, N.J., N.Y., & PA.
Appalachian tectonics; metamorphic petrology

MABERRY, ANDREA L. **John**
International Program Assistant
5905 Old Sawmill Road
Fairfax, Virginia 22030
Office of International Geology
Reston **(703) 860-6410**

MABERRY, JOHN O. **Andrea**
Geologist **(703) 830-1712**
5905 Old Sawmill Road
Fairfax, Virginia 22030
Office of Environmental Geology
Reston **(703) 860-6412**
U. of Colorado-BA; Colorado Schl. of Mines-MS
Energy land program
Stratigraphy; engineering geology; energy resource geology

MABEY, DON R.
Geophysicist
Br. of Regional Geophysics
Salt Lake City **(801) 524-5640**
U. of Utah-BS
Geothermal resources; mineral resources; exploration geophysics

MacDONALD, EDWARD F.
PST **(202) 659-0393**
2030 F. Street, N.W., Apt. 1007
Washington, D.C. 20006
Br. of Paleontology & Stratigraphy
Washington, D.C. **(202) 343-3495**
U. of Maine-BS
Conodonts

MACHETTE, MICHAEL N. **Nancy**
Geologist **(303) 744-3014**
1190 S. Emerson Street
Denver, Colorado 80210
Br. of Central Environmental Geology
Denver **(303) 234-5167**
San Jose State U.-BS; U. of Colorado-MS
Neotectonics & Quaternary dating
Soils; Cenozoic stratigraphy; tectonics

MACKE, DAVID L.
Geologist **(303) 279-5641**
1411 Ulysses #8
Golden, Colorado 80401
Br. of Uranium-Thorium Resources
Denver **(303) 234-5811**
U. of Illinois-BS; Colorado State U.-MS
Tertiary of northern Great Plains; thorium resources of Florida's raised beaches
Fluvial geomorphology; sedimentology; stratigraphy

MACLACHLAN, MARJORIE E. **James**
Geologist **(303) 986-7192**
2540 S. Brentwood
Lakewood, Colorado 80227
Office of Scientific Publications
Denver **(303) 234-3212**
Mount Holyoke-BA; U. of Colorado
Upper Triassic rocks of northeast New Mexico
Stratigraphy & stratigraphic nomenclature; Triassic; sedimentation

MACLEOD, NORMAN S. **Caroline**
Geologist **(408) 739-9461**
1037 Enderby Way
Sunnyvale, California 94087
Br. of Western Mineral Resources
Menlo Park **(415) 323-8111 x2359**
U. of Cal. (Riverside)-BA; U. of Cal. (Santa Barbara)-PhD
Oregon geothermal, Cascade wilderness
Volcanology; regional geology of Pacific NW; igneous petrology

MACQUEEN, LAURA M.
Librarian **(703) 860-5317**
12035 Greywing Square, B-2
Reston, Virginia 22091
Library
Reston **(703) 860-6679**
Wayne State U.-BA; U. of Michigan-AMLS

MADOLE, RICHARD F.
Geologist **(303) 494-8693**
737 Glen Haven Court
Boulder, Colorado 80303
Br. of Central Environmental Geology
Denver **(303) 234-5258**
Case-Western Reserve U.-BS; Ohio State U.-MSc, PhD
Energy lands: NW Colorado & central Utah
Quaternary stratigraphy; glacial geology; soil science

MADRID, RAUL J.
Geologist **(415) 327-8249**
1275 Harriet Street
Palo Alto, California 94301
Br. of Western Minerals Resources
Menlo Park **(415) 323-2214**
U. of Cal. (Berkeley)-BS; San Jose State U.; Stanford U.
Kinematics of emplacement & styles of deformation of the Roberts Mtn. allochthon.
Structural geology

MADSEN, BETH M.
Geologist **(303) 988-0827**
740 S. Youngfield Court
Lakewood, Colorado 80228
Br. of Sedimentary Mineral Resources
Denver **(303) 234-3402**
Cornell College-BA; U. of Wisconsin
Mineralogy of non-metallic deposits; petrology of chemical sediments

MAGNER, JERRY E.
PST
Br. of Special Projects
Denver **(303) 234-2391**
Denver Community College; Metropolitan State College-AS
Rock mechanics; hydrofracturing

MAHRT, LOUIS R., JR.
PST
Br. of Sedimentary Mineral Resour.
Denver **(303) 234-3402**
U. of Dayton; Northern Arizona U.-BS
Heavy liquid mineral separations

MAJOR, VIRGINIA L.
Geologist
3301 5th Street, S.E.
Washington, D.C. 20032
Office of Scientific Publications
Reston **(703) 860-6517**
Virginia Union U.-BS; Virginia State Coll.-MS
Geologic Inquiries Group

MALCOLM, MOLLIE J. **Ronald**
PST **(303) 697-8874**
Route 1, Box 612
Morrison, Colorado 80465
Br. of Analytical Laboratories
Denver **(303) 234-6404**
Colorado State U.-BS

MALDE, HAROLD E. **Caroline**
Geologist **(303) 442-3628**
842 Grant Place
Boulder, Colorado 80302
Br. of Central Environmental Geology
Denver **(303) 234-2864**
Willamette U.-AB; Harvard U.; U. of Colorado
Geologic factors for evaluating reclamation potential
Geomorphology; Quaternary & Cenozoic geology; environmental geology

MALDONADO, FLORIAN **Chris**
Geologist
9017 Dover
Broomfield, Colorado
Br. of Special Projects
Denver **(303) 234-2261**
U. of New Mexico-BS, MS
Nevada nuclear waste investigations
Volcanology & granite studies

MALLIS, ROBERT R.
Geophysicist **(415) 325-7932**
1033 Del Norte Avenue
Menlo Park, California 94025
Br. of Field Geochemistry & Petrology
Menlo Park **(415) 323-8111 x2551**
Virginia Polytechnic Inst.-BS
Geothermal research program
Seismology

MALLOY, MARY J.
Secretary
Sunnyvale, California 94087
Br. of Alaskan Geology
Menlo Park **(415) 323-8111**

MAMAY, SERGIUS H. **Hermoise**
Botanist **(301) 365-3168**
8001 Carita Court
Bethesda, Maryland 20034
Br. of Paleontology & Stratigraphy
Washington, D.C. **(202) 343-2435**
U. of Akron-BS; U. of Washington-MA, PhD
Paleozoic paleobotany; Permian floras, SW United States

MANGUM, JOHN H. **Bessie**
PST **(202) 399-0610**
4258 E. Capitol Street, N.E., Apt. 201
Washington, D.C. 20019
Br. of Isotope Geology
Reston **(703) 860-6593**
Bloomfield Coll.; Howard U.
Geochemistry & geophysics; mineral deposits

MANHEIM, FRANK T. **Ose**
Geologist **(617) 548-6426**
9 Broken Bow Lane
Teaticket, Massachusetts 02536
Br. of Atlantic-Gulf of Mex. Geology
Woods Hole **(617) 548-8700**
Harvard-BA; U. of Minn.-MS; U. of Stockholm-PhD, DSc
Geochemistry & hydrochemistry of marine sediments, mineral resour. & groundwater
Formation fluids; resource policy

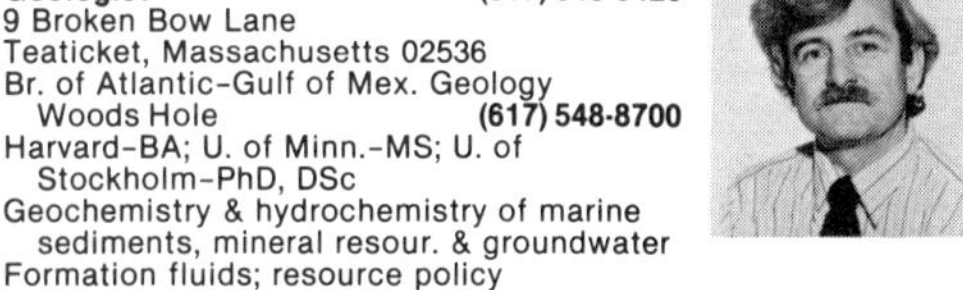

MANKINEN, EDWARD A. **Jeanne**
Geologist **(415) 969-0693**
1594 Dennis Lane
Mountain View, California 94040
Br. of Petrophysics & Remote Sens.
Menlo Park **(415) 323-8111 x2418**
San Jose State Coll.-BS, MS
Geomagnetic polarity time scale & paleosecular variation
Paleomagnetism

MANKINEN, JEANNE C. **Edward**
Administrative Clerk **(415) 969-0693**
1594 Dennis Lane
Mountain View, California 94040
Office of the Chief Geologist
Menlo Park **(415) 323-8111**
San Francisco State U.; San Jose State U.

MANLEY, DOROTHY J. **Eugene**
Secretary **(703) 378-4923**
13518 Ellendale Drive
Chantilly, Virginia 22021
Office of Resource Analysis
Reston **(703) 860-6451**

MANLEY, KIM
Geologist **(303) 232-3281**
145 S. Garrison Street
Lakewood, Colorado 80226
Br. of Central Environmental Geology
Denver **(303) 234-5168**
U. of Colorado-BA, PhD; U. of Texas-MEd
Mapping in Aztec, N.M. 1 x 2° quad.
Tertiary stratigraphy

MANN, DENNIS M.
Geologist
Br. of Pacific-Arctic Geology
Menlo Park **(415) 323-8111 x2072**
San Jose State U.-BS, MS
Marine seismic data analysis & processing
Seismic geology; sedimentology; paleontology

MANN, DOROTHY J. — John
Administrative Officer (303) 936-6395
2345 S. Newton Street
Denver, Colorado 80219
Br. of Isotope Geology
Denver (303) 234-5531

MAPEL, WILLIAM J. — Mary
Geologist (303) 237-7012
11290 Benthaven Drive
Lakewood, Colorado 80215
Br. of Coal Resources
Denver (303) 234-3530
Stanford U.-BA
Stratigraphy; sedimentation

MARCHAND, DENIS E.
Geologist (415) 328-7985
1004 Almanor Avenue
Menlo Park, California 94025
Br. of Western Environmental Geology
Menlo Park (415) 323-8111 x2009
Colorado Schl. of Mines-GE; U. of Cal. (Berkeley)-PhD
Soil correlation & dating, Western region; tectonic studies, California
Quaternary stratigraphy; pedology

MARGERUM, RICHARD — Helen
PST
200A G. Street, S.W.
Washington, D.C. 20024
Br. of Paleontology & Stratigraphy
Washington, D.C. (202) 343-3424
Cornell U.
Fusulinid biostratigraphy

MARINCOVICH, LOUIE N., JR.
Geologist
Br. of Paleontology & Stratigraphy
Menlo Park (415) 323-8111 x2264
UCLA-BA; U. of Southern Cal.-MS, PhD
Cenozoic mollusks, especially of Alaska

MARINENKO, JOHN W. — Olga
Chemist (703) 471-5375
1530 Stuart Road
Herndon, Virginia 22070
Br. of Analytical Laboratories
Reston (703) 860-7247
American U.-BS, MS
Micro mineral analysis; trace elements analysis

MARINENKO, OLGA H. — John
Chief. Instit. & Prof. Dev. Sec. (703) 471-5375
1530 Stuart Road
Herndon, Virginia 22070
Office of International Geology
Reston (703) 860-6441
American U.-BA

MARK, ROBERT K. — Mary
Physical Scientist (415) 321-7028
725 Cowper Street
Palo Alto, California 94301
Br. of Western Environmental Geology
Menlo Park (415) 323-8111 x2059
CCNY-BS; Stanford-MS, PhD
Statistics of reservoir-induced seismicity; statistics of surface faulting
Computer statistics applications to geologic problems

MARKEWICH, HELAINE WALSH
Geologist
Br. of Eastern Environmental Geology
Reston (703) 860-6421

MARKOCHICK, DENNIS J.
Geologist (303) 989-7669
12621 W. Mississippi Avenue
Lakewood, Colorado 80228
Br. of Oil & Gas Resources
Denver (303) 234-6288
Waynesburg Coll.-BS; Texas Christian U.-MS

MARLOW, MICHAEL S. — Elaine
Geophysicist
3893 Corina Way
Palo Alto, California 94303
Br. of Pacific-Arctic Geology
Menlo Park (415) 856-7092
Stanford-BS, MS, PhD
Geologic framework & resource assessment of the Bering Sea
Tectonics of continental margins

MARRANZINO, ALBERT P. — JoAnn
Chemist
12805 W. Alameda Drive
Lakewood, Colorado 80228
Office of the Chief Geologist
Denver (303) 234-6764
Regis Coll.-BS; U. of Colorado; U. of Denver
Nuclear reactor administration
Geochemical studies; facilities management

MARSH, SHERMAN P.
Geologist (303) 986-0939
891 S. Kline Way
Lakewood, Colorado
Br. of Exploration Research
Denver (303) 234-6400
U. of New Mexico-BS
Resource appraisal
Titanium

MARSHALL, B. VAUGHN — Yvonne
Geophysicist
Br. of Tectonophysics
Menlo Park (415) 323-8111 x2270
San Jose State U.-BA
Geothermal studies
Heat flow/computer software

MARTIN, E. ANN — Ray
Chemist (512) 992-7969
5417 Sugar Creek
Corpus Christi, Texas 78413
Br. of Atlantic-Gulf of Mex. Geology
Corpus Christi (512) 888-3241
Sam Houston State U.-BS, MA
Coastal plain estuaries
Geochronology; trace metals; geochemistry

MARTIN, JANE H.
Cartographic Technician
Office of Scientific Publications
Reston (703) 860-6495
Johns Hopkins; Maryland Inst. of Fine Arts

MARTIN, PETER L. — Gail
Technical Editor (303) 733-0924
1287 S. Fillmore Street
Denver, Colorado 80210
Office of Scientific Publications
Denver (303) 234-5723
U. of Denver-BA

MARTIN, RAY G. — Ann
Geologist (512) 992-7969
5417 Sugar Creek
Corpus Christi, Texas 78413
Br. of Atlantic-Gulf of Mex. Geology
Corpus Christi (512) 888-3294
Vanderbilt U.-BA; U. of Tennessee-MS
Gulf of Mexico tectonics
Marine geophysics; energy resources assessment

MARTIN, RONNY A. — **Janice**
Geophysicist — **(303) 278-3589**
135 S. Flora Way
Golden, Colorado 80401
Br. of Electromag. & Geomagnetism
Denver — **(303) 234-5491**
Texas Tech. U.-BS
Wilderness area studies
Magnetics

MARTIN, WAYNE E.
PST
100 E. Oak Street
Alexandria, Virginia 22301
Br. of Paleontology & Stratigraphy
Washington, D.C. — **(202) 343-5368**

MARTINEZ, ROBERTO J.
PST
180 S. 11th Street
Brighton, Colorado 80601
Br. of Oil & Gas Resources
Denver — **(303) 234-5026**
Metro College.
Bore hole gravity project

MARTNA, MARET H.
Librarian — **(703) 532-6264**
210 East Fairfax Street #303
Falls Church, Virginia 22046
Library
Reston — **(703) 860-6612**
Stockholm U.; Uppsala U.; U de Montreal-MA; McGill U.-BLS, MLS

MARVIN, RICHARD F. — **Lillian**
Geologist — **(303) 233-0920**
2470 Miller Street
Lakewood, Colorado 80215
Br. of Isotope Geology
Denver — **(303) 234-5531**
Montana Schl. of Mines-BS, MS
K-Ar geochronology; radiometric age data bank
Geochronology

MASCARDO, JUANITA S.
Secretary
Br. of Pacific-Arctic Geology
Menlo Park — **(415) 856-7050**

MASE, CHARLES W.
Geophysicist
Br. of Tectonophysics
Menlo Park — **(415) 323-8111 x2524**
San Diego State U.-BS; U. of Utah-MS
Geothermal studies
Heat flow & geothermal studies

MASON, CHARLES E. — **Linda**
PST — **(703) 823-2676**
179 Normandy Hill Drive
Alexandria, Virginia 22304
Br. of Paleontology & Stratigraphy
Washington, D.C. — **(202) 343-8620**
Morehead State U.-BS; George Washington U.-MS
Carboniferous biostratigraphy of northeastern Kentucky
Carboniferous cephalopods; trace fossils

MASON, DAVID H.
Engineering Technician — **(617) 548-6053**
68 Sippewissett Road, P.O. #526
Woods Hole, Massachusetts 02543
Br. of Atlantic-Gulf of Mex. Geology
Woods Hole — **(617) 548-8700 x147**
Marietta College
Marine seismic operations
Mechanical engineering, marine seismic technology, submersible operations

MASON, GEORGE T. JR.
Programmer
Office of Resource Analysis
Reston — **(703) 860-6455**
Langston College-BS
CRIB

MASSINGILL, LINDA M. — **Gary**
Geologist — **(512) 643-4352**
1005 Espana
Portland, Texas 78374
Br. of Atlantic-Gulf of Mex. Geology
Corpus Christi — **(512) 888-3294**
St. Mary's U.-BS; U. of Texas (El Paso)-MS
Mississippi Delta project
Mineral deposits; geochemistry; marine geology

MASSONI, CAMILLO J. — **Minnie**
Engineering Technician
1701 N. Nelson Street
Arlington, Virginia 22207
Br. of Analytical Laboratories
Reston — **(703) 860-7442**

MAST, RICHARD F. — **Joyce**
Geologist — **(303) 232-3632**
8 Skyline Drive
Lakewood, Colorado
Br. of Oil & Gas Resources
Denver — **(303) 234-4750**
U. of Illinois-BS, MS
Petroleum reservoir rocks, Western Interior; petrophysics & petroleum recovery

MASTERS, CHARLES D. — **Sandy**
Geologist — **(703) 759-3489**
1028 Walker Road
Great Falls, Virginia 22066
Office of Energy Resources
Reston — **(703) 860-6431**
Yale-BS, PhD; Colorado U.-MS
Physical stratigraphy

MASURSKY, HAROLD
Geologist
Br. of Astrogeolgic Studies
Flagstaff — **(602) 779-3311 x1454**
Yale U.-BS, MS
Viking Mars, Venus Pioneer, Voyager Saturn
Planetary geology, remote sensing, general geology

MATHEWS, SUSAN K.
Physical Scientist
552 Military Way
Palo Alto, California 94306
Br. of Pacific-Arctic Geology
Menlo Park — **(415) 856-7068**
Harvard U.-AB
Shelf sediment dynamics

MATHIESON, SCOTT A. — **Elizabeth L.**
Geologist — **(408) 293-5852**
82 Pierce Avenue
San Jose, California 95110
Br. of Ground Motion & Faulting
Menlo Park — **(415) 323-8111 x2064**
California State U. (Hayward)-BS
Coastal tectonics, western U.S.
Pleistocene glacial stratigraphy & landform geomorphology, Quaternary dating

MATTI, JONATHAN C. — **Susan T. Miller**
Geologist
Br. of Western Environmental Geology
Menlo Park — **(415) 323-8111 x2031**
U. of Calif. (Riverside)-BS, MS; Stanford-PhD
Cenozoic tectonics & wilderness program; southern California
Tectonic evolution of eastern Transverse Ranges; hazards, strike-slip faults; Quaternary of southern California

MATTICK, ROBERT E. **Barbara**
Geophysicist **(703) 368-1702**
9250 Stonewall Court
Manassas, Virginia 22110
Br. of Oil & Gas Resources
Reston **(703) 860-7250**
San Diego State U.-BS; U. of Colorado-MS
Resources of the Atlantic continental margin
Geophysics

MAUGHAN, EDWIN K.
Geologist **(303) 238-8750**
1317 Yank Street
Golden, Colorado 80401
Br. of Oil & Gas Resources
Denver **(303) 234-3465**
Utah State U.-BS; UCLA; U. of Colorado
Pennsylvanian & Permian petroleum source rocks
Upper Paleozoic stratigraphy, paleogeography

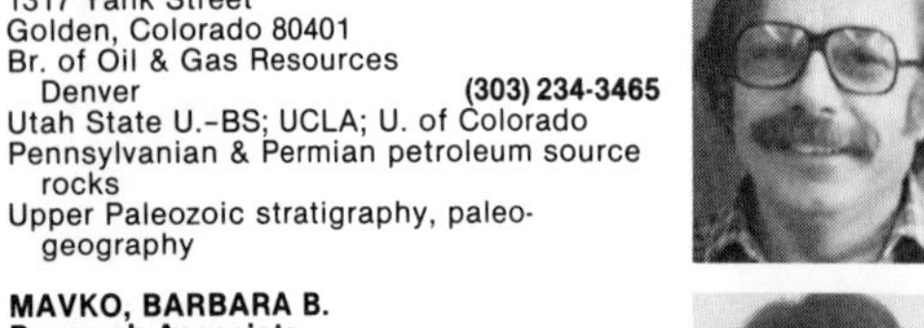

MAVKO, BARBARA B.
Research Associate
Br. of Seismology
Menlo Park **(415) 323-8111 x2024**
U. of Utah-BS; Stanford-MS, PhD
Earthquake studies

MAVKO, GERALD M.
Geophysicist
Br. of Tectonophysics
Menlo Park **(415) 323-8111 x2038**
Cornell U.-BS; Stanford-MS, PhD
Rock mechanics; tectonophysics

MAXWELL, CHARLES H.
Geologist
Br. of Central Mineral Resources
Denver **(303) 234-3226**
U. of New Mexico-BS, MS
Ore deposits of New Mexico
Mineralization; structural geology

MAY, FRED E. **Linda**
Geologist **(303) 979-6201**
5894 W. Alder Avenue
Littleton, Colorado 80123
Br. of Paleontology & Stratigraphy
Denver **(303) 234-3624**
Weber State Coll.-BS; Penn State U.-MS; Virginia Polytech. Inst.-PhD
Marine palynology, northern Alaska
Dinoflagellates, acritarchs, pollen & spores

MAY, IRVING **Florence**
Chemist **(301) 649-1431**
917 Brentwood Lane
Silver Spring, Maryland 20902
Office of the Chief Geologist
Reston **(703) 860-6531**
CCNY-BS; George Washington U.-MS; U. of Maryland
Geochemical analysis

MAY, RODD J. **Jeanne**
Geologist **(415) 364-2763**
3617 Oak Drive
Menlo Park, California 94025
Br. of Isotope Geology
Menlo Park **(415) 323-8111 x2036**
San Jose State U.-BS; Stanford-PhD
Geochronology; thermoluminescence dating

MAY, STEVEN D.
Geophysicist
Br. of Pacific-Arctic Geology
Menlo Park **(415) 323-8111 x2695**
W. Washington State U.-BS; Stanford-MS
Beaufort-Chukchi Seas resource assessment
Seismic & gravity data processing & interpretation

MAYFIELD, CHARLES F.
Geologist
Br. of Alaskan Geology
Menlo Park **(415) 323-8111 x2147**
Ohio U.-BA; U. of Idaho-MS
Regional geology of western Brooks Range, Alaska
Field mapping; stratigraphy; petrology

MAYFIELD, SUSAN E.
Scientific Illustrator
P.O. Box 284
Menlo Park, California 94025
Office of Scientific Publications
Menlo Park **(415) 323-8111**
San Jose State U.-BA

MAYS, ROBERT E.
Chemist
Br. of Analytical Laboratories
Menlo Park **(415) 323-2214**
City College of S.F.-AA; U. of California (Berkeley)-BA
Spectrographic analysis

McBROOME, LISA A.
Geologist **(303) 499-4031**
2955 Carnegie
Boulder, Colorado 80303
Br. of Central Environmental Geology
Denver **(303) 234-5381**
U. of Missouri-BS; U. of Colorado
Snake River Plain (eastern)
Volcanology; geologic field mapping; mineralogy; petrology

McCALL, BENJAMIN A.
PST **(202) 396-4791**
565-23rd Place, NE
Washington, D.C. 20002
Br. of Analytical Laboratories
Reston **(703) 860-7543**
X-ray laboratory

McCAMMON, RICHARD B. **Helen**
Geologist **(301) 469-7465**
8430 Bradley Boulevard
Bethesda, Maryland 20034
Office of Resource Analysis
Reston **(703) 860-6446**
MIT-BSc; U. of Michigan-MSc; U. of Indiana-PhD
Uranium resource assessment
Resource analysis; computer applications; mathematical geology

McCARTHY, GERARD V.
Electronics Technician
24 Eastland Road
Boston, Massachusetts 02130
Br. of Atlantic-Gulf of Mex. Geology
Woods Hole **(617) 548-8700**
Northeastern U.; Southern Methodist U.-BS
Underwater acoustic navigation system
Underwater acoustics; data processing; computer analysis of geophysical systems

McCARTHY, J. HOWARD, JR. **Lucy**
Chemist **(303) 674-6650**
Route 7, Box 573H
Evergreen, Colorado 80439
Br. of Exploration Research
Denver **(303) 234-5350**
U. of Denver-BA
Geochemical exploration; volatile elements & gases
Measurement of gases for exploration

McCARTHY, ROBERT P. **Barbara**
Geophysicist
Box 578
Broomfield, Colorado 80020
Br. of Global Seismology
Denver **(303) 234-5080**
RPI-BS; New Mex. Instit. Mining & Tech-BS
Seismic data review & data services

McCAULEY, JOHN F. **Camilla**
Geologist
60 Wilson Lane
Flagstaff, Arizona 86001
Br. of Astrogeologic Studies
Flagstaff **(602) 779-3311**
Fordham U.-BS; Columbia-PhD
Eolian processes
Geology of Moon, Mars, Mercury; desert studies

McCLELLAN, PATRICK H.
Geologist
3252 South Court
Palo Alto, California 94306
Br. of Pacific-Arctic Geology
Menlo Park **(415) 856-7100**
Utah State U.-BS; U. of Cal. (Berkeley)-MA
Marine biostratigraphy of the western Gulf of Alaska
Vertebrate paleontology; earthquake prediction

McCLEARN, ROBERT W.
Electronics Technician
Br. of Network Operations
Menlo Park **(415) 323-8111 x2704**

McCULLOH, THANE H. **Mary Ann**
Research Geologist **(206) 524-2795**
4046 NE 86th Street
Seattle, Washington 98115
Br. of Pacific-Arctic Geology
Seattle **(206) 442-1995**
Pomona Coll.-BA; UCLA-PhD
Geology of petroleum; sedimentary petrology; petrophysics

McCUNE, DOROTHY G.
Clerk-Typist
1490 S. Ammons
Lakewood, Colorado 80226
Br. of Earthquake Tectonics & Risk
Denver **(303) 234-5087**
Mesa College

McDADE JOHNNIE
Chemist **(303) 333-9627**
2636 Colorado Boulevard
Denver, Colorado 80207
Br. of Analytical Laboratories
Denver **(303) 234-6401**
U. of Colorado-BS

McDANAL, STEVEN K.
Computer Technician
Br. of Exploration Research
Denver **(303) 234-2361**

McDANIEL, SUSAN E. **Randall**
PST **(512) 854-6053**
534 Hoffman
Corpus Christi, Texas 78404
Br. of Atlantic-Gulf of Mex. Geology
Corpus Christi **(512) 888-3241**
U. of Texas-BS
Trace metals; uranium-thorium dating

McDOUGALL, KRISTIN
Geologist
Br. of Paleontology & Stratigraphy
Menlo Park **(415) 323-8111 x2185**
U. of Washington-BS, MS; U. of Southern California-PhD
Cenozoic benthic foraminifers; Mesozoic calcareous nannoplankton

McDOWELL, ROBERT C.
Geologist **(703) 554-8256**
Route 1, Box 209
Bluemont, Virginia 22012
Br. of Eastern Environmental Geology
Reston **(703) 860-6503**
Virginia Polytech. Inst.-BS, MS, PhD; Harvard U.
Structural analysis of Giles County area, Virginia—West Virginia
Structure; stratigraphy; mapping

McGAHA, ANA C.
Clerk-Typist
805 W. Nettle Tree Road
Sterling, Virginia 22170
Office of the Chief Geologist
Reston **(703) 860-6544**

McGEE, KENNETH A. **Linda**
Geologist **(703) 430-2730**
116 W. Brighton Avenue
Sterling, Virginia 22170
Br. of Exper. Geochem. & Mineralogy
Reston **(703) 860-6620**
U. of Missouri-BS, MA
Chemical speciation in aqueous systems
Stability of minerals & dissolved ions; techniques for remote monitoring of hazards

McGEE, LINDA C.
Geologist
Br. of Petrophysics & Remote Sens.
Denver **(303) 234-4897**
Illinois State U.-BS; Colorado Schl. of Mines
Remote sensing applications; economic geology

McGILL, JOHN T. **Carol**
Geologist **(303) 233-8211**
11705 West 24th Place Circle
Lakewood, Colorado 80215
Br. of Engineering Geology
Denver **(303) 234-4760**
UCLA-AB, MA, PhD
Pacific Palisades landslide area, Los Angeles, California
Engineering geology; landslides; coastal geomorphology

McGREGOR, BONNIE A. **William Stubblefield**
Geologist
4110 Woodridge Road
Miami, Florida 33133
Br. of Atlantic-Gulf of Mex. Geology
Miami **(305) 350-4239**
Tufts-BS; U. of Rhode Island-MS; U. of Miami-PhD
Marine geology & geophysics; sea floor stability

McGREGOR, EDWARD E. **Karen**
Geologist **(303) 333-9834**
1649 Race Street
Denver, Colorado 80206
Br. of Engineering Geology
Denver **(303) 234-2841**
Wayne State U.-BA
Engineering geology of Price Utah area
Environmental impact statements; mapping.

McGREGOR, JOSEPH K., JR.
Library Technician
Library
Denver **(303) 234-4004**
U. of Colorado

McGREGOR, ROBERT E.
PST
Br. of Analytical Laboratories
Denver **(303) 234-3245**

McGUIRE, VIRGINIA A. Tom
Clerk-Typist
Office of Environmental Geology
Reston (703) 860-6411

McHENDRIE, A. GRAIG
System Analyst
Br. of Pacific-Arctic Geology
Menlo Park (415) 856-7147
Williams Coll.-BA; Stanford U.-MS
Applications programming & systems analysis
Computer applications for marine geologic & geophysical data

McHUGH, JOHN B.
Chemist (303) 985-0475
12347 W. Ohio Circle
Lakewood, Colorado 80228
Br. of Exploration Research
Denver (303) 234-2684
Colorado State U.-BS
Geochemistry of the Richfield 1° x 2° quadrangle, Utah
Geochemistry methods; geochemistry of wilderness

McINTYRE, DAVID H.
Geologist.
Br. of Central Mineral Resources
Denver (303) 234-3392
U. of Washington-BS, MS; Washington State U.-PhD
Challis 1° x 2° quadrangle
Volcanic rocks

McKEE, EDWIN D. Barbara
Geologist (303) 794-2854
4845 Redwood Drive
Littleton, Colorado 80123
Br. of Oil & Gas Resources
Denver (303) 234-5010
Cornell; U. of Arizona; U. of Cal. (Berkeley); Yale
Director of sedimentology structures laboratory
Sedimentology; stratigraphy

McKELVEY, VINCENT E. Genevieve
Geologist
510 Runnymeade Road
St. Cloud, Florida 32769
Office of the Director & Chief Geologist
Reston (703) 860-6531
Syracuse-BS; U. of Wisconsin-MA, PhD
Phosphate & uranium deposits; energy resources, seabed resources
Scientist; administrator; Law of the Sea advisor

McKEOWN, FRANK A. Helen
Geologist
Br. of Earthquake Tectonics & Risk
Denver (303) 234-5086
U. of Pennsylvania-BA; Johns Hopkins-MA
Seismic source zones of U.S.
Tectonics

McKEOWN, HELEN L. Frank
Administrative Technician (303) 237-7092
9025 West 2nd Avenue
Lakewood, Colorado 80226
Office of Geochemistry & Geophysics
Denver (303) 234-5461

McKINNEY, ROBERT H. Mary
Photographer
Br. of Paleontology & Stratigraphy
Washington, D.C. (202) 343-3841

McKOWN, DAVID M. Carolyn
Chemist (303) 988-6828
1704 S. Vancouver Court
Lakewood, Colorado 80228
Br. of Analytical Laboratories
Denver (303) 234-3624
West Virginia U.-BS; U. of Kentucky-PhD
Radiochemistry, Denver
Radioanalytical chemistry; chemical analysis

McLAUGHLIN, PHILIP V.
PST
315 Bryant Court
Palo Alto, California 94301
Br. of Engineering Geology
Menlo Park (415) 856-7119
U. of California-BA
Laboratory soil testing; soil drilling

McLAUGHLIN, ROBERT J. Judith
Geologist (408) 252-0589
Br. of Western Environmental Geology
Menlo Park (415) 323-2214 x2475
San Jose State U.-BS, MS
Geysers-Clear Lake geothermal area; RARE 2
Structural geology & tectonics of W. North American continental margin

McLEAN, HUGH
Geologist (415) 494-2622
930 Los Robles Avenue
Palo Alto, California 94306
Br. of Oil & Gas Resources
Menlo Park (415) 856-7012
San Diego State U.-BA; U. of Washington-MS, PhD
Geologic characteristics of deep water sandstones
Sedimentary petrology; sedimentology

McLELLAN, MARGUERITE W. Bruce
Geologist
474 Spruce Road, Lookout Mountain
Golden, Colorado 80401
Br. of Coal Resources
Denver (303) 234-3578
U. of Oklahoma-BS; U. of New Mexico
Zuni-Salt Lake Coal Field, New Mexico
Coal & petroleum geology; stratigraphy

McMASTERS, CATHERINE R.
PST (415) 328-5387
782 Coleman Avenue, Apt. B
Menlo Park, California 94025
Br. of Western Environmental Geology
Menlo Park (415) 323-8111 x2512
Stanford-BS
Tectonics of the San Francisco Bay region
Quaternary faulting

McNAIR, DONALD W.
Electronics Technician (303) 569-3125
Box 182
Empire, Colorado 80438
Br. of Isotope Geology
Denver (303) 234-5531

McNEAL, JAMES M. Ilene
Geologist (303) 989-4571
10267 W. Arkansas Avenue
Lakewood, Colorado 80226
Br. of Regional Geochemistry
Denver (303) 234-5629
College of Wooster-BA; Penn State U.-PhD
Trace element occurrence in soils, rocks, sediments; geochemical exploration; geochemical mapping

McNUTT, MARCIA K.
Geophysicist (415) 941-4046
546 University Avenue
Los Altos, California 94022
Br. of Tectonophysics
Menlo Park (415) 323-8111 x2479
Scripps Inst. of Oceanography-PhD
Longterm lithospheric properties
Gravity

McPHILLIPS, MAUREEN
PST
Br. of Coal Resources
Denver **(303) 234-3519**
Wellesley-BA
Coal exploratory drilling—drill support group

McQUEEN, DAVID R. **Shirley**
Geologist **(703) 435-1521**
1132 Shannon Place
Herndon, Virginia 22070
Office of Resource Analysis
Reston **(703) 860-6456**
Southeastern Christian-AS; U. of Tennessee-BA; U. of Michigan-MS
Metallogenic studies; CRIB
Economic geology; structural geology; general geology

MEAD, JENNIFER R. **Stephen**
Administrative Clerk **(303) 278-3612**
1116 13th Street
Golden, Colorado 80401
Office of Earthquake Studies
Denver **(303) 234-5085**
U. of Colorado-BA

MEDINA, EDWARD S. **Carlotta**
Electronics Technician
9206 Lagrima De Oro, NE
Albuquerque, New Mexico 87111
Br. of Global Seismology
Albuquerque **(505) 844-4637**
NATI-ASEE
Digital telemetry

MEDLIN, ANTOINETTE L.
Computer Specialist
Office of Resource Analysis
Reston **(703) 860-7307**
U. of Illinois-BS
National coal resources data system
Applications of computers to geologic problems

MEDLIN, JACK H.
Geologist
Br. of Coal Resources
Reston **(703) 860-7734**
U. of Georgia-BS, MS; Penn State U.-PhD
Energy & mineral resources; regional geology

MEEHAN, RICHARD H.
Administrative Officer
Br. of Alaskan Geology
Anchorage **(907) 271-4150**
U. of Montana-BS, MBA

MEGEATH, JOE D.
Computer Scientist
4577 S. Hannibal
Aurora, Colorado 80015
Br. of Oil & Gas Resources
Denver **(303) 234-3435**
Colorado Schl. of Mines-PhD
Estimation of undiscovered resources
Probability models; computer systems

MEHNERT, HARALD H. **Brigitte**
Geologist **(303) 985-7956**
12285 W. Wisconsin Drive
Lakewood, Colorado 80228
Br. of Isotope Geology
Denver **(303) 234-3876**
U. of Colorado-BA
Geochronology; mass spectrometry
K-Ar dating

MEI, LEUNG
Chemist
Br. of Analytical Laboratories
Reston **(703) 860-7653**
U. of Arkansas-BS; American U.
Spectrographic analysis

MEIER, ALLEN L. **Margaret**
Chemist **(303) 421-8729**
6742 Urban Court
Arvada, Colorado 80004
Br. of Exploration Research
Denver **(303) 234-6146**
Colorado State U.-BS
Research of analytical methods
Analytical chemistry

MEISSNER, CHARLES R., JR. **Irene**
Geologist
2711 Soapstone Drive
Reston, Virginia 22091
Br. of Coal Resources
Reston **(703) 860-7734**
LeHigh U.-BA; U. of Maine
Coal geology and resources, Virginia & West Virginia
Coal exploration—geologic mapping; stratigraphy; resource evaluation

MENARD, H. WILLIAM **Gifford**
Geologist
Office of the Director
Reston **(703) 860-6531**
Cal Tech-BS, MS; Harvard-PhD
Marine geology; tectonics; mineral resources & environment
Scientist; administrator

MENDENHALL, WALTER C. **Alice**
Geologist
Office of the Director
Reston **(703) 860-6531**
Ohio Normal U.-BS; Harvard; Heidelberg U.
Appalachian coals; geology of Alaska, ground water & mineral resource studies; stratigraphy
Scientist; administrator; explorer

MENDES, ROY V. **Johanna**
Computer Specialist
770 Cole Drive
Golden, Colorado 80401
Br. of Regional Geochemistry
Denver **(303) 234-2438**
Colorado Schl. of Mines-GE
Geochemical data systems
Storage & retrieval of geochemical data

MENDOZA, CARLOS
Geophysicist
Br. of Global Seismology
Denver **(303) 234-3994**
U. of Wisconsin (Milwaukee)-BS, MS
Global seismology; earthquake tectonics

MENZIE, W. DAVID **Carolyn**
Geologist **(415) 858-0807**
836 Driftwood Drive
Palo Alto, California 94303
Office of Resource Analysis
Menlo Park **(415) 323-8111 x2906**
Dickinson Coll.-BS; Penn State U.-MS, MA, PhD
Mineral resource assessment; applied statistics & modeling

MERCILLIOTT, BEVERLY A.
Secretary **(703) 471-7383**
1522 Northgate Square
Reston, Virginia 22090
Office of the Chief Geologist
Reston **(703) 860-6532**

MEREWETHER, E. ALLEN — **Helen**
Geologist — **(303) 422-3203**
6319 Iris Way
Arvada, Colorado 80004
Br. of Oil & Gas Resources
Denver — **(303) 234-4314**
U. of Oregon–BS, MS
Lower Upper Cretaceous strata–stratigraphy & petroleum potential
Mid-Cretaceous stratigraphy of Wyoming & eastern South Dakota

MERRITT, VIOLET M.
PST
Br. of Analytical Laboratories
Denver — **(303) 234-6401**

MEYER, CHARLES E. — **Lynda**
Geologist — **(415) 964-4972**
992 Sleeper Avenue
Mountain View, California 94040
Br. of Western Environmental Geology
Menlo Park — **(415) 323-8111 x2696**
Humboldt State U.–BA
Tephrochronology of the western United States
Geochronology of Tertiary volcanics; fission track dating; electron microprobe analysis

MEYER, RICHARD F.
Geologist
P.O. Box 227
Warrenton, Virginia 22186
Office of Resource Analysis
Reston — **(703) 860-6446**
Dartmouth–BA; Harvard–MA; U. of Kansas–PhD
Energy resources

M'GONIGLE, JOHN W. — **Georgia L.**
Geologist — **(303) 237-8610**
12185 Applewood Knolls Drive
Lakewood, Colorado 80215
Br. of Coal Resources
Denver — (303) 234-3578
U. of New Mexico–BS, MS; Penn State U.–PhD
Coal geology of Hams Fork region, Wyoming
Sedimentary and structural geology

MICHALSKI, DANIEL C.
Cartographic Technician
P.O. Box 623
Westminster, Colorado 80030
Br. of Regional Geophysics
Denver — **(303) 234-4772**
U. of Colorado
Drafting; photography

MICHALSKI, THOMAS C. — **Janet**
Geologist — **(303) 320-5386**
1651 Adams
Denver, Colorado 80206
Br. of Oil & Gas Resources
Denver — **(303) 234-5105**
Wayne State U.–BS
Curator of core library
Core preservation; descriptive mineralogy

MICHELITCH, ARLEEN V. — **Harry**
Secretary
9817 Mill Run Drive
Great Falls, Virginia 22066
Office of Resource Analysis
Reston — **(703) 860-6455**

MIESCH, ALFRED T. — **Norma**
Research Geologist — **(303) 233-2525**
10300 West 23rd Avenue
Lakewood, Colorado 80215
Br. of Regional Geochemistry
Denver — **(303) 234-4672**
St. Joseph's Coll.–BS; Indiana U.–MA; Northwestern U.–PhD
Statistical geochemistry & petrology
Environmental geochemistry; exploration geochemistry; petrology

MIHALYI, DALE L.
Geologist
Br. of Atlantic-Gulf of Mex. Geology
Corpus Christi — **(512) 888-3294**
U. of Rochester–BS, MS
Texas-Louisiana outer continental shelf seismic stratigraphy
Seismic stratigraphy; carbonate petrology

MIKESELL, JON L. — **Betty**
Physicist — **(301) 552-3682**
8681 Greenbelt Road
Greenbelt, Maryland 20770
Br. of Isotope Geology
Reston — **(703) 860-7662**
MIT–BS; Howard U.–MS
Nuclear borehole logging
Neutron activation; natural radioactivity

MIKUNI, DIANE E. — **Alan**
Foreign Participant Coordinator
Office of International Geology
Menlo Park — **(415) 323-8111 x2926**

MILAN, MARY P.
Secretary — **(415) 322-5004**
303 Chester Street
Menlo Park, California 94025
Br. of Western Environmental Geology
Menlo Park — **(415) 323-8111 x2474**

MILES, CHERYL — **Bot**
Administrative Aid
Office of Earthquake Studies
Reston — **(703) 860-6475**
Tomlinson Coll.–AA

MILLARD, HUGH T., JR.
Chemist
Br. of Analytical Laboratories
Denver — **(303) 234-4201**
Coe Coll.–BA; Cal Tech–PhD
Radiochemistry, Denver
Activation analysis; geochemistry

MILLER, C. DAN — **Lois**
Geologist — **(303) 238-9903**
3030 Upham Court
Denver, Colorado 80215
Br. of Engineering Geology
Denver — **(303) 234-5681**
U. of Washington–BS, MS; U. of Colorado–PhD
Volcanic hazards
Volcanology; Quat. stratigraphy; glacial geology

MILLER, CARTER H. — **Willie**
Geophysicist — **(303) 237-8227**
158 S. Teller Street
Lakewood, Colorado 80226
Br. of Engineering Geology
Denver — **(303) 234-5577 (3721)**
Colorado U.–BA; Colorado Sch. Mines–ME
Geophysics for engineering geology
Engineering geophysics; engineering geology; rock & soil mechanics

MILLER, DANNY R. — **Judy**
Engineering Technician
11551 West Tennessee Place
Lakewood, Colorado 80226
Br. of Engineering Geology
Denver — **(303) 234-2529**
Western State Coll.; Northern Colorado U.; Metro. State College.
Testing & evaluation of stress & rock property instrumentation
In situ stress-strain; rock mechanics

MILLER, DAVID M.
Geologist
Br. of Western Environmental Geology
Menlo Park **(415) 323-8111 x2956**
SUNY (Binghamton)-BS; UCLA-PhD
Tectonics of northern Utah & eastern Mojave Desert
Structure & tectonics of U.S. Cordillera

MILLER, FRED K. **Catherine**
Geologist **(509) 466-4463**
N15504 Fircrest Circle
Spokane, Washington 99218
Br. of Western Environmental Geology
Spokane **(509) 456-4677**
U. of Cal. (Riverside)-BA; Stanford-PhD
Relation of geologic framework to uranium resources in Sandpoint 2° quadrangle
Regional & structural geology; petrology

MILLER, HARLEY P.
Electronics Technician
Office of the Chief Geologist
Denver **(303) 234-3771**

MILLER, HOPE D. **Charles**
Staff Assistant **(703) 450-4247**
802 N. Watford Street
Sterling, Virginia 22170
Office of the Chief Geologist
Reston **(703) 860-7420**

MILLER, JANET L.
Secretary
831 Cortez Lane
Foster City, California 94404
Office of the Chief Geologist
Menlo Park **(415) 467-2214**

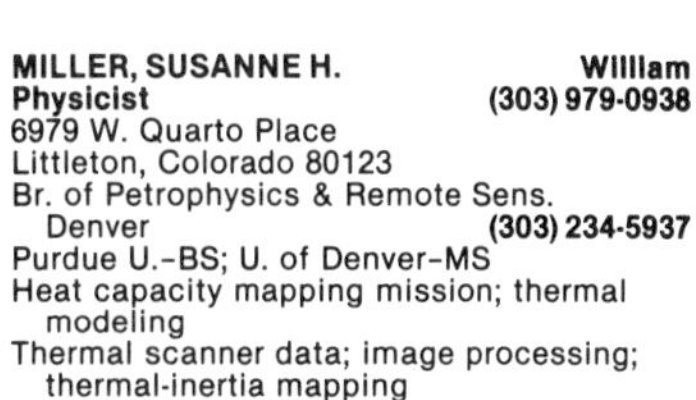

MILLER, JOHN W. **Dorris**
Museum Specialist **(415) 493-0388**
3736 Cass Way
Palo Alto, California 94306
Br. of Paleontology & Stratigraphy
Menlo Park **(415) 323-8111 x2540**
New Mexico State U.-BA; UCLA; San Jose State U.
Alaska Pennisula paleontology & stratigraphy
Upper Mesozoic mollusks

MILLER, LAURIE D.
Clerk Typist
Br. of Paleontology & Stratigraphy
Reston **(703) 860-7745**

MILLER, MARY H.
Geologist
453 S. Independence Court
Lakewood, Colorado 80226
Br. of Central Mineral Resources
Denver **(303) 234-2937**
U. of Missouri-BA, MA
Mineral resource evaluation, Rolla Mo., 2° quadrangle
Mineral resources; antimony-bismuth commodity studies

MILLER, M. MEGHAN
GFA
Br. of Pacific-Arctic Geology
Menlo Park **(415) 856-7051**
Yale-BS

MILLER, RALPH L. **Ansel**
Geologist **(301) 229-7841**
5215 Abingdon Road
Washington, D.C. 20016
Office of International Geology
Reston **(703) 860-6531**
Haverford Coll.-BS; Columbia-PhD
Coal resources of South Korea—mapping, San Juan Basin, New Mexico
Geology of fuels

MILLER, ROBERT D. **Judy A.**
Geologist
291 Old Y Road
Golden, Colorado 80401
Br. of Engineering Geology
Denver **(303) 234-2960**
Purdue U.; U. of Illinois; U. of Colorado-AB
Physical properties & earthquake hazards of surficial deposits
Pleistocene geology; physical properties of surficial materials; geologic hazards

MILLER, ROBUT E. **Barbara**
Geologist
2 Middleton Lane
Sterling, Virginia 22170
Br. of Atlantic-Gulf of Mex. Geology
Reston **(703) 860-7164**
U. of Tulsa-BG, MS; Texas A&M U.-PhD
Marine organic geochemistry; resource & environmental assessment
Organic geochem. of petroleum & sedimentary minerals; environmental geochem.

MILLER, RONALD J.
Oceanographer
5808 Hinman
Corpus Christi, Texas 78412
Br. of Atlantic-Gulf of Mex. Geology
Corpus Christi **(512) 888-3294**
Corpus Christi State U.-MS
Carbonate geology; nearshore processes

MILLER, SUSANNE H. **William**
Physicist **(303) 979-0938**
6979 W. Quarto Place
Littleton, Colorado 80123
Br. of Petrophysics & Remote Sens.
Denver **(303) 234-5937**
Purdue U.-BS; U. of Denver-MS
Heat capacity mapping mission; thermal modeling
Thermal scanner data; image processing; thermal-inertia mapping

MILLER, THERON E. **Lois**
Electronics Technician **(303) 776-1026**
1803 Meadow Street
Longmont, Colorado 80501
Br. of Engineering Geology
Denver **(303) 234-4477**
Western Radio Institute
Instrumentation research

MILLER, THOMAS P. **Shirla**
Chief, Br. of Alaskan Geology **(907) 333-5651**
4962 Barat Circle
Anchorage, Alaska 99504
Br. of Alaskan Geology
Anchorage **(907) 271-4150**
U. of Minnesota-BA, MS; Stanford-PhD
Alaska geothermal; western Alaska uranium
Volcanology; geochemistry; economic geology

MILLER, WILLIAM R. **Jody**
Geologist
12429 W. 17th Avenue
Lakewood, Colorado 80215
Br. of Exploration Research
Denver **(303) 234-6189**
Ohio State-BCE; U. of Wyoming-MS, PhD
Geochemical mapping of the Richfield, Utah 1° x 2° quadrangle
Hydrogeochemical prospecting; mineral-solution equilibria; statistics

MILLER-HOARE, MARTI L. **Michael Hoare**
PST **(408) 851-7062**
Star Route, Box 68
Woodside, California 94062
Br. of Alaskan Geology
Menlo Park **(415) 323-8111 x2614**
U. of Cal. (Berkeley); Stanford-BS, MS; U. of Cal. (Santa Cruz)
Chugach Nat'l Forest, RARE II
Geologic mapping; petrography

MILLS, FRANCIS R.
Cartographic Technician **(415) 494-7198**
629 Keats Court
Palo Alto, California 94303
Office of International Geology
Menlo Park **(415) 323-8111 x2927**
Circum-Pacific map project

MILLS, HUGH H.
Geologist **(615) 528-3121**
Br. of Eastern Environmental Geology
U. of North Carolina-MS; U. of Washington
Geomorphology of Giles County & vicinity, Virginia
Relative dating techniques; periglacial features in the southeast U.S.

MILTON, CHARLES **Leona**
Geologist **(301) 565-2330**
Beechbank Road, Forest Glen
Silver Spring, Maryland 20910
Br. of Chemical Resources
Washington, D.C. **(202) 676-7118**
Johns Hopkins-PhD
Green River mineralogy; vanadium mineralogy
Mineralogy; petrology

MILTON, DANIEL J.
Geologist **(703) 476-4322**
11557 Rolling Green
Reston, Virginia 22091
Br. of Eastern Environmental Geology
Reston **(703) 860-6503**
Harvard-AB, PhD; Cal Tech-MS
Charlotte 2° sheet
Appalachian geology; impact cratering; lunar & planetary geology

MILTON, NANCY
Botanist **(703) 437-4926**
907 Monroe Street
Herndon, Virginia 22070
Br. of Petrophysics & Remote Sens.
Reston **(703) 860-7233**
Howard U.-BS; Johns Hopkins-PhD
Spectral reflectance characteristics of plants
Plant ecology; geobotany; remote sensing

MINARD, JAMES P. **Velma**
Geologist **(206) 334-7776**
9207-60th Street, NE
Everett, Washington 98205
Br. of Western Environmental Geology
Seattle **(206) 442-7300**
Upsala Coll.-AB; Rutgers U.-MS
Puget Sound ESA Project
Atlantic Coastal Plain stratigraphy; Puget Sound glacial geology

MINKIN, JEAN A. **Max**
Physicist **(703) 560-4861**
3440 Round Table Court
Annandale, Virginia 22003
Br. of Coal Resources
Reston **(703) 860-6788**
Bryn Mawr Coll.-BA
Coal petrochemistry & related investigations
Elemental abundances in coal macerals & minerals; coal petrology; mineralogy

MINKLER, PETER W. **Trudy**
PST **(415) 328-2716**
Br. of Pacific-Arctic Geology
Menlo Park **(415) 856-7005**
Harvard-BA
Beaufort Sea

MINSCH, JOHN H. **Supatra**
Geophysicist **(303) 278-3029**
492 Deframe Court
Golden, Colorado 80401
Br. of Global Seismology
Denver **(303) 234-3994**
Kansas State U.-BS
U.S. earthquakes

MIRANDA, SANDRA L.
Computer Programmer
Br. of Tectonophysics
Menlo Park **(415) 323-2214**
San Jose State U.-BA
Computerized geophysical laboratory
Computer-aided education; earth tide studies

MITCHELL, CHARLES M.
Electronics Technician
Br. of Electromag. & Geomagnetism
Denver **(303) 234-2589**

MIXON, ROBERT B. **Krista**
Geologist
528 Springvale Road
Great Falls, Virginia 22066
Br. of Eastern Environmental Geology
Reston **(703) 860-6406**
Louisiana State U.-PhD
Inner Coastal Plain tectonics; middle Atlantic States
Tectonics; sedimentology & stratigraphy

MIYAOKA, RONNY T.
PST **(408) 294-9675**
601-A N. 3rd Street
San Jose, California 95112
Br. of Alaskan Geology
Menlo Park **(415) 323-8111 x2406**
U. of Cal. (Santa Cruz)-BS
Valdez & Mt. Hayes quadrangle, Alaska

MOCHIZUKI, HARUO E.
Photographer
2224 F. Street, NW
Washington, D.C. 20037
Br. of Paleontology & Stratigraphy
Washington, D.C. **(202) 343-3841**
Photography

MODRESKI, PETER J.
Chemist
8075 W. Fremont Drive
Littleton, Colorado 80123
Br. of Central Mineral Resources
Denver **(303) 234-6808**
Rutgers Coll.-BA; Penn State-PhD
Challis, Idaho 2° quad
Phase equilibria; igneous petrology; mineralogy

MOENCH, ROBERT H. **Sarah**
Geologist **(303) 526-0428**
593 Mt. Evans Road
Golden, Colorado 80401
Br. of Central Mineral Resources
Denver **(303) 234-3424**
Boston U.-PhD
Coordinator, Sherbrook-Lewiston CUSMAP (Me, NH, Vt)
Metamorphic stratigraphy, mineral resource applications

MOHR, PAMELA J.
GFA
Br. of Petrophysics & Remote Sens.
Denver **(303) 234-3624**
SUNY (Buffalo)-BA

MOLENAAR, CORNELIUS M. **Barbara**
Geologist **(303) 278-3596**
2570 Eldridge Street
Golden, Colorado 80401
Br. of Oil & Gas Resources
Denver **(303) 234-4642**
U. of Washington-BS
Cretaceous studies; North Slope of Alaska
Cretaceous stratigraphy; physical stratigraphy

MOLL, ELIZABETH J.
Geologist (415) 327-6970
170 Hawthorne Avenue
Palo Alto, California 94301
Br. of Alaskan Geology
Menlo Park (415) 323-8111 x2025
U. of Cal. (Santa Cruz)-BA, MS
Alaska mineral resources
Regional geology of western Alaska; volcanic rocks of western Alaska

MOLNIA, BRUCE F. Mary
Geologist
1066 Muir Way
Los Altos, California 94022
Br. of Pacific-Arctic Geology
Menlo Park (415) 856-7019
Harpur Coll.-BA; Duke U.-MA; U. of S. Carolina-PhD
Eastern gulf of Alaska environmental studies
Marine geology; coastal processes; glacial-marine sedimentation

MOLNIA, CAROL L.
Geologist
Br. of Coal Resources
Denver (303) 234-3548
Princeton U.-BS
Sheridan, Wyoming 1° x ½° quadrangle—coal folio project
Coal mining; coal leasing; environmental regulation

MONS-WENGLER, MARGARET CLARE
Clerk-Typist
Seagrove Road
South Dennis, Massachusetts 02660
Br. of Atlantic-Gulf of Mex. Geology
Woods Hole (617) 548-8700

MONTAGUE, NANCY L. Nelson
Administrative Assistant
1324 Westhills Lane
Reston, Virginia 22090
Office of Mineral Resources
Reston (703) 860-6571

MONTIJO, ELOUISE V. Frank
Administrative Technician
Br. of Coal Resources
Denver (303) 234-3578
U. of Denver-AA

MOORE, DAVID W.
Geologist
Br. of Central Environmental Geology
Denver (303) 234-5373
College of Wooster-AB; U. of N. Carolina-MS; U. of Illinois-PhD
Environmental aspects of energy resources, San Juan Basin
Pleistocene stratigraphy; glacial deposits; reclamation of surface-mined areas

MOORE, ELLEN J. George W.
Geologist (415) 493-7895
828 La Jennifer Way
Palo Alto, California 94306
Br. of Paleontology & Stratigraphy
Menlo Park (415) 323-8111
Oregon State U.-BA; U. of Oregon-MS
Tertiary biostratigraphy, Pacific northwest
Cenozoic biostratigraphy; ecology & taxonomy of Cenozoic marine mollusks

MOORE, GEORGE W. Ellen
Geologist (415) 493-7895
828 La Jennifer Way
Palo Alto, California 94306
Br. of Pacific-Arctic Geology
Menlo Park (415) 856-7159
Stanford U.-BS, MS; Yale U.-PhD
Western Gulf of Alaska
Structure; stratigraphy; sedimentation

MOORE, HENRY J. Patsy Ann
Geologist
Br. of Astrogeologic Studies
Menlo Park (415) 323-8111 x2361
Stanford U.-PhD
Mars—volcanic processes; physical properties of surface materials
Planetary geology

MOORE, JAMES G. Deborah
Geologist (415) 494-0356
4045 Ben Lomand Drive
Palo Alto, California 94306
Br. of Field Geochemistry & Petrology
Menlo Park (415) 323-8111
Stanford-BS; U. of Washington-MS; Johns Hopkins-PhD
Geology of Sequoia & Kings Canyon National Parks
Igneous petrology; volcanology

MOORE, KEITH R.
PST (703) 536-2546
1404 North Tuckahoe Street
Arlington, Virginia 22205
Br. of Paleontology & Stratigraphy
Washington, D.C. (202) 343-4300
Marshall U.-BS

MOORE, RICHARD B. Kristine
Geologist (808) 967-7388
Box 94
Hawaii National Park, Hawaii 96718
Br. of Field Geochem. & Petrology
Hawaiian Volcano Observ. (808) 967-7328
Tufts U.-BS; U. of N. Dakota-MS; U. of New Mexico-PhD
Geology & petrochemistry of Kilauea & Hualalai volcanoes, Hawaii
Volcanology; igneous petrology

MOORE, SAMUEL L. Virginia
Geologist (303) 986-3086
962 S. Quail Way
Lakewood, Colorado 80226
Br. of Central Mineral Resources
Denver (303) 234-6297
Franklin & Marshall Coll.-BS; U. of Colorado-MS
Resource appraisal & mapping of Mescalero-Apache reservation
Petrology & geochem.-mineral occurrences

MOORE, SHARON T.
Secretary
Office of the Chief Geologist
Reston (703) 860-6711

MOORE, SUSAN L. Steve
Administrative Clerk (415) 967-3193
500 W. Middlefield, #63
Mt. View, California 94043
Br. of Western Environmental Geology
Menlo Park (415) 323-8111 x2001
Calif. Poly.-BS; U. of Cal. (Davis)

MOORE, THOMAS E.
PST (415) 328-3149
Br. of Alaskan Geology
Menlo Park (415) 323-8111 x2470
U. of Cal. (Santa Barabara)-BA; San Diego State U.-MS; Stanford U.
Devonian clastic rocks, Brooks Range, Alaska
Field geology; petrology; ophiolites

MOORE, WILLIAM J.
Geologist
127 Dunsmuir Way
Menlo Park, California 94025
Br. of Western Mineral Resources
Menlo Park (415) 323-8111 x2547
Iowa State U.-BS, MS; Stanford U.-PhD
Walker Lake 1° x 2° quadrangle; plutonic rocks of central Wasatch Mts., Utah
Igneous petrology; prophyry copper deposits

MORGAN, BENJAMIN A., III
Geologist
Br. of Field Geochemistry & Petrology
Reston **(703) 860-7453**
U. of N. Carolina-BS, MA; Princeton-PhD
Metamorphic petrology; microprobe; field mapping

MORGAN, IDA M.
Geologist
2065 Royal Fern Court
Reston, Virginia 22091
Office of Scientific Publications
Reston **(703) 860-6517**
George Washington U.-BA
Geologic Inquiries Group

MORGAN, JOSEPH O. **Pitaya**
Remote Sensing Specialist
Office of International Geology
Reston **(703) 860-6555**
Purdue U.-MS
Satellite remote sensing, international technology exchange

MORGAN, JOHN W.
Chemist **(703) 437-6351**
11430 Orchard Lane
Reston, Virginia 22090
Br. of Analytical Laboratories
Reston **(703) 860-6852**
Birmingham U.-BSc; Australian National U.-PhD
Radiochemistry; geochemistry; cosmochemistry

MORGENSTERN, JOSEPH C. **Louise**
Engineering Technician **(408) 739-5573**
630 Barnsley Way
Sunnyvale, California 94087
Br. of Field Geochemistry & Petrology
Menlo Park **(415) 323-8111 x2644**

MORGENSTERN, KAREN M.
Administrative Officer
261 Encina Avenue
Redwood City, California 94061
Br. of Alaskan Geology
Menlo Park **(415) 323-8111 x2181**

MORIN, ROBERT L. **Edith Q.**
PST
420 San Luis Avenue
Los Altos, California 94022
Br. of Regional Geophysics
Menlo Park **(415) 323-8111 x2607**
San Jose State U.-BA

MORLEY, JAMES M.
Cartographer
160 Waverley Street
Palo Alto, California 94301
Br. of Pacific-Arctic Geology
Menlo Park **(415) 856-7102**
U. of California-BA
Oblique diagrams; base maps

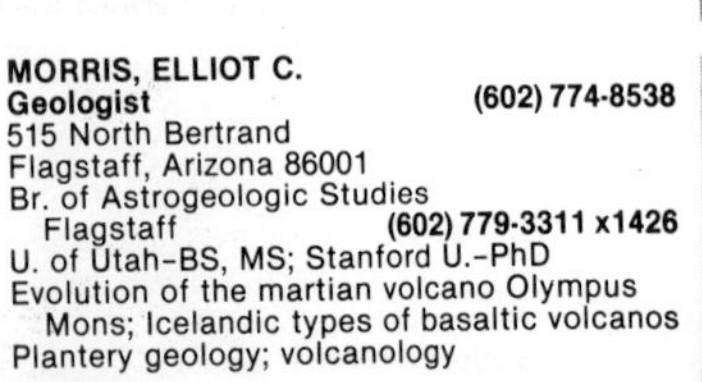

MORRIS, ELLIOT C.
Geologist **(602) 774-8538**
515 North Bertrand
Flagstaff, Arizona 86001
Br. of Astrogeologic Studies
Flagstaff **(602) 779-3311 x1426**
U. of Utah-BS, MS; Stanford U.-PhD
Evolution of the martian volcano Olympus Mons; Icelandic types of basaltic volcanos
Plantery geology; volcanology

MORRIS, HAL T. **Elizabeth**
Geologist **(415) 854-6569**
2220 Camino A Los Cerros
Menlo Park, California 94025
Br. of Western Mineral Resources
Menlo Park **(415) 323-8111**
U. of Utah-BS, MS
Richfield 2° quadrangle
Mineral deposits; mineral resources

MORRIS, LYNN G. **Woody**
Management Technician
822½ Freeman Avenue, NW
Albuquerque, New Mexico 87107
Br. of Global Seismology
Albuquerque **(505) 844-4637**
U. of New Mexico

MORRIS, ROBERT H. **Mary Nell**
Geologist **(703) 476-5496**
2521 Brofferton Court
Herndon, Virginia 22070
Office of Environmental Geology
Reston **(703) 860-6411**
Upsala Coll.-BA; Rutgers U.-MS
Manager, reactor hazards research program
Engineering geology

MORTENSEN, CARL E. **Lynn**
Geophysicist
Br. of Tectonophysics
Menlo Park **(415) 323-8111 x2583**
U. of Cal. (Berkeley)-BA; Colorado Schl. of Mines; Stanford U.
Tilt operations, instrumental strain
Instrumental monitoring of crustal deformation

MORTON, DOUGLAS M. **Robyn**
Geologist
Office of Environmental Geology
Reston **(703) 860-6411**
U. of Cal. (Riverside)-BA; UCLA-PhD
Earthquake hazards reduction program

MOSELY, RODERICK C.
Physicist
2727 Midtown Court, #28
Palo Alto, California 94303
Br. of Isotope Geology
Denver **(303) 234-3624**
U. of Cal. (Santa Barbara)-BS
Radiocarbon dating lab
Dating equipment, research & development

MOSER, JEANETTE M. **Howard**
Secretary
Br. of Special Projects
Denver **(303) 234-2112**

MOSES, THOMAS H. **Pat**
Engineer
Woodside, California 94062
Br. of Tectonophysics
Menlo Park **(415) 323-8111 x2235**
U. of Oklahoma-BS
Geothermal studies
Heat flow, well logging & research drilling

MOSIER, DAN L.
PST
Office of Reource Analysis
Menlo Park **(415) 323-8111 x2906**
San Francisco State U.-BA

MOSIER, ELWIN L. **Lou**
Chemist
Br. of Exploration Research
Denver (303) 234-6166
U. of Nebraska-BS
Emission spectroscopy; analytical chemistry

MOSSOTTI, VICTOR G.
Chemist
Br. of Analytical Laboratories
Reston (703) 323-8111 x2946
Iowa State U.-PhD
X-ray fluorescence
X-ray fluorescence; time-series analysis

MOWINCKEL, PENELOPE K.
Geologist
Office of Scientific Publications
Menlo Park (415) 323-8111 x2301
San Francisco State U.-BS

MOYER, FRANCES V. **Wilson**
Secretary (415) 593-7094
3220 Brittan Avenue
San Carlos, California 94070
Br. of Pacific-Arctic Geology
Menlo Park (415) 856-7143
Droughon's Business College

MROSE, MARY E.
Geologist (703) 528-6448
114 N. Wayne Street, Apt. 2
Arlington, Virginia 22201
Br. of Exper. Geochem. & Mineralogy
Reston (703) 860-6670
Salem State Coll.-BS; Boston U.-MA
Mineralogical investigations
Minerals of the nitrate deposits of Chile; determinative mineralogy

MUELLER, ROBERT J. JR.
PST (415) 321-2173
Br. of Tectonophysics
Menlo Park (415) 323-8111 x2533
Humboldt State U.-BA

MUFFLER, L.J. PATRICK **Pat**
Geologist (415) 493-6439
961 Ilima Way
Palo Alto, California 94306
Br. of Field Geochemistry & Petrology
Menlo Park (415) 323-8111 x2398
Pomona Coll.-BA; Princeton-MA, PhD
Volcanic rocks, tectonics & geothermal resources of the Lassen Region, California
Geothermal resources; hydrothermal systems; Quaternary volcanic rocks

MULHAUSEN, LORRAINE E.
Secretary
525 Florida Avenue, #101
Herndon, Virginia 22070
Office of Mineral Resources
Reston (703) 860-6567
Wilderness program

MULL, CHARLES G.
Geologist
SRA, Box 371-X
Anchorage, Alaska 99507
Br. of Alaskan Geology
Anchorage (907) 271-4150
U. of Colorado-BA, MS
Brooks Range and Arctic Slope, Alaska

MULLER, DOUGLAS C. **Patricia**
Geophysicist (303) 989-0348
Lakewood, Colorado 80226
Br. of Special Projects
Denver (303) 234-2371
Colorado Schl. of Mines-BS
Computer applications for well log analysis

MULLINEAUX, DONAL R. **Diana**
Geologist
2245 Lee Street
Lakewood, Colorado
Br. of Engineering Geology
Denver (303) 234-3312
U. of Washington-BS, MS, PhD
Volcanic hazards
Surficial & engineering geology

MUNROE, ROBERT J. **Maura**
Geophysicist
2037 Sharon Road
Menlo Park, California 94025
Br. of Tectonophysics
Menlo Park (415) 323-8111 x2381
San Jose State U.-BA; Stanford U.-MS
Geothermal studies
Heat flow, thermal conductivity

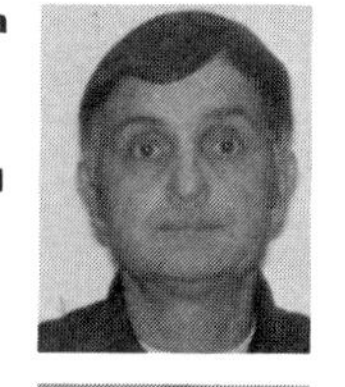

MURDOCK, JAMES N.
Geophysicist (505) 299-5254
611 Green Valley Drive, SE
Albuquerque, New Mexico 87123
Br. of Global Seismology
Albuquerque (505) 844-4638
U. of N. Carolina
Seismic research observatories
Crustal structure; earthquake studies

MURPHY, JACK F. **Eleanor**
Geologist (303) 279-1899
2011 Mt. Zion Drive
Golden, Colorado 80401
Office of Energy Resources
Denver (303) 234-4735
Dartmouth-BA, MA; Cal. Tech.
Stratigraphy & sedimentation

MURRAY, THOMAS L.
PST
Br. of Tectonophysics
Menlo Park (415) 323-8111 x2705
Stanford U.-BS
Tilt operations
Instrumentation for measuring crustal tilting

MURREY, DONALD G. **Mary Lou**
PST
Lakewood, Colorado 80226
Br. of Uranium-Thorium Resources
Denver (303) 234-5531
Helium

MUSIALOWSKI, FRANK R. **Josephine**
Oceanographer (703) 569-5185
7129 Lamar Drive
Springfield, Virginia 22150
Br. of Atlantic-Gulf of Mex. Geology
Reston (703) 860-6489
U. of Alabama-BS
Nearshore sediment transport

MYERS, ALFRED T. **Edna**
Chemist (303) 237-5667
11675 W. 31st Place
Lakewood, Colorado 80215
Br. of Analytical Laboratories
Denver (303) 234-6401
U. of Maryland-BS, MS
Analytical methods & trace elements in environmental health
Geochemistry of rhenium & germanium

MYERS, DONALD A. **Margaret**
Geologist **(303) 237-4595**
1240 Cody Street
Lakewood, Colorado 80215
Br. of Central Environmental Geology
Denver **(303) 234-6198**
Stanford-BA; Johns Hopkins U.
Manzano Los Pinos Mts., New Mexico
Pennsylvanian Permian stratigraphy; fusulinid foraminifera; geologic mapping

MYLES, PERCY R. **Dawnnita**
PST **(408) 988-4169**
3711 Lafayette Street, #3
Santa Clara, California 95050
Office of the Chief Geologist
Menlo Park **(415) 323-8111 x2253**

MYREN, GLENN D.
PST
474 O'Connor Street
Menlo Park, California 94025
Br. of Tectonophysics
Menlo Park **(415) 323-8111 x2705**
Marietta Coll.; San Jose State U.-BA
Instrumental strain

MYTTON, JAMES W. **Jeanne**
Geologist **(303) 237-5649**
2720 Simms Street
Lakewood, Colorado 80215
Br. of Coal Resources
Denver **(303) 234-3570**
Dartmouth-AB; U. of Wyoming-MA
San Juan Basin, New Mexico coal resource study
Sedimentary mineral deposits, energy resources

NAESER, CHARLES W.
Geologist **(303) 986-6947**
445 S. Wright Street, #214
Lakewood, Colorado 80228
Br. of Isotope Geology
Denver **(303) 234-4201**
Dartmouth-BA, MA; Southern Methodist U.-PhD
Geochronology; fission track dating

NAKAGAWA, HARRY M. **Frances**
Chemist **(303) 773-1054**
7696 E. Napa Place
Denver, Colorado 80237
Br. of Exploration Research
Denver **(303) 234-2849**
U. of Colorado-BS
Methods development
Electrochemistry; atomic absorption

NAKATA, JENNIFER S.
PSA
Br. of Field Geochem. & Petrology
Hawaiian Volcano Observ. **(808) 967-7328**
Seismic reading

NASH, J. THOMAS **Marti**
Geologist **(303) 674-7653**
32059 Quarterhorse Road
Evergreen, Colorado 80439
Br. of Uranium-Thorium Resources
Denver **(303) 234-2954**
Amherst Coll.-BA; Columbia-MA, PhD
Geology of uranium deposits
Geology & geochemistry of uranium and base metal deposits

NASON, ROBERT D.
Geophysicist
Br. of Ground Motion & Faulting
Menlo Park **(415) 323-8111 x2760**
Cal. Tech-BS; U. of Cal. (San Diego)-PhD
California earthquakes & seismic intensity
Earthquake damage processes; tectonic fault creep

NATHENSON, MANUEL **Judith H. Windt**
Mechanical Engineer **(415) 328-4173**
239 Elliot Drive
Menlo Park, California 94025
Br. of Field Geochemistry & Petrology
Menlo Park **(415) 323-8111 x2730**
Carnegie Mellon U.-BS; Stanford-MS, PhD
Geothermal reservoirs
Geothermal physics; geothermal resources

NAVARRO, RICHARD **Ana**
Geophysicist
Br. of Ground Motion & Faulting
Menlo Park **(415) 323-8111 x2161**
U. of Texas-BS, ME; Baylor U.-BA
Ground motion prediction; instrumentation; seismic documentation

NEALEY, L. DAVID **Myrna**
Geologist **(602) 526-9511**
3923 N. Geneva Circle
Flagstaff, Arizona 86001
Br. of Central Environmental Geology
Flagstaff **(602) 779-3311 x1538**
U. of Florida-BS; Northern Arizona U.-MS; U. of New Mexico
San Francisco & Springerville volcanic fields
Igneous petrology; volcanology; remote sensing

NEEDELL, SALLY W.
Physicial Scientist
43 Althea Road
North Falmouth, Massachusetts
Br. of Atlantic-Gulf of Mex. Geology
Woods Hole **(617) 548-8700**
Middlebury College-BA

NEEDHAM, RUSSELL E. **Charlene**
Geophysicist **(303) 421-7336**
6464 Welch Street
Arvada, Colorado 80004
Br. of Global Seismology
Denver **(303) 234-3994**
Ohio U.-BS
NEIS operations
Analysis of digital recorded seismograms; epicenter determination; seismicity

NEELEY, CATHY L.
PST
Br. of Eastern Mineral Resources
Reston **(703) 860-6913**
Eastern Kentucky U.-BS

NEGRI, JOHN C. **Jennie**
Chemist **(303) 477-2506**
5262 Zuni Street
Denver, Colorado 80221
Br. of Exploration Research
Denver **(303) 234-6183**
Regis College-BS

NEIL, SARAH T.
Chemist
4055 Ben Lomond Drive
Palo Alto, California 94306
Br. of Analytical Laboratories
Menlo Park **(415) 323-8111**
Memphis State U.-BS
Chemical analysis of rocks & minerals; classical chemical analysis

NEIMAN, HARRIETT G.
Chemist **(303) 278-3806**
474 Gladiola Street
Golden, Colorado 80401
Br. of Analytical Laboratories
Denver **(303) 234-6401**
U. of Michigan-BS

NELMS, CHARLES A. **Estella**
Geophysicist **(303) 371-0792**
12973 E. 47th Avenue
Denver, Colorado 80239
Br. of Petrophysics & Remote Sens.
Denver **(303) 234-5388**
U. of Texas (El Paso)-BS, MS
Laboratory studies of soil moisture
Thermal properties of rocks; borehole geophysics

NELSON, C. HANS
Geologist
26360 Calle del Sol
Los Altos Hills, California 94022
Br. of Pacific-Arctic Geology
Menlo Park **(415) 856-7023**
Carleton Coll.; U. of Minnesota; Oregon State
Environmental geology of the Bering Shelf
Marine geology; sedimentology; environmental geology

NELSON, CLIFFORD M.
Geologist **(703) 437-0345**
1412 Northgate Square, Apt. 11B
Reston, Virginia 22090
Office of Scientific Publications
Reston **(703) 860-6575**
U. of Illinois-BS; Michigan State-MS; U. of California (Berkeley)-PhD
Marine mollusk biostratigraphy & paleontology (Neogene, Pacific—Arctic); history of geology

NELSON, GLENN D.
Geophysicist
Office of Earthquake Studies
Menlo Park **(415) 323-8111 x2839**
MIT-BS; Brown U.-MS
Dry lake earthquakes
Digital filtering

NELSON, KAREN A.
Administrative Assistant
Br. of Central Mineral Resources
Denver **(303) 234-4842**
El Camino Coll.-AA; San Francisco State U.-BA

NELSON, MARTHA J.
Secretary **(415) 345-5202**
29 Tollridge Court
San Mateo, California 94402
Br. of Pacific-Arctic Geology
Menlo Park **(415) 856-7142**

NELSON, MAURICE M. **Louise**
Engineering Technician **(907) 456-6115**
833 4th Avenue
Fairbanks, Alaska 99701
Br. of Electromag. & Geomagnetism
College, Alaska **(907) 479-6146**

NELSON, WILLIS H. **Evelyn**
Geologist **(415) 856-6987**
2692 Louis Road
Palo Alto, California 94303
Br. of Alaskan Geology
Menlo Park **(415) 323-8111**
Montana State Coll.-BS; U. of Washington
Structural geology; petrology & petrography

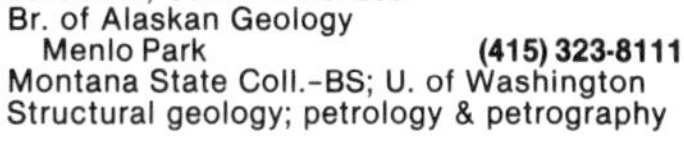

NERVICK, KEVIN H.
PST **(303) 988-0832**
10555 W. Jewell Avenue 4-303
Lakewood, Colorado 80226
Br. of Electromag. & Geomagnetism
Denver **(303) 234-5157**
Macaiester Coll.-BA
Geothermal resources
Geothermal, mineral resources; radiation waste disposal

NEUERBURG, GEORGE J. **Faye**
Geologist **(303) 986-3577**
872 S. Newcombe Way
Lakewood, Colorado 80226
Br. of Exploration Research
Denver **(303) 234-2932**
UCLA-BA, MA, PhD
Geochemical characterization of metallogenic provinces & mineralized areas
Prospecting—theory & methods

NEUMAN, ROBERT B. **Arline**
Geologist **(202) 362-7147**
5027 Klingle Street, NW
Washington, D.C. 20016
Br. of Paleontology & Stratigraphy
Washington, D.C. **(202) 343-3319**
U. of N. Carolina-BS; Johns Hopkins-PhD
Caledonide orogen
Ordovician brachiopods, stratigraphy & paleogeography, eastern U.S.

NEUVILLE, ARNETT K.
PST **(703) 435-1334**
11717 North Shore Drive
Reston, Virginia 22090
Br. of Analytical Laboratories
Reston **(703) 860-6964**
West Hampton Coll.-BA

NEVIN, JAN M.
Editorial Assistant **(303) 279-0882**
2569 Alkire
Golden, Colorado 80401
Br. of Global Seismology
Denver **(303) 234-3994**
U. of Colorado; Kansas State U.-BA

NEWELL, MARCIA F.
Geologist **(703) 364-2722**
P.O. Box 144
Rectortown, Virginia 22140
Br. of Isotope Geology
Reston **(703) 860-6592**
Oberlin Coll.-BA
U-Th-Pb zircon geochronology
Geochronology

NEWELL, WAYNE L. **Mary Jane**
Geologist **(703) 777-3595**
135 Belmont Drive
Leesburg, Virginia 22075
Br. of Eastern Environmental Geology
Reston **(703) 860-6406**
Franklin & Marshall-BA; Dartmouth-MA; Johns Hopkins-PhD
Cenozoic stratigraphy of tidewater Va. & Md
Geomorphology; Quaternary stratigraphy; neotectonics

NEWMAN, EVELYN B. **Fred**
Mathematician **(408) 248-8206**
1547 Sandpiper Court
Sunnyvale, California 94087
Br. of Western Environmental Geology
Menlo Park **(415) 323-2914**
New Mexico State U.-BS
Reservoir induced seismicity
Computer science; vegetative recovery rates

NEWMAN, WILLIAM L. **Lucille**
Geologist **(301) 299-4724**
11125 Hunt Club Drive
Potomac, Maryland 20854
Office of Scientific Publications
Reston **(703) 860-6575**
Beloit Coll.-BS; Montana Schl. of Mines; U. of Montana
Mineral deposits & their economics

NEWTON, SALLY K. **Matthew**
GFA
Br. of Paleontology & Stratigraphy
Menlo Park **(415) 323-8111 x2489**
Foothill Jr. Coll.; Stanford
Radiolaria processing techniques
Radiolaria; shoreline processes

NICCOLLS, LINDA J.
Administrative Assistant
Br. of Eastern Environmental Geology
Reston **(703) 860-6406**

NICHOLS, DONALD R. **Louise**
Geologist **(303) 526-1304**
978 Coneflower Drive
Golden, Colorado 80401
Br. of Engineering Geology
Denver **(303) 234-3624**
U. of Virginia; U. of Nebraska-BS; Yale
Engineering, environmental & Pleistocene geology

NICHOLS, DOUGLAS J. **Jan**
Geologist **(303) 979-2825**
6939 West Quarto Place
Littleton, Colorado 80123
Br. of Paleontology & Stratigraphy
Denver **(303) 234-5871**
New York U.-BA, MS; Penn State-PhD
Tertiary nonmarine basins; thrust belt biostratigraphy; mid-Cretaceous events
Mesozoic & Cenozoic palynology

NICHOLS, FREDERIC H. **Kirstin**
Oceanographer **(415) 328-1684**
1189 Harker Avenue
Palo Alto, California 94301
Br. of Pacific-Arctic Geology
Menlo Park **(415) 856-7196**
Hamilton Coll.-BA; U. of Washington-MS, PhD
San Francisco Bay
Biological oceanography; marine & estuarine ecology

NICHOLS, KATHRYN M.
Geologist
Br. of Oil & Gas Resources
Denver **(303) 234-4109**
U. of Cal. (Riverside)-BS; Stanford-PhD
Paleozoic carbonate reservoir rocks, Disturbed Belt, Montana
Carbonate petrography & diagenesis; stratigraphy

NICHOLS, LINDSAY R. **Ken**
Secretary
1412 Sadlers Wells Drive
Herndon, Virginia 22070
Office of Environmental Geology
Reston **(703) 860-6411**

NICHOLS, ROGER W. **Marilyn**
Engineering Technician **(303) 421-9729**
6569 Van Gordon Street
Arvada, Colorado 80004
Br. of Engineering Geology
Denver **(303) 234-5187**
Fort Hays State Coll.; Colorado U.
Geotechnical measurements & services project

NICHOLS, THOMAS C., JR. **Joey**
Geologist **(303) 979-7546**
5091 W. Ottawa Avenue
Littleton, Colorado 80123
Br. of Engineering Geology
Denver **(303) 234-5682**
U. of Colorado-BA; Texas A&M-MS
In-situ stress in shales
Rock mechanics; engineering geology

NILSEN, TOR H. **Sharon**
Geologist **(415) 368-5997**
751 Canyon Road
Redwood City, California 94062
Br. of Western Environmental Geology
Menlo Park **(415) 323-8111 x2470**
CCNY-BS; U. of Wisconsin-MS, PhD
Devonian clastics of Brooks Range, Alaska
Sedimentology; tectonics; marine geology

NISHI, JAMES M.
Physicist **(303) 989-2700**
465 S. Wright Street, # 116
Lakewood, Colorado 80228
Br. of Central Mineral Resources
Denver **(303) 234-5746**
U. of Hawaii-BA
Scanning electron microscopy

NOBLE, E.A. **Polly**
Geologist **(703) 860-1011**
11407 Great Meadow Drive
Reston, Virginia 22091
Office of Energy Resources
Reston **(703) 860-6432**
Tufts-BA; U. of New Mexico-MS; U. of Wyoming-PhD
Ore deposits, especially those in sedimentary rocks

NOBLE, MARLENE A.
Oceanographer
Woods Hole, Massachusetts
Br. of Atlantic-Gulf of Mex. Geology
Woods Hole **(617) 548-8700**
Princeton-MS; MIT-MS
Continental shelf circulation, transport of pollutants & sediments
Physical oceanography of continental margins

NOKLEBERG, WARREN J.
Geologist **(415) 797-7341**
4315 Norris Road
Fremont, California 94536
Br. of Alaskan Geology
Menlo Park **(415) 323-8111 x2277**
UCLA-BA; U. of Cal. (Santa Barbara)-PhD
Mt. Hayes quadrangle
Tectonics; geochemistry; economic geology

NOLAN, THOMAS B. **Pete**
Geologist
2219 California Street, NW
Washington, D.C. 20008
Office of the Director & Chief Geologist
Reston
Yale-PhB, PhD
Geology & mineral resources of the Great Basin
Scientist, administrator

NORD, GORDON L., JR.
Geologist **(703) 860-4522**
2220 White Cornus Lane
Reston, Virginia 22091
Br. of Exper. Geochem. & Mineralogy
Reston **(703) 860-6660**
U. of Wisconsin-BS; U. of Idaho-MS; U. of California (Berkeley)-PhD
Microstructure analysis
Mineralogy; mechanism & kinetics of phase transformations; electron microscopy

NORMAN, MEADE B., II
PST
Sunnyvale, California 94086
Br. of Western Environmental Geology
Menlo Park **(415) 323-2214**
Michigan Coll. of Mining & Tech.; Wayne State U.; Florida State U.-BS; San Jose State U.
X-ray diffraction & spectrography; mineral identification & staining; quartz crystallinity studies

NORMARK, WILLIAM R. **DJ**
Geologist **(408) 735-9937**
1225 Rembrandt Drive
Sunnyvale, California 94087
Br. of Pacific-Arctic Geology
Menlo Park **(415) 856-7045**
Stanford-BS; U. of Cal. (San Diego)-PhD
Submarine fans off California
Marine geology & geophysics; deep-water instrumentation; geology of spreading ctrs.

NORTH, LESTER D. **Barbara**
Computer Specialist **(617) 540-1115**
122 Barrows Road
E. Falmouth, Massachusetts 02536
Br. of Atlantic-Gulf of Mex. Geology
Woods Hole **(617) 548-8700**
Virginia Polytechnic Inst.-BSc
Computer applications

NORTON, DANIEL R. Rachel
Chemist (303) 674-5150
29611 Fairway Drive
Evergreen, Colorado 80439
Br. of Analytical Laboratories
Denver (303) 234-2521
Antioch Coll.-BS; Princeton-MS, PhD
Conventional rock analysis
Electroanalytical studies; spectrophotometry; thermal analysis

NORTON, JAMES J. Katharine
Geologist (303) 420-3365
3602 Taft Court
Wheat Ridge, Colorado 80033
Br. of Central Mineral Resources
Denver (303) 234-5740
Princeton-BA; Northwestern U.-MS; Columbia-PhD
Black Hills, South Dakota
Pegmatites; metamorphic rocks; lithium

NOVAK, STEVEN W. Ruth
Geologist
Br. of Eastern Environmental Geology
Reston (703) 860-6406
Virginia Polytech. Institute-MS
Appalachian volcanics
Petrology; geochemistry

NOWLAN, GARY A. Connie
Geologist (303) 499-6892
1470 Brown Circle
Boulder, Colorado 80303
Br. of Exploration Research
Denver (303) 234-6188
Union Coll.-BA; U. of Colorado-PhD
Surface & ground water in geochemical exploration
Exploration geochemistry; low temperature geochemistry

OBERMEIER, STEPHEN F.
Civil Engineer (703) 338-6235
8 Country Club Drive
Purcellville, Virginia 22132
Br. of Engineering Geology
Reston (703) 860-6406
Purdue-PhD
Safe mine waste disposal—Appalachia
Regional engineering geology; soil mechanics

OBI, CURTIS M.
PST
1122 Whipple Avenue, #19
Redwood City, California 94062
Br. of Alaskan Geology
Menlo Park (415) 323-8111 x2342
U. of California (Santa Cruz)-BS

OBRADOVICH, JOHN D. Jan
Geophysicist (303) 986-2055
13428 W. Exposition Drive
Lakewood, Colorado 80228
Br. of Isotope Geology
Denver (303) 234-3876
U. of California (Berkeley)-PhD
Potassium—argon geochronology

O'CONNELL, EVERETT M. Doris
PST (303) 758-4497
1832 S. Garfield Street
Denver, Colorado 80210
Br. of Central Environmental Geology
Denver (303) 234-5378
U. of Kentucky-BS

ODELL, LYNDON A. Elfrieda
Geophysicist (509) 447-4690
Route 4, Box 475
Newport, Washington 99156
Br. of Global Seismology
Newport (509) 447-3195
Southwestern Oklahoma State Coll.-BS
Seismology; geomagnetism; tsunami warning system

ODLAND, SARAH K.
Geologist
1445 S. Pennsylvania
Denver, Colorado 80210
Br. of Exploration Research
Denver (303) 234-3891
Colorado Coll.-BA
Rolla 2° sheet

O'DONNELL, CLAUDINE S. Robert
Secretary (303) 234-3625
2793 S. Vrain Street
Denver, Colorado 80236
Office of the Chief Geologist
Denver (303) 234-3625

O'DONNELL, JIM E. Janice
Geophysicist (303) 988-3071
2284 S. Youngfield Court
Lakewood, Colorado 80228
Br. of Electromag. & Geomagnetism
Denver (303) 234-5159
U. of California (Berkeley)-BA, MA
Development of magnetotelluric & telluric methods
Magnetotelluric surveys & geological interpretation

O'DONNELL, ROBERT W. Claudine
PST (303) 936-4146
2793 S. Vrain Street
Denver, Colorado 80236
Br. of Paleontology & Stratigraphy
Denver (303) 234-5919
Preparation of invertebrate & vertebrate fossils

ODUM, JACK K. Lynda
Geologist
1713 N. College Avenue
Ft. Collins, Colorado
Br. of Engineering Geology
Denver (303) 234-5349
Colorado State U.-BS
Engineering geology; engineering geophysics; geomorphology

OFFIELD, TERRY W.
Geologist
Br. of Uranium-Thorium Resources
Denver (303) 234-3697
Virginia Polytech. Inst.-BS; U. of Illinois-MS; Yale-PhD
Remote sensing; uranium geology; structural geology

OFTEDAHL, ORRIN G. Soledad
Chemist (703) 430-6087
1050-B Margate Court
Sterling, Virginia 22170
Br. of Paleontology & Stratigraphy
Reston (703) 860-7745
U. of Wichita-BS
Palynologic processes

OGLE, HELEN M. Malcolm
Clerk-Typist (415) 326-2326
70 Jordan Place
Palo Alto, California 94303
Br. of Pacific-Arctic Geology
Menlo Park (415) 856-7006
U. of Minnesota-BA

O'HARA, CHARLES J. Mary Kay
Geologist (617) 540-0833
22 Maravista Avenue Ext.
Teaticket, Massachusetts 02536
Br. of Atlantic-Gulf of Mex. Geology
Woods Hole (617) 548-8700
U. of Buffalo-BA
Massachusetts cooperative marine geology project
Marine geology & geophysics; glacial geology

OHLMACHER, GREGORY C.
Geologist
12301 Brookhaven Drive
Wheaton, Maryland 20902
Br. of Engineering Geology
Reston **(703) 860-6421**
U. of Maryland-BS
Safe mine waste disposal—Appalachians
Engineering geology; geomorphology; remote sensing

OKAMURA, ARNOLD T. **Patricia**
Geologist **(808) 959-8019**
212 Pohakulani Street
Hilo, Hawaii 96720
Br. of Field Geochem. & Petrology
Hawaiian Volcano Observ. **(808) 967-7328**
U. of Hawaii-BA; San Jose State U.-MS
Surface deformation
Structural geology; computer modeling; volcanic studies

OKAMURA, REGINALD T. **Jane**
PST **(808) 959-8020**
60 Kapualani Street
Hilo, Hawaii 96720
Br. of Field Geochem. & Petrology
Hawaiian Volcano Observ. **(808) 967-7328**
U. of Hawaii-BA
Hawaii volcanic hazard studies
Geodetic measurements; Lava lake; geochemical studies

OLDALE, ROBERT N. **Gail**
Geologist **(617) 563-6450**
53 Ocean View Avenue
North Falmouth, Massachusetts 02556
Br. of Atlantic-Gulf of Mex. Geology
Woods Hole **(617) 548-8700**
St. Lawrence-BS
Massachusetts marine geologic cooperative
Quaternary geology; marine geology & geophysics

O'LEARY, DENNIS W.
Geologist
Br. of Atlantic-Gulf of Mex. Geology
Woods Hole **(617) 548-8700 x160**
Boston Coll.-BS; U. of Missouri (Rolla)-MS; Penn State-PhD
Environmental geology, Georges Bank segment of Atlantic continental shelf
Geomorphology; remote sensing; tectonics

O'LEARY, RICHARD M.
Chemist **(303) 237-4720**
12895 W. 14th Place
Golden, Colorado 80401
Br. of Exploration Research
Denver **(303) 234-6151**
College of Santa Fe-BS; Colorado Schl. of Mines
Alaskan mineral resource appraisal program
Geochemical analysis

OLHOEFT, GARY R. **Jean**
Geophysicist **(303) 279-6345**
226 S. Holman Way
Golden, Colorado 80401
Br. of Petrophysics & Remote Sens.
Denver **(303) 234-5393**
MIT-BS, MS; U. of Toronto-PhD
Physical properties of rocks and minerals, nonlinear complex resistivity logging
Petrophysics; borehole geophysics; radar sounding

OLIVE, WILDS W. **Ruth**
Geologist
Jl. Rancabentang 5
Bandung, Indonesia
Br. of Middle Eastern & Asian Geology
Bandung, Indonesia
U. of North Carolina-BS; Louisiana State U.-PhD
Science & technology project with USAID/Indonesia

OLIVER, HOWARD W. **Betty**
Geophysicist **(415) 322-7863**
550 St. Francis Court
Menlo Park, California 94025
Br. of Regional Geophysics
Menlo Park **(415) 323-8111 x2255**
U. of Cal. (Berkeley)-BA; Harvard-MA
Radioactive waste disposal
Gravity; isostasy; batholiths

OLIVER, WILLIAM A., JR. **Johanna**
Geologist **(301) 946-0387**
4203 McCain Court
Kensington, Maryland 20795
Br. of Paleontology & Stratigraphy
Washington, D.C. **(202) 343-3523**
U. of Illinois-BS; Cornell-MA, PhD
Pre-Carboniferous corals; Devonian biostratigraphy

OLSEN, ALICE L.
Secretary
Br. of Ground Motion & Faulting
Menlo Park **(415) 323-8111 x2720**

OLSEN, HAROLD W.
Civil Engineer
2307 Ivy
Denver, Colorado 80207
Br. of Engineering Geology
Denver **(303) 234-4717**
MIT-BS, MS, ScD
Marine geotechnical investigations
Soil mechanics; transport phenomena; marine geotechnics

OLSON, ANNABEL B. **Richard**
Geologist
5207 Wilson Lane
Bethesda, Maryland 20014
Br. of Coal Resources
Reston **(703) 860-7734**
U. of Chicago-BS; Stanford
Photogeologic mapping & stratigraphy

OLSON, JANE CIENER **Allison**
Geologist **(703) 437-0664**
409 Cavendish Street
Herndon, Virginia 22070
Office of Resource Analysis
Reston **(703) 860-6451**
U. of Florida-BS, MS
Mineral data system reference data base
Bibliography; data base management

OLSON, RITA J.
Administrative Clerk
Office of Environmental Geology
Reston **(703) 860-6416**

OMAN, CHARLES L. **Joanne**
Geologist **(703) 430-9603**
200 Fir Court
Sterling, Virginia 22170
Br. of Coal Resources
Reston **(703) 860-7426**
American U.-BS; Texas Tech.-MS
Geochemistry of coals of the eastern U.S.
Geochemistry; coal

OMAN, JOANNE K. **Charles**
PST **(703) 430-9603**
200 Fir Court
Sterling, Virginia 22170
Br. of Oil & Gas Resources
Reston **(703) 860-6652**
American U.-BS
Gulf coast lignites
Lignites

O'NEAL, REBECCA K.
Secretary
10555 W. Jewell
Lakewood, Colorado 80226
Br. of Coal Resources
Denver **(303) 234-3624**

O'NEIL, JAMES R.
Chemist **(415) 323-5251**
674 Webster Street
Palo Alto, California 94301
Br. of Isotope Geology
Menlo Park **(415) 323-8111 x2598**
Loyola U.-BS; Carnegie-Mellon-MS; U. of Chicago-PhD
Oxygen isotopes
Stable isotope geochemistry; mantle processes; paleoclimatology

O'NEILL, J. MICHAEL
Geologist **(303) 494-4831**
4270 Prado Drive
Boulder, Colorado 80303
Br. of Central Environmental Geology
Denver **(303) 234-6285**
University of Colorado-PhD
Geologic framework, RARE II, SW Montana Precambrian geology; Neogene tectonics & volcanism, N. New Mexico
Structural geology & tectonics

O'NEILL, MARY E. **Rex V. Allen**
Mathematician
P.O. Box 915
Menlo Park, California 94025
Office of Earthquake Studies
Menlo Park **(415) 323-8111 x2761**
U. of Cal. (Davis)-BA; U. of Cal. (Berkeley)-MA
Mathematical represenation of seismic instrument response; earth attentuation of seismic waves

ORIEL, STEVEN S. **Esther**
Geologist **(303) 279-2384**
14195 Crabapple Road
Golden, Colorado 80401
Br. of Central Environmental Geology
Denver **(303) 234-3337**
Columbia-BA; Yale-MS, PhD
Idaho—Wyoming thrust belt — Snake River plain studies
Structural & areal geology

ORKILD, PAUL P. **Barbara**
Geologist **(303) 237-7607**
130 Flower Street
Lakewood, Colorado 80226
Br. of Special Projects
Denver **(303) 234-2391**
U. of Illinois-BS; GWU; U. of Colorado
Tectonics of S. Nevada, geology of Nevada Test Site
Volcanology, structural geology, areal mapping

ORLANDO, ROBERT C.
PST
Br. of Pacific-Arctic Geology
Menlo Park **(415) 856-7064**
San Francisco State U.-BA; San Jose State U.-MA
Bedform studies in lower Cook Inlet, AK
Sedimentology; geophysics; photographic applications

ORRIS, GRETA J.
Geologist
Office of Resource Analysis
Denver **(303) 234-6281**
San Jose State U.-BA, MS
CRIB; geothermal studies

ORSINI, NICHOLAS A.
Physical Scientist
725-306 Mariposa Avenue
Mt. View, California 94041
Office of Earthquake Studies
Menlo Park **(415) 323-8111**
Colgate U.-BS; St. Louis U.
Global aftershock project
Technical requirements for contracting

OSBAKKEN, WILLIS E. **Sylvia**
PST **(907) 747-3901**
1719 Sawmill Creek Road
Sitka, Alaska 99835
Br. of Electromag. & Geomagnetism
Sitka, Alaska **(907) 747-3332**
Washington State U.-BS
Geomagnetism

OSBERG, PHILIP H. **Priscilla**
Geologist **(207) 866-4126**
Star Route
Orono, Maine 04473
Br. of Eastern Environmental Geology
Orono, Maine **(207) 581-2730**
Dartmouth-BA; Harvard-MA, PhD
Geology of Bangor 2° sheet
Petrology, structural geology, regional tectonics

OSCARSON, ROBERT L. **Sheri**
PST **(415) 328-5287**
Office of the Chief Geologist
Menlo Park **(415) 323-8111 x2385**
San Jose State U.
Scanning electron microscope laboratory

OSTERWALD, FRANK W. **Doris**
Geologist **(303) 237-0583**
40 South Dover Street
Lakewood, Colorado 80226
Br. of Engineering Geology
Denver **(303) 234-3818**
U. of Wyoming-BA, MA; U. of Chicago-PhD
Geotechnical research, western energy lands
Mine deformation; structural geology

OTA, YOSHIKO P. **Kurt**
Secretary **(415) 341-6086**
733-27th Avenue
San Mateo, California 94403
Br. of Field Geochemistry & Petrology
Menlo Park **(415) 323-8111**

OTTON, JAMES K. **Sue**
Geologist **(303) 989-4214**
S. Van Gordon Street
Lakewood, Colorado 80228
Br. of Uranium-Thorium Resources
Denver **(303) 234-3481**
Penn State-BS, PhD
Uranium in Arizona Tertiary basins
Basin & Range tectonics; uranium in volcanic sedimentary rocks

OUTERBRIDGE, WILLIAM F. **Dorothy**
Geologist **(703) 860-8537**
11969 Greywing Court
Reston, Virginia 22091
Br. of Eastern Environmental Geology
Reston **(703) 860-6595**
Brown-BA; American U.-MS
Safe coal waste disposal
Areal geology; engineering geology

OVENSHINE, A. THOMAS **Elinor**
Geologist
Office of Mineral Resources
Reston **(703) 860-6561**
Yale-BS; Virginia Polytech. Inst.-MS; UCLA-PhD

OVERTURF, DEE E. **Jacqueline**
Electronics Technician **(303) 838-7389**
25 Pinto Lane
Bailey, Colorado 80421
Br. of Earthquake Tectonics & Risk
Denver **(303) 234-6352**
Murray State Coll; Del Mar College

OWEN, DOUGLASS E. **Debbie**
Geologist **(512) 855-6986**
4802 Alma
Corpus Christi, Texas 78411
Br. of Atlantic-Gulf of Mex. Geology
Corpus Christi **(512) 888-3294**
Kent State U.-BS
Marine geology; sedimentology; environmental geology

OWEN, MARY J. Kenneth
Geologic Inquiries Asst. (703) 788-4824
Route 1, Box 199B
Catlett, Virginia 22019
Office of Scientific Publications
Reston (703) 860-6575

OWENS, JAMES P.
Geologist (301) 460-9154
14528 Bauer Drive
Rockville, Maryland 20853
Br. of Eastern Environmental Geology
Reston (703) 860-6421
U. of Buffalo-BA, MA
Cape Fear project
Stratigraphy; sedimentology; regional geology

PAGE, NORMAN J. Nina
Geologist (415) 325-6401
122 East Creek, Apt. 5B
Menlo Park, California 94025
Br. of Western Mineral Resources
Menlo Park (415) 323-8111 x2671
Dartmouth-BA; U. of California (Berkeley)-PhD
Platinum-group metals
Economic geology; petrology; geochemistry

PAGE, ROBERT A. Marna
Geophysicist
Br. of Ground Motion & Faulting
Menlo Park (415) 323-8111 x2881
Harvard-BA; Columbia-PhD
Alaska earthquake studies
Seismology & tectonics

PAIDAKOVICH, MATTHEW E. Mary
Geologist (703) 471-7816
13315 Schwenger Place
Herndon, Virginia 22070
Office of Resource Analysis
Reston (703) 860-6604
Northern Arizona U.-BS
Metallogenic studies
Economic geology; metallogeny; mineral deposit data processing

PALACAS, JAMES G. Joanna
Geologist (303) 238-0109
11300 W. 27th Place
Lakewood, Colorado 80215
Br. of Oil & Gas Resources
Denver (303) 234-3332
Harvard-BA; Penn State-MS; U. of Minnesota-PhD
Petroleum geochemistry of carbonate rocks
Organic geochemistry of recent & ancient sediments; source rock studies

PALLISTER, JOHN S. Jane
Geologist
Br. of Field Geochemistry & Petrology
Menlo Park (415) 323-8111
Emory U.-BS; U. of California (Santa Barbara)-PhD
Igneous petrology; ophiolites; volcanology

PAMPEYAN, EARL H. Joan
Geologist (415) 948-5937
747 Los Altos Avenue
Los Altos, California 94022
Br. of Engineering Geology
Menlo Park (415) 323-8111 x2340
Pomona Coll.-BA; Claremont Coll.-MA; Stanford
San Mateo & Montana mountain 7.5' quads
Areal mapping; environmental geology; mineralogy

PANKRATZ, LEROY W. Judith
Geophysicist (303) 988-3032
14274 W. Center Drive
Lakewood, Colorado 80228
Br. of Regional Geology
Denver (303) 234-5463
Mankato State Coll.-BA
Central Arizona fissure investigations
Applications of seismic refraction techniques

PANTEA, MICHAEL P.
PST
Br. of Sedimentary Mineral Resources
Denver (303) 234-5915

PAPP, CLARA S.E.
Chemist (303) 985-9382
549 S. Swadley
Lakewood, Colorado 80228
Br. of Regional Geochemistry
Denver (303) 234-3453
U. of Exact & Natural Sciences (Argentina)-BS, MS
Geochemistry of plants

PAPP, JOHN E. Jo
Geophysicist (907) 479-6757
SR3, Box 30470
Fairbanks, Alaska 99701
Br. of Electromag. & Geomagnetism
College, Alaska (907) 479-6146
U. of Massachusetts-BS
Magnetic & sesimic observatory project
Geomagnetism & seismology; data collection; archiving & dissemination

PARCEL, ALICE A.
Technical Publications Editor
Office of Scientific Publications
Menlo Park (415) 323-8111 x2302
Columbia; U. of Edinburgh

PARK, ROBERT B.
Computer Specialist (303) 279-1015
1608 Utah Street
Golden, Colorado 80401
Br. of Ground Motion & Faulting
Denver (303) 234-3624
Baylor U.-BA; U. of Texas (Arlington)-BS
Golden computer center

PARMENTER, CAROL M.
Geologist
407 Quaker Road
N. Falmouth, Massachusetts
Br. of Atlantic-Gulf of Mex. Geology
Woods Hole (617) 548-8700
Middlebury Coll.-BA
Sediment transport & suspended sediments, Georges Bank
Suspended sediments; marine diatoms

PAROLSKI, KENNETH F.
PST
Br. of Atlantic-Gulf of Mex. Geology
Woods Hole (617) 548-8700
Northeastern U.-BA

PASTORE, EILEEN R. Raymond
Administrative Assistant (703) 860-3532
2046 Durand Drive
Reston, Virginia 22091
Br. of Coal Resources
Reston (703) 860-7734

PASTRANA, LOURDES N. Felix
Research Assistant
1305 N. Argonne Avenue
Sterling Park, Virginia 22170
Office of Energy Resources
Reston (703) 860-6531

PATTERSON, SAM H. **Barbara**
Geologist
Br. of Eastern Mineral Resources
Reston **(703) 860-6913**
Coe Coll.-BA; U. of Iowa-MS; U. of Illinois-PhD
Florida wilderness projects
Bauxite; clays

PATTON, WILLIAM W., JR. **Peggy**
Geologist **(415) 326-2033**
951 Coleman Avenue
Menlo Park, California 94025
Br. of Alaskan Geology
Menlo Park **(415) 323-8111 x2248**
Cornell U.-BA, MS; Stanford-PhD
AMRAP-Medfra quadrangle
Regional geology & tectonics of western Alaska & Bering Sea; ophiolites of Alaska

PAUL, LAWRENCE E.
PST
Br. of Electromag. & Geomagnetism
Denver **(303) 234-2588**
Western State Coll. of Colorado-BA
Magnetic field surveys

PAULL, CHARLES K.
Geologist
P.O. Box 652
Woods Hole, Massachusetts 02543
Br. of Atlantic-Gulf of Mex. Geology
Woods Hole **(617) 548-8700**
Harvard-BA; U. of Miami-MS
Acoustic stratigraphy; deep sea erosion; continental margin devleopment

PAULSON, WILLIAM C.
Clerk-Typist **(202) 362-5674**
4409 Faraday Place, N.W.
Washington, D.C. 20016
Br. of Paleontology & Stratigraphy
Washington, D.C. **(202) 343-3206**
Bowdoin Coll.-BA; George Washington U.-MA

PAVICH, MILAN J. **Deborah**
Geologist
112 Ayrlee Avenue
Leesburg, Virginia 22075
Br. of Eastern Environmental Geology
Reston **(703) 860-6595**
Franklin & Marshall-BA; Johns Hopkins-PhD
Soil & saprolite stratigraphy, SE U.S.
Soil & saprolite geochemistry; geomorphology; denudation rates

PAVLIDES, LOUIS **Evangeline**
Geologist **(301) 474-9158**
7518 Creighton Drive
College Park, Maryland 20740
Br. of Eastern Environmental Geology
Reston **(703) 860-6503**
Brooklyn Coll.-BS; Columbia-MA
Tectonics of Virginia piedmont
Structure; stratigraphy; metamorphism

PAWLEWICZ, MARK J.
PST
Br. of Oil & Gas Resources
Denver **(303) 234-4640**
Penn State-BS; Colorado Schl. of Mines
Vitrinite reflectance in oil & gas fields & geothermal fields in S. California

PEAKE, LOREN G.
Computer Programmer
1208 Mariposa Avenue
San Jose, California 95126
Br. of Pacific-Arctic Geology
Menlo Park **(415) 323-2214 x2073**
San Jose State U.
Digital data processing laboratory

PEARL, JAMES E. **Rosa**
Geologist **(408) 733-9172**
10490 Madera Drive
Cupertino, California 95014
Br. of Pacific-Arctic Geology
Menlo Park **(415) 856-7015**
U. of California (Berkeley)-BA; San Jose State U.-MS
Oregon-Washington onshore-offshore
Sedimentary petrology

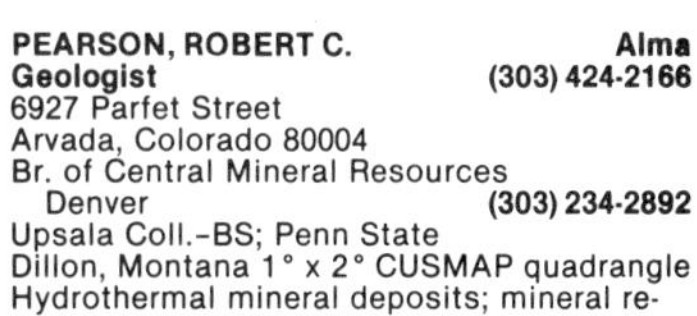

PEARSON, ROBERT C. **Alma**
Geologist **(303) 424-2166**
6927 Parfet Street
Arvada, Colorado 80004
Br. of Central Mineral Resources
Denver **(303) 234-2892**
Upsala Coll.-BS; Penn State
Dillon, Montana 1° x 2° CUSMAP quadrangle
Hydrothermal mineral deposits; mineral resource assessment

PECK, DALLAS L. **Sue**
Geologist **(703) 620-3706**
2524 Hearthcliff Lane
Reston, Virginia 22091
Office of the Chief Geologist
Reston **(703) 860-6531**
Harvard-PhD
Merced Peak, California
Igneous petrology; volcanology; mineral deposits

PECORA, WILLIAM T. **Wyn**
Geologist
Office of the Director
Reston
Princeton-BS; Harvard-MA, PhD
Petrology & mineralogy of igneous rocks; carbonatites; geology of Montana
Scientist; administrator

PEDDIE, NORMAN W. **Lori**
Geophysicist **(303) 499-3597**
5051 E. Euclid Avenue
Boulder, Colorado 80303
Br. of Electromag. & Geomagnetism
Denver **(303) 234-5497**
Hope Coll.-BA
Geomagnetic analysis & modeling
Geomagnetism; mathematics

PELTON, JOHN R.
Geophysicist
Br. of Seismology
Menlo Park **(415) 323-8111 x2791**
U. of Utah-PhD
Source mechanism studies

PENMAN, MARTHA K.
Illustrator
Office of Scientific Publications
Reston **(703) 860-6495**
U. of Virginia; Northern Virginia Community Coll.; American U.-BA

PEPER, JOHN D.
Geologist **(703) 437-9223**
11605 Vantage-Hill Road
Reston, Virginia 22090
Br. of Eastern Environmental Geology
Reston **(703) 860-6406**
CCNY-BS; U. of Massachusetts-MS; U. of Rochester-PhD
Massachusetts Co-operative project
Igneous & metamorphic rocks; stratigraphy & structure; Pleistocene deposits

PERKINS, DAVID M. **Marie**
Geophysicist
Boulder, Colorado
Br. of Earthquake Tectonics & Risk
Denver **(303) 234-2832**
George Washington U.-BA; Brown; U. of California (Berkeley)-MS
Seismic risk mapping
Seismicity; probability; strong ground motion

PERNOKAS, MARTHA A.
PST
2711 Emerson Street
Palo Alto, California 94603
Br. of Western Environmental Geology
Menlo Park **(415) 323-8111 x2282**
U. of California (Santa Cruz)-BS
Economic geology

PERRY, JANE W. **Bill**
Librarian **(703) 620-2625**
11906 Barrel Cooper Court
Reston, Virginia 22091
Library
Reston **(703) 860-6612**
Frostburg State Coll.-BA; U. of Maryland-MLS
Cataloging; OCLC specialist

PERRY, WILLIAM J., JR. **Diane**
Geologist
5700 15th Road, North
Arlington, Virginia 22205
Br. of Oil & Gas Resources
Denver **(303) 234-4750**
Johns Hopkins-BA; U. of Michigan-MS; Yale-MPhil, PhD
Appalachian structural patterns
Structural geology; tectonophysics

PESELNICK, LOUIS
Physicist
319 Waverly Street
Menlo Park, California 94025
Br. of Tectonophysics
Menlo Park **(415) 323-8111 x2394**
Catholic U.-PhD
Anelastic properties of rocks
Elastic & anelastic properties of rocks

PETERMAN, ZELL E. **Gladys**
Geologist
9795 W. Ohio Drive
Lakewood, Colorado 80226
Br. of Isotope Geology
Denver **(303) 234-5531**
Colorado Schl. of Mines-BS; U. of Minnesota-MS; U. of Alberta-PhD
Geochronology; isotope geochemistry; Precambrian geology

PETERS, DOUGLAS C.
Geologist **(303) 278-1540**
17844 W. Lunnonhaus Drive, #11
Golden, Colorado 80401
Br. of Petrophysics & Remote Sens.
Denver **(303) 234-4897**
U. of Pitt.-BS; Colorado Schl. of Mines-MS
Airborne multispectral scanner for uranium alteration in Wyoming
Remote sensing; economic geology; uranium studies

PETERSON, DAVID M.
Geologist
Br. of Western Environmental Geology
Menlo Park **(415) 323-2601**
San Jose State U.-BA, MS
Mapping Quaternary deposits, Mojave Desert
Hillslope erosional processes; geomorphology; Quaternary geology

PETERSON, DONALD W. **Betty**
Geologist **(415) 494-8896**
250 Davenport Way
Palo Alto, California 94306
Br. of Western Mineral Resources
Menlo Park **(415) 323-8111 x2291**
Cal. Tech.-BS; Washington State U.-MS; Stanford-PhD
Papago-Ajo, Arizona
Volcanology; petrology of volcanic rocks; field geology

PETERSON, FRED **Christine**
Geologist
P.O. Box 471
Morrison, Colorado 80465
Br. of Uranium-Thorium Resources
Denver **(303) 234-5813**
San Diego State U.-BS; Stanford-PhD
Uranium-bearing Jurassic rocks, Colorado Plateau
Mesozoic stratigraphy; sedimentology & tectonism, Colorado Plateau

PETERSON, JAMES A. **Gladys**
Geologist **(406) 542-2087**
301 Pattee Canyon Drive
Missoula, Montana 59801
Br. of Oil & Gas Resources
Missoula, Montana **(406) 542-2087**
Northwestern U.; U. of Wisconsin; St. Louis U.-BS; U. of Minnesota-MS, PhD
Stratigraphy; petroleum geology; carbonate geology

PETERSON, JOCELYN A.
Geologist
Palo Alto, California 94306
Br. of Western Mineral Resources
Menlo Park **(415) 323-8111 x2549**
Hope Coll.-BA; Stanford-MS
Ishi, Polk Springs, Mill Creek, & Butt Mtn. wilderness areas
CRIB; STATPAC; volcanic rocks

PETERSON, JON R. **Joan**
Geophysicist **(505) 299-7398**
3056 Ole Court, N.E.
Albuquerque, New Mexico 87111
Br. of Global Seismology
Albuquerque **(505) 844-4637**
U. of Minnesota-BS
Global seismograph networks

PETERSON, WARREN L.
Geologist
1816 Golf View Court
Reston, Virginia 22090
Br. of Eastern Environmental Geology
Reston **(703) 860-6503**
Augustana Coll.-BA; U. of Chicago-MS
Glacial geology of Iron River 2° quadrangle
Geologic mapping; petrology of sedimentary rocks; glacial geology

PETRAFESO, FRANK A. **Bonnie**
Cartographic Technician **(303) 233-1657**
8700 W. 1st Avenue
Lakewood, Colorado 80226
Br. of Regional Geophysics
Denver **(303) 234-5504**
Mesa Coll.-AA
Aeromagnetic data processing

PFLAUM, BERNARD H.
Engineering Technician **(415) 629-9256**
341 Greenpark Way
San Jose, California 95136
Br. of Field Geochemistry & Petrology
Menlo Park **(415) 323-2644**
U. of North Dakota
Geochemical & geophysical apparatus
Research & development

PFLUKE, JOHN H. **Sybil**
Geophysicist **(415) 321-6316**
221 Kingsley Avenue
Palo Alto, California 94301
Br. of Seismology
Menlo Park **(415) 323-8111 x2764**
St. Louis U.-BS; Penn State-PhD
Contract monitor: all external research in earthquake prediction
Slow earth-deformation; earthquake statistics; earthquake prediction

PHAIR, GEORGE **Cecie**
Geologist **(301) 926-8944**
14700 River Road
Potomac, Maryland 20854
Br. of Field Geochemistry & Petrology
Reston **(703) 860-7421**
Hamilton Coll.-BS; Rutgers-MS; Princeton-MA, PhD
Behavior of U & Th in igneous, metamorphic, & hydrothermal processes
Geochemistry; petrology; economic geology

PHELPS, WILLIAM E., JR. **Nancy**
Administrative Officer
Office of Earthquake Studies
Reston **(703) 860-6475**
U. of Maryland

PHILLIPS, CAROL A.
Computer Aid
6188 Vivian Court
Arvada, Colorado 80004
Br. of Uranium-Thorium Resources
Denver **(303) 234-5149**
U. of Illinois

PHILLIPS, JEFFREY D. **Susan**
Geophysicist **(703) 471-5390**
1629 Apricot Court
Reston, Virginia 22090
Br. of Regional Geology
Reston **(703) 860-7454**
U. of California-BA; Stanford-MS, PhD
Eastern overthrust geophysics; MAGSAT
Exploration geophysics; geomagnetism

PHIPPS, CAROLYN G.
Travel Clerk
Office of International Geology
Reston **(703) 860-6348**

PICKERING, MICHAEL J.
PST **(703) 527-7090**
2321 N. Richmond Street
Arlington, Virginia 22207
Br. of Analytical Laboratories
Reston **(703) 860-7543**
Virginia Polytech. Inst.-BS

PIERCE, CHARLOTTE J. **Gerald**
Administrative Officer **(408) 257-6854**
10086 Lamplighter Square
Cupertino, California 95014
Br. of Pacific-Arctic Geology
Menlo Park **(415) 856-7139**
San Jose State U.-BA

PIERCE, KENNETH L. **Linda**
Geologist **(303) 526-0466**
451 Woodland Lane
Golden, Colorado 80401
Br. of Central Environmental Geology
Denver **(303) 234-2737**
Stanford-BS; Yale-PhD
Quaternary dating & neotectonics
Quaternary geology; neotectonics; geomorphology

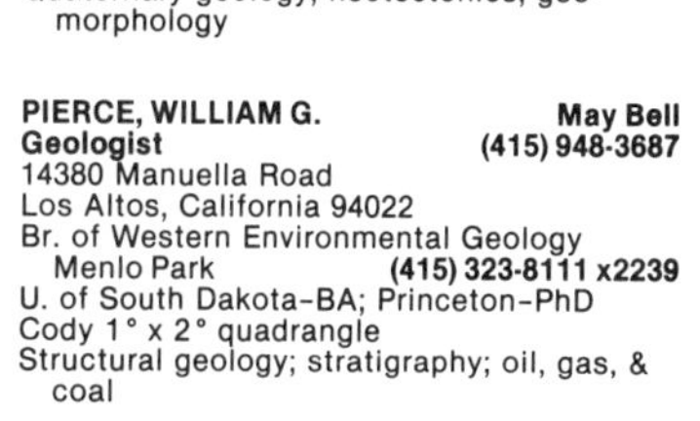

PIERCE, WILLIAM G. **May Bell**
Geologist **(415) 948-3687**
14380 Manuella Road
Los Altos, California 94022
Br. of Western Environmental Geology
Menlo Park **(415) 323-8111 x2239**
U. of South Dakota-BA; Princeton-PhD
Cody 1° x 2° quadrangle
Structural geology; stratigraphy; oil, gas, & coal

PIKE, JANE E.
Geologist
Br. of Western Environmental Geology
Menlo Park **(415) 323-8111 x2092**
George Washington U.-BA; U. of Michigan-MS; Stanford-PhD
Igneous & metamorphic petrology; volcanology; geochemistry

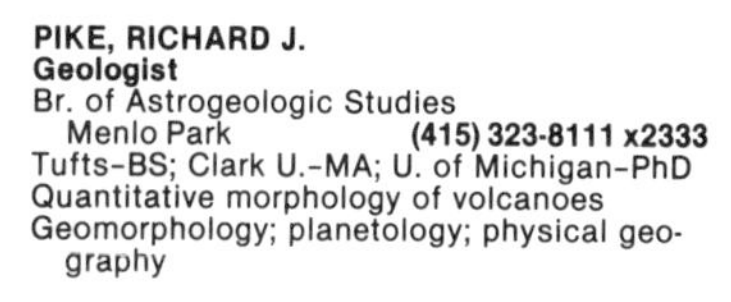

PIKE, RICHARD J.
Geologist
Br. of Astrogeologic Studies
Menlo Park **(415) 323-8111 x2333**
Tufts-BS; Clark U.-MA; U. of Michigan-PhD
Quantitative morphology of volcanoes
Geomorphology; planetology; physical geography

PIKE, ROBERT S. **Cecile**
Geologist **(303) 978-9714**
7059 West Walker Avenue
Littleton, Colorado 80123
Br. of Oil & Gas Resources
Denver **(303) 234-5235**
Oklahoma City U.
Resource appraisal
Logging interpretation; petroleum geology; resource appraisals

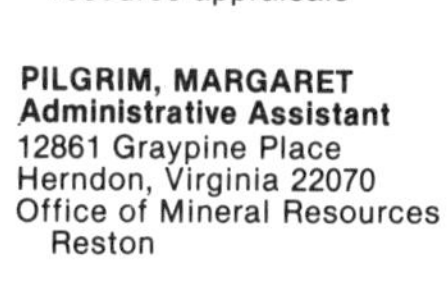

PILGRIM, MARGARET
Administrative Assistant
12861 Graypine Place
Herndon, Virginia 22070
Office of Mineral Resources
Reston **(703) 860-6571**

PILLERA, JOSEPH S. **Diane**
Administrative Officer
8901 W. Stanford Avenue
Littleton, Colorado 80123
Office of the Chief Geologist
Denver **(303) 234-3622**
U. of Maryland-BS

PILLMORE, CHARLES L. **Arlene**
Geologist **(303) 985-0676**
682 South Beech Street
Lakewood, Colorado 80228
Br. of Central Environmental Geology
Denver **(303) 234-3475**
U. of Colorado-BA, MS
Southern Raton Basin energy lands, Raton 2° sheet
Coal; publications; application of photogrammetry to geology

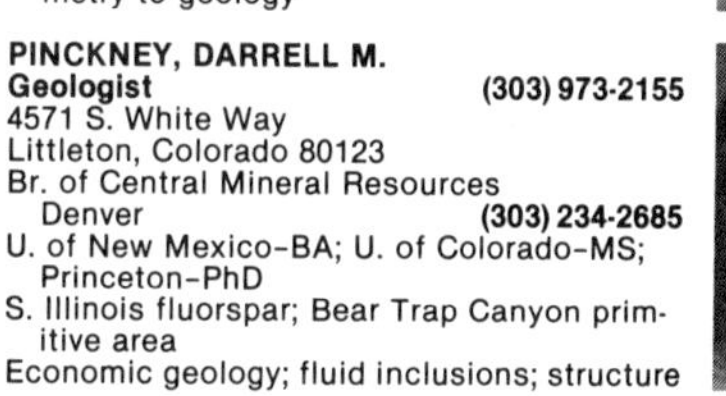

PINCKNEY, DARRELL M.
Geologist **(303) 973-2155**
4571 S. White Way
Littleton, Colorado 80123
Br. of Central Mineral Resources
Denver **(303) 234-2685**
U. of New Mexico-BA; U. of Colorado-MS; Princeton-PhD
S. Illinois fluorspar; Bear Trap Canyon primitive area
Economic geology; fluid inclusions; structure

PINCKNEY, WILLIAM C., JR.
PST **(202) 829-6450**
1446 Tuckerman Street, N.W., Apt. 107
Washington, D.C. 20011
Br. of Paleontology & Stratigraphy
Washington, D.C. **(202) 343-3523**
Morehouse College-BS
Paleozoic corals

PINKERTON, JAMES B. **Virginia**
Geologist **(408) 378-0362**
105 Calle Marguerita
Los Gatos, California 95030
Office of Scientific Publications
Menlo Park **(415) 323-8111 x2301**
U. of California (Berkeley)-BA; San Jose State U.-MS
Structural geology; stratigraphy

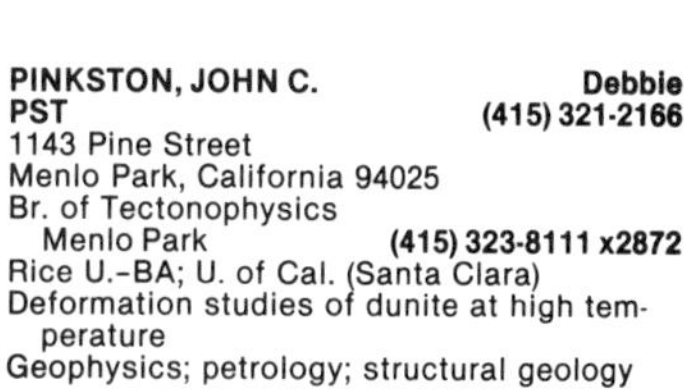

PINKSTON, JOHN C. **Debbie**
PST **(415) 321-2166**
1143 Pine Street
Menlo Park, California 94025
Br. of Tectonophysics
Menlo Park **(415) 323-8111 x2872**
Rice U.-BA; U. of Cal. (Santa Clara)
Deformation studies of dunite at high temperature
Geophysics; petrology; structural geology

PIPER, DAVID Z.
Geologist **(415) 499-8261**
3649 Ramona
Palo Alto, California 94302
Br. of Pacific-Arctic Geology
Menlo Park **(415) 856-7158**
Scripps Inst. of Oceanography-PhD
Deep-ocean minerals
Marine geology

PIPPIN, MARJORIE F.
Editorial Assistant
799 S. Robb Way
Lakewood, Colorado 80226
Br. of Engineering Geology
Denver **(303) 234-3819**

PITKIN, JAMES A. **Patricia**
Geologist **(303) 989-2541**
529 S. Alkire Street
Lakewood, Colorado 80228
Br. of Petrophysics & Remote Sens.
Denver **(303) 234-5228**
S. Ill. U.; Southern Methodist U.-BS, MS
Gamma-ray spectrometry for uranium exploration in crystalline terrane
Gamma-ray spectrometry; airborne geophysics; applied geophysics

PITT, ANDREW M.
Geophysicist
Br. of Seismology
Menlo Park **(415) 323-8111 x2571**
Colorado Schl. of Mines-GE
Yellowstone seismic analysis
Local earthquake studies

PITTS, JOANN K. **Rolfe**
Administrative Clerk
9659 Lindenbrook Street
Fairfax, Virginia 22031
Br. of Coal Resources
Reston **(703) 860-7734**

PITTS, SUSAN E. **Frank**
Clerk-Typist
15029 Red Fox Court
Haymarket, Virginia 22069
Office of the Chief Geologist
Reston **(703) 860-7420**

PLAFKER, GEORGE **Ruth**
Geologist
Br. of Alaskan Geology
Menlo Park **(415) 323-8111 x2201**
Bklyn Coll.-BS; U. of Cal.-MS; Stanford-PhD
Alaska geologic earthquake hazards; Gulf of Alaska resources
Neotectonic studies; Tertiary stratigraphy; petroleum resources

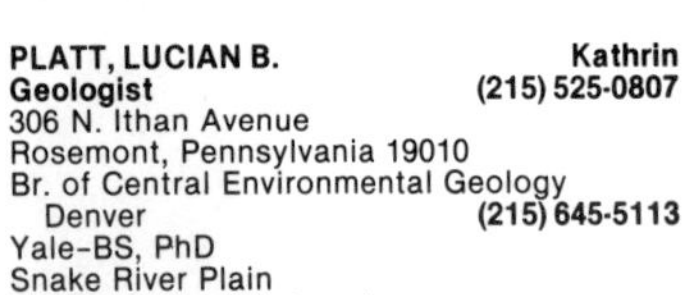

PLATT, LUCIAN B. **Kathrin**
Geologist **(215) 525-0807**
306 N. Ithan Avenue
Rosemont, Pennsylvania 19010
Br. of Central Environmental Geology
Denver **(215) 645-5113**
Yale-BS, PhD
Snake River Plain
Regional structural geology

PLESHA, JOSEPH L.
PST **(303) 494-2207**
4820 Thunderbird Circle, #319
Boulder, Colorado 80303
Br. of Regional Geophysics
Denver **(303) 234-4691**
Humboldt State U.-BS
Geophysics

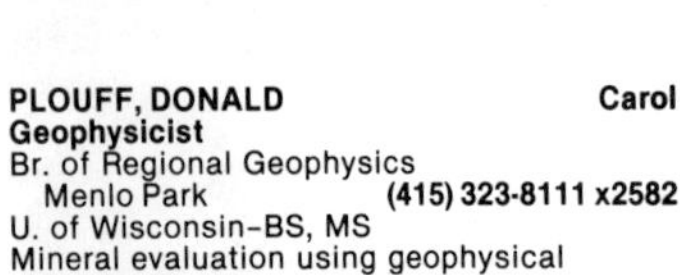

PLOUFF, DONALD **Carol**
Geophysicist
Br. of Regional Geophysics
Menlo Park **(415) 323-8111 x2582**
U. of Wisconsin-BS, MS
Mineral evaluation using geophysical methods
Computer applications to geophysical analysis & modeling

POAG, C. WYLIE **Martha**
Geologist **(617) 548-4407**
14 Sandpiper Circle
E. Falmouth, Massachusetts 02536
Br. of Paleontology & Stratigraphy
Woods Hole **(617) 548-8700**
Florida State U.-BS; Louisiana State U.-MS; Tulane U.-PhD
Stratigraphy & paleoenvironments of U.S. Atlantic margin
Micropaleontology-Foraminifera

PODWYSOCKI, MELVIN H. **Patricia**
Geologist
12868 Whitefur Lane
Herndon, Virginia 22070
Br. of Petrophysics & Remote Sens.
Reston **(703) 860-7403**
Wayne State U.-BS; Penn State-MS, PhD
Remote sensing for uranium exploration; remote sensing, Richfield, UT 1° x 2° quad.
Multivariate analysis applied to remote sensing data; structural geology

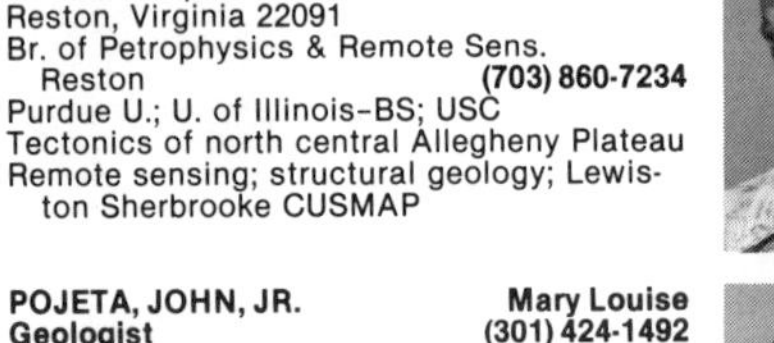

POHN, HOWARD A. **Sheila**
Geologist **(703) 476-4632**
12142 Stirrup Road
Reston, Virginia 22091
Br. of Petrophysics & Remote Sens.
Reston **(703) 860-7234**
Purdue U.; U. of Illinois-BS; USC
Tectonics of north central Allegheny Plateau
Remote sensing; structural geology; Lewiston Sherbrooke CUSMAP

POJETA, JOHN, JR. **Mary Louise**
Geologist **(301) 424-1492**
1492 Dunster Lane
Rockville, Maryland 20854
Br. of Paleontology & Stratigraphy
Washington, D.C. **(202) 343-5097**
Capital U.-BS; U. of Cincinnati-MS, PhD
Cambrian mollusks of Great Basin
Paleozoic pelecypods; Cambrian mollusks; Ordovician stratigraphy

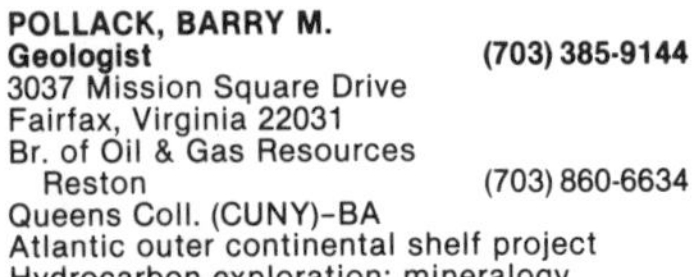

POLLACK, BARRY M.
Geologist **(703) 385-9144**
3037 Mission Square Drive
Fairfax, Virginia 22031
Br. of Oil & Gas Resources
Reston (703) 860-6634
Queens Coll. (CUNY)-BA
Atlantic outer continental shelf project
Hydrocarbon exploration; mineralogy

POLLASTRO, RICHARD M.
Geologist
4902 W. 35th Avenue
Denver, Colorado 80212
Br. of Oil & Gas Resources
Denver **(303) 234-6470**
SUNY Coll. at Buffalo-BA; SUNY at Buffalo-MA
Reservoir rocks-carbonates
Clay mineralogy; mineralogy; sedimentary petrology

POLOVTZOFF, OLEG C. **Xenia**
PST **(415) 493-7491**
4203 Park Boulevard
Palo Alto, California 94306
Br. of Field Geochemistry & Petrology
Menlo Park **(415) 323-8111**
U. of Southern California-BA
Sierra Nevada batholith project
Granitic rocks

POMEROY, JOHN S. **Sally**
Geologist **(301) 424-3711**
711 Fordham Street
Rockville, Maryland 20850
Br. of Eastern Environmental Geology
Reston **(703) 860-6503**
Lehigh U.-BA; U. of Utah
Landslides, western Pennsylvania
Mass movement; environmental geology; remote sensing

PONCE, DAVID A.
PST **(408) 257-1245**
20800 Homestead Road
Cupertino, California 95014
Br. of Regional Geophysics
Menlo Park **(415) 323-8111 x2607**
San Jose State U.-BS
Gravity investigations for nuclear waste isolation program
Geophysics, geology

PONTI, DANIEL J.
Geologist
Br. of Ground Motion & Faulting
Menlo Park **(415) 323-8111 x2233**
Stanford-BS, MS
Quaternary stratigraphy & deformation, Antelope Valley, California
Quaternary geology; engineering geology

POOLE, FORREST G. **Patricia**
Geologist **(303) 237-3154**
11760 West 30th Place
Lakewood, Colorado 80215
Br. of Central Mineral Resources
Denver **(303) 234-2979**
Ohio U.-BS; Colorado U.-MS
Basin-Range studies
Regional stratigraphy, structure & mineral deposits

POOLE, PATRICIA C. **Forrest**
Editorial Assistant **(303) 237-3154**
11760 West 30th Place
Lakewood, Colorado 80215
Br. of Engineering Geology
Denver **(303) 234-4697**
Mesa College

POORE, RICHARD Z.
Geologist
Br. of Paleontology & Stratigraphy
Menlo Park **(415) 323-8111 x2807**
Brown-PhD
Micropaleontology; paleoclimates

POPENOE, PETER **Anna**
Geologist
15 Greengate Road
Falmouth, Massachusetts 02540
Br. of Atlantic-Gulf of Mex. Geology
Woods Hole **(617) 548-8700**
Ohio State-BS
Southeastern Atlantic environmental studies
Marine & Appalachian geology & geophysics

POPPE, BARBARA B. **Herbert**
Technical Editor **(303) 258-7617**
Magnolia Star Route
Nederland, Colorado 80466
Office of Scientific Publications
Denver **(303) 234-2445**
Radcliffe-BA
Computer software documentation

POPPE, LAWRENCE J. **Ruth**
Geologist
48 Green Pond Road
East Falmouth, Massachusetts 02536
Br. of Atlantic-Gulf of Mex. Geology
Woods Hole **(617) 548-8700**
Boston Coll.-BS; U. of Connecticut-MS
Clay mineralogy

PORCELLA, RONALD L.
Geophysicist
Br. of Seismic Engineering
Menlo Park **(415) 323-8111 x2881**

PORTER, PEARL B. **O.B.**
Computer Programmer
Herndon, Virginia 22070
Office of Scientific Publications
Reston **(703) 860-6738**
Midway Coll.-AA; Berea Coll.-BS
Geomap Index

POTTS, ROSEMARIE
Secretary
Office of Earthquake Studies
Reston **(703) 860-6473**
Northern Virginia Community College

POWELL, JOHN W. **Emma**
Geologist
Office of the Director
Reston
Illinois Coll.; Illinois Institute; Oberlin
Arid western lands; Colorado River & Plateau studies; ethnic cultures
Teacher; scientist; explorer, lobbyist

POWELL, YVONNE J. **Jeff**
Administrative Clerk **(703) 860-8580**
11819 Breton Court, 2C
Reston, Virginia 22091
Office of Scientific Publications
Reston **(703) 860-6784**

POWERS, M. SUSANN W. **John**
Library Technician **(303) 237-5120**
1845 Yarrow
Lakewood, Colorado 80215
Library
Denver **(303) 234-4133**
U. of Colorado

POWERS, PHILIP S. **LaDonna**
Engineering Technician
Br. of Engineering Geology
Denver **(303) 234-2529**
Colorado State U.; U. of Colorado; MSC-BS
Geotechnical measurements laboratory
Applications programming; subsurface investigations

POWERS, RICHARD B.
Geologist **(303) 936-0918**
1155 South Otis Place
Lakewood, Colorado 80226
Br. of Oil & Gas Resources
Denver **(303) 234-3435**
Augustana Coll.-BA; U. of Missouri-MA; U. of Colorado
Oil & gas studies—west overthrust belt
Oil & gas appraisal studies of Rocky Mtn. region; Atlantic offshore & US public lands

PRATT, WALDEN P. **Jan**
Geologist **(303) 421-4746**
7048 Parfet Court
Arvada, Colorado 80004
Br. of Central Mineral Resources
Denver **(303) 234-2756**
U. of Rochester-BA; Yale; Stanford-MS, PhD
Rolla & Springfield 2° quads, Missouri
Mineral resource appraisal

PRESCOTT, WILLIAM H. **Judith**
Geophysicist
190 Los Palmos Drive
San Francisco, California 94127
Br. of Tectonophysics
Menlo Park **(415) 323-8111 x2967**
Middlebury Coll.-BA; U. of California (Berkeley)-MA; Stanford
Crustal strain project
Geodetic determination of crustal strain; computer modeling of plate boundaries

PRESTON, MARGARET W.
Administrative Assistant
Office of Geochemistry & Geophysics
Reston **(703) 860-6585**

PRICHARD, GEORGE E. **Jane**
Geologist **(303) 934-4548**
3665 W. Floyd Avenue
Denver, Colorado 80236
Br. of Central Environmental Geology
Denver **(303) 234-3952**
U. of Kentucky-BS; U. of Denver; U. of Minnesota
Geologic mapping, Nebraska
Areal mapping; stratigraphy; sedimentology

PRINZ, WILLIAM C. **Barbara**
Geologist
Box 6
Garrett Park, Maryland 20766
Br. of Eastern Mineral Resources
Reston **(703) 860-6561**
Ohio State-BS, MS; Yale-PhD
Michigan Precambrian; Saudi Arabia
Metallic mineral deposits; Precambrian geology

PROWELL, DAVID C. **Glenda**
Geologist
Route 1, Box 368
Leesburg, Virginia 22075
Br. of Eastern Environmental Geology
Reston **(703) 860-6421**
Emory U.-BS, MS; U. of California-PhD
Brittle tectonics in the southeastern U.S.
Cenozoic faulting

PRZEDPELSKI, MARY ANN
Geologist
Office of Scientific Publications
Reston **(703) 860-6517**
SUNY (Buffalo)-BA

PUGH, YOUNG-JA
Administrative Officer
Br. of Coal Resources
Reston **(703) 860-7734**
Pusan Jr. Women's College (Korea)

PUNIWAI, GARY S. **Judy**
PST
Br. of Field Geochemistry & Petrology
Hawaiian Volcano Observ. **(808) 967-7328**
Hilo College-BA
Ground tilt deformation studies

PURDY, SUSAN J.
Secretary
Box 358
Cataumet, Massachusetts 02534
Br. of Atlantic-Gulf of Mex. Geology
Woods Hole **(617) 548-8700**
Mary Washington College

PURDY, TERRI L.
Geologist
Br. of Petrophysics & Remote Sens.
Reston **(703) 860-7407**
George Washington U.-BS
Remote sensing; sedimentary petrology

QUEEN, DONALD G.
Driller
Br. of Eastern Environmental Geology
Reston **(703) 860-7689**

QUIGLEY, MARGARET R.
Clerk-Typist **(703) 435-4721**
13225 Keach Place
Herndon, Virginia 22070
Office of Scientific Publications
Reston **(703) 860-6517**

QUINTERNO, PAULA J.
Geologist
Br. of Pacific-Arctic Geology
Menlo Park **(415) 856-7165**
San Jose State U.-BA, MS
Micropaleontology

RACHLIN, JACK **Marjorie**
Geologist **(202) 362-9486**
2919 Brandywine Street, NW
Washington, D.C. 20001
Br. of Eastern Environmental Geology
Reston **(703) 860-7256**
City College of New York-BS
Geology & environmental analysis, USSR

RADBRUCH-HALL, DOROTHY H. **Wayne Hall**
Geologist **(415) 366-1633**
1028 Wilmington Way
Redwood City, California 94062
Br. of Engineering Geology
Menlo Park **(415) 856-2545**
U. of Colorado-BA
Engineering geologic map of the conterminous United States
Engineering geologic & environmental geologic mapping; landslides

RAINES, GARY L. **Ewa Katarzyna**
Geologist **(303) 278-1420**
13557 W. 22nd Place
Golden, Colorado 80401
Br. of Petrophysics & Remote Sens.
Denver **(303) 234-4898**
UCLA-BA; Colorado Schl. of Mines-MS, PhD
Remote sensing; economic geology; regional tectonics

RAIT, NORMA
Chemist
Br. of Analytical Chemistry
Reston **(703) 860-7653**
CUNY-BS; New York U.-MS
Atomic emission spectrography

RALEIGH, C. BARRY **Gail**
Geophysicist
44 Mesa Court
Atherton, California 94025
Office of Earthquake Studies
Menlo Park **(415) 323-8111 x2893**
Pomona Coll.-BA; Claremont Coll.-MA; UCLA-PhD
Earthquake prediction program

RAMBO, WILLIAM L. **Esther**
Geologist **(408) 257-9534**
5823 Randewood Court
San Jose, California 95129
Office of the Chief Geologist
Menlo Park **(415) 323-8111 x2480**
San Jose State U.-BS
Geologic information
Western region of the United States

RANKIN, DOUGLAS W. **Mary**
Geologist **(202) 337-8659**
1614 44th Street, NW
Washington, D.C. 20007
Br. of Eastern Environmental Geology
Reston **(703) 860-6404**
Colgate U.-BA; Harvard-MA, PhD
Volcanic rocks of Appalachian orogen
Appalachian orogenic belt; volcanic rocks; geochemistry

RAPP, JOHN B.
Chemist
Br. of Pacific-Arctic Geology
Menlo Park **(415) 856-7146**
U. of California (Berkeley)-BS
Organic geochemistry
Geochemistry of hydrocarbons & amino acids (geochronology)

RAPPEPORT, MEL
Geophysicist **(415) 856-1396**
870 Bruce
Palo Alto, California 94303
Br. of Pacific-Arctic Geology
Menlo Park **(415) 856-7064**
Lehigh U.-BS, MS; Stanford-PhD
Marine sediment transport & bedforms
Shallow surface stratigraphy; sediment transport; basin evaluation

RAPPORT, AMY L.
PST **(415) 325-8009**
840 Coleman Avenue, #8
Menlo Park, California 94025
Br. of Seismology
Menlo Park **(415) 323-8111 x2192**
Sonoma State Coll.-BS

RASPET, RUDOLPH **Ann**
Engineering Technician **(301) 593-2077**
804 Kerwin Road
Silver Spring, Maryland 20901
Br. of Analytical Laboratories
Reston **(703) 860-7472**
Carnegie Tech.; Johns Hopkins U.

RATCLIFFE, NICHOLAS M. **Katherine**
Geologist
Box 189
Waterford, Virginia 22190
Br. of Eastern Environmental Geology
Reston **(703) 860-6406**
Williams Coll.-BA; Penn State-PhD
Northeastern U.S. seismicity & tectonics
Structural geology; petrology; tectonics

RATLIFF, MARY L. **Eugene**
Administrative Officer
928 Park Avenue
Herndon, Virginia 22070
Office of Scientific Publications
Reston **(703) 860-6784**

RAUGH, MIKE R.
Mathematician
Br. of Ground Motion & Faulting
Menlo Park **(415) 323-8111 x2727**
UCLA-BA; Stanford-PhD
Source mechanics; time series analysis; digital filtering

RAUP, OMER B. **Phyllis**
Geologist **(303) 233-0879**
12295 Applewood Knolls Drive
Lakewood, Colorado 80215
Br. of Sedimentary Mineral Resources
Denver **(303) 234-3507**
American U.-BS; Colorado U.-PhD
Paradox basin salt studies
Petrology & geochemistry; marine evaporites

RAUP, ROBERT B., JR. **Carol**
Geologist **(303) 697-9037**
6570 S. Crestbrook Drive
Morrison, Colorado 80465
Br. of Central Environmental Geology
Denver **(303) 234-2650**
Columbia-AB; U. of Michigan-MA
Environmental geology; NW Colorado
Environmental geology; mineral deposits; stratigraphy

RAY, JAMES D. **Joyce**
Warehouseman **(512) 937-0207**
305 Oak Ridge Drive
Corpus Christi, Texas 78418
Br. of Atlantic-Gulf of Mex. Geology
Corpus Christi **(512) 888-3294**

RAYMOND, WILLIAM H. **Shari**
Geologist **(303) 697-4602**
P.O. Box 323
Morrison, Colorado 80465
Br. of Central Mineral Resources
Denver **(303) 234-3454**
Miami U.-BA, MS
Precambrian sulfide deposits of Colorado
Economic geology; mineralogy; placer gold

REAGOR, BOBBY G.
Geophysicist
Br. of Global Seismology
Denver **(303) 234-3994**
Texas Technological College-BS
United States earthquakes
Historical catalog of U.S. earthquakes

REARIC, DOUGLAS M.
PST
470 N. 3rd Street
San Jose, California 95112
Br. of Pacific-Arctic Geology
Menlo Park **(415) 856-7003**
San Jose State U.-BS
Ice gouge processes of the north slope shelf
Marine geology

REED, BETTY S.
Editorial Assistant (512) 855-7531
4113 Kevin Drive
Corpus Christi, Texas 78413
Br. of Atlantic-Gulf of Mex. Geology
Corpus Christi **(512) 888-3241**

REED, BOBBI J. **Mike**
Personnel Technician
P.O. Box 955
Rockport, Texas 78382
Br. of Atlantic-Gulf of Mex. Geology
Corpus Christi **(512) 888-3294**

REED, BOBBY M.
Librarian **(703) 471-5490**
11524 Links Drive
Reston, Virginia 22090
Library
Reston **(703) 860-6611**
Adams State Coll. (Colorado)-BA

REED, BRUCE L.
Geologist
Br. of Alaskan Geology
Anchorage **(907) 271-4150**
U. of Maine-BS; Washington State U.-MS; Harvard-PhD
Tin commodity research
Geochemistry of tin; granites; economic geology

REED, JOHN C., JR. **Linda**
Geologist **(303) 526-1511**
26756 Columbine Glen
Golden, Colorado 80401
Br. of Central Environmental Geology
Denver **(303) 234-4857**
Johns Hopkins U.-PhD
Precambrian rocks, Sangre de Cristo range, New Mexico
Structural geology; metamorphic petrology; Quaternary geology

REED, MARSHALL J. **Catherine**
Geologist
1630 Manitoba Drive
Sunnyvale, California 94087
Br. of Field Geochemistry & Petrology
Menlo Park **(415) 323-8111 x2151**
U. of California (Berkeley)-BA; U. of California (Riverside)-MA
Geothermal assessment
Hydrothermal metamorphism; fluid geochemistry

REEVES, JESSIE F.
Management Assistant (415) 327-8635
425 Grant Avenue, #30
Palo Alto, California 94306
Office of Earthquake Studies
Menlo Park **(415) 323-8111 x2765**
Coker College-BA

REGA, NOREEN H.
Coal Data Assistant
4105 Port Rae Lane
Fairfax, Virginia 22030
Br. of Coal Resources
Reston **(703) 860-7306**

REHEIS, MARITH C.
Geologist **(303) 277-1843**
3955 Douglas Mountain Drive
Golden, Colorado 80401
Br. of Engineering Geology
Denver **(303) 234-5380**
U. of Georgia-BS; U. of Colorado-MS
Surficial geology, Powder River & Bighorn Basins
Quaternary geology; soils

REIF, LOUISE M. **Jack**
Clerk-Typist **(303) 986-6984**
1760 S. Carr Street
Lakewood, Colorado 80226
Br. of Regional Geochemistry
Denver **(303) 234-6566**
Colorado State U.

REIMER, G. MICHAEL
Geologist
122 S. Devinney Street
Golden, Colorado 80401
Br. of Uranium-Thorium Resources
Denver **(303) 234-5146**
Alfred U.-BA; U. of Pennsylvania-PhD
Helium detection
Use of gases in exploration; fission-track geochronology

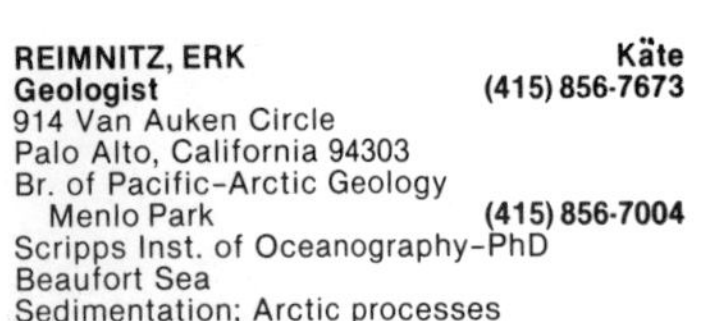

REIMNITZ, ERK **Käte**
Geologist **(415) 856-7673**
914 Van Auken Circle
Palo Alto, California 94303
Br. of Pacific-Arctic Geology
Menlo Park **(415) 856-7004**
Scripps Inst. of Oceanography-PhD
Beaufort Sea
Sedimentation; Arctic processes

REINEMUND, JOHN A. **Ruth**
Geologist **(703) 777-1491**
P.O. Box 890
Leesburg, Virginia 22075
Office of International Geology
Reston **(703) 860-6418**
Augustana College; U. of Chicago
Circum-Pacific map project; treasurer of International Union of Geological Sciences
Structural geology; sedimentation; coal geology

REINHARDT, JUERGEN **Judy**
Geologist **(703) 777-5374**
158 Edwards Ferry Road
Leesburg, Virginia 22075
Br. of Eastern Environmental Geology
Reston **(703) 860-6595**
Brown U.-BA; Johns Hopkins-PhD
Western Georgia coastal plain
Sedimentology; stratigraphy

REISER, HILLARD N.
Geologist
P.O. Box 751
Menlo Park, California 94025
Br. of Alaskan Geology
Menlo Park **(415) 323-8111**
CCNY-BS; Johns Hopkins U.; Stanford U.
Eastern Brooks Range, Alaska
Geologic mapping, Brooks Range

RENDIGS, RICHARD R.
Geologist
Br. of Atlantic-Gulf of Mex. Geology
Woods Hole **(617) 548-8700**
West. Conn. State Coll.-BS; U. of Waterloo-MSc
Modern sediment accumulation off southern New England coast
Lead 210 isotope analysis; heavy minerals

REPENNING, CHARLES A. **Nancy**
Geologist **(415) 968-0458**
2440 Villanueva Way
Mountain View, California 94040
Br. of Paleontology & Stratigraphy
Menlo Park **(415) 323-8111 x2366**
U. of New Mexico-BS; U. of Cal. (Berkeley)-MA
Neogene mammalian biochronology
Neogene mammalian paleontology

REPETSKI, JOHN E. **Donna**
Geologist **(703) 698-5277**
7217 Wayne Drive
Annandale, Virginia 22003
Br. of Paleontology & Stratigraphy
Washington, D.C. **(202) 343-2409**
U. of Indiana; U. of Pennsylvania-BS; U. of Missouri-MA, PhD
Cambrian-Ordovician & Lower-Middle Ordovician boundary studies
Conodonts & biostratigraphy

REYNOLDS, MITCHELL W. **Sandra**
Geologist
Office of the Chief Geologist
Reston **(703) 860-6544**
Harvard-BA; U. of California-PhD
Program manager, geologic framework prgm.
Structural geology; tectonics; stratigraphy-sedimentology

REYNOLDS, RICHARD L.
Geologist
4331 Eldorado Springs Drive
Boulder, Colorado 80303
Br. of Uranium-Thorium Resources
Denver **(303) 234-5482**
Princeton-BA; U. of Colorado-PhD
Rock magnetism & paleomagnetism

REYNOLDS, ROBERT D. **Karin**
Computer Specialist **(505) 298-1785**
12604 Turquoise, NE
Albuquerque, New Mexico 87123
Br. of Global Seismology
Albuquerque **(505) 844-4637**
East Texas State U.-BS
Seismic instrumentation & related computer systems

RHEA, B. SUSAN
Geophysicist
Flagstaff Star Route
Boulder, Colorado 80302
Br. of Earthquake Tectonics & Risk
Denver **(303) 234-4091**
U. of Colorado-BA
South Carolina Seismic Program

RHODEHAMEL, EDWARD C. **Rosemary**
Geologist **(703) 830-2949**
14621 Baugher Drive
Centreville, Virginia 22020
Br. of Oil & Gas Resources
Reston **(703) 860-6634**
Michigan State U.-BS, MS
Atlantic outer continental shelf oil & gas resources
Atlantic coastal plain geology & hydrology; Pleistocene geology

RICE, CHARLES L. **Lois**
Geologist
12356 Lima Lane
Reston, Virginia 22091
Br. of Coal Resources
Reston **(703) 860-7426**
U. of Cal. (Berkeley)-BA
Carboniferous stratigraphy; sedimentary petrology

RICE, CYNTHIA A.
Oceanographer
Br. of Atlantic-Gulf of Mex. Geology
Corpus Christi **(512) 888-3241 x27**
Texas A&I-BS
Trace metal analysis; isotope dating

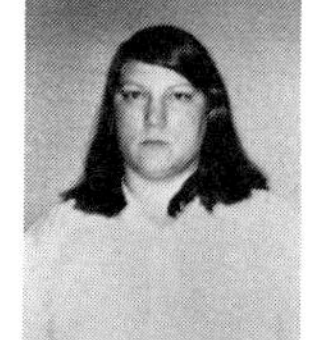

RICE, DUDLEY D. **Thelma**
Geologist **(303) 988-9671**
2395 S. Hoyt Street
Lakewood, Colorado 80227
Br. of Oil & Gas Resources
Denver **(303) 234-4167**
Cornell-BA; U. of Alberta-MSc
Natural gas resources of northern Great Plains
Sedimentology; stratigraphy & geochemistry of natural gases

RICE, THOMAS L. **Melissa**
Geologist
Br. of Atlantic-Gulf of Mex. Geology
Corpus Christi **(512) 888-3241**
West Virginia U.-BS
Geotechnical properties of marine soils

RICHARDS, EARL, JR. **Ellen**
Maintenance man **(303) 433-2844**
4845 Bryant Street
Denver, Colorado 80221
Office of the Chief Geologist
Denver **(303) 234-3771**
Northwestern State U. (Oklahoma)

RICHMOND, BRUCE M. **Karen**
PST **(408) 462-1623**
698 Olive Springs Road
Santa Cruz, California 95065
Br. of Pacific-Arctic Geology
Menlo Park **(415) 856-7073**
U. of Cal. (Santa Cruz)-BS; U. of Waikato (New Zealand)-MSc
Coastal
Nearshore dynamics & facies; eolian processes; estuarine deposits

RICHMOND, DEANNE L.
Drilling Operations Clerk **(303) 935-5374**
3014 S. Zurich Court
Denver, Colorado 80236
Br. of Coal Resources
Denver **(303) 234-4230**
Metropolitan State College

RICHMOND, GERALD M. **Amelie**
Geologist **(303) 758-0992**
3950 S. Hillcrest Drive
Denver, Colorado 80237
Br. of Central Environmental Geology
Denver **(303) 234-3372**
Brown-BA; Harvard-MA; U. of Colorado-PhD
Quaternary map of the United States & adjacent parts of Canada
Quaternary stratigraphy; chronology; climate

RICOTTA, RALPH J. **Sylvia**
Cartographic Technician **(303) 421-4513**
8248 Harlan Court
Arvada, Colorado 80003
Br. of Earthquake Tectonics & Risk
Denver **(303) 234-5074**
Engineering Drafting School

RIDDLE, GEORGE O.
PST **(303) 988-0899**
9582 W. Alameda Place
Lakewood, Colorado 80226
Br. of Analytical Laboratories
Denver **(303) 234-6406**
Panhandle A&M Coll.; U. of Colorado

RIDGLEY, JENNIE L. **Vic**
Geologist
12860 W. 75th Avenue
Arvada, Colorado 80005
Br. of Uranium-Thorium Resources
Denver **(303) 234-5814**
Penn State-BS; U. of Wyoming-MS
Uranium studies—Chama & N. San Juan Basins
Mesozoic stratigraphy, W. Interior

RINEHART, C. DEAN **Susan**
Geologist **(415) 327-3604**
626 Coleridge Avenue
Palo Alto, California 94301
Br. of Western Mineral Resources
Menlo Park **(415) 323-8111 x2349**
Coll. of Wooster-BA; U. of Colorado
Geol. map—Colville Indian Reservation, Washington
Igneous & metamorphic geology

RIVERS, WILLIE C.
PST
P.O. Box 15367
Lakewood, Colorado 80215
Br. of Special Projects
Denver **(303) 234-2371**
Denver Community Coll.-AS

ROBB, JAMES M. **Cheryl**
Geologist **(617) 540-1997**
Box 402, Shapquit Bars Road
West Falmouth, Massachusetts 02574
Br. of Atlantic-Gulf of Mex. Geology
Woods Hole **(617) 548-8700**
Cornell-BA; U. of Rhode Island-MS
Environmental assessment of Mid-Atlantic continental margin
Marine geology

ROBBIN, DANIEL M.
PST
730 Palermo
Coral Gables, Florida 33134
Br. of Oil & Gas Resources
Miami Beach **(305) 672-1784**
U. of Cal. (Berkeley)-BA
Compaction of carbonate sediments; caliche accumulation; studies of coral banding

ROBBINS, ELEANORA I. **Brian**
Geologist **(703) 354-0574**
6304 Lachine Lane
Alexandria, Virginia 22312
Br. of Coal Resources
Reston **(703) 860-6697**
Ohio State U.-BS; U. of Arizona-MS; Penn State
Coal in Triassic rift basins
Paleoecology; palynology

ROBBINS, STEPHEN L. **Heather**
Geophysicist **(303) 232-0440**
Lakewood, Colorado 80215
Br. of Oil & Gas Resources
Denver **(303) 234-4595**
San Francisco State U.-AB; San Jose State U.-MS
Borehole gravity
Applied gravity & magnetics; well log analysis

ROBERTS, ALAN A.
Chemist **(303) 447-9641**
1999 Bluebell
Boulder, Colorado 80302
Br. of Oil & Gas Resources
Denver **(303) 234-3624**
Earlham Coll.-BA; U. of Cal. (Santa Barbara)-PhD
Helium detection in petroleum exploration
Geochemical prospecting; organic geochemistry

ROBERTS, ALBERT E. — **Evelyn**
Geologist — **(408) 732-2629**
22343 Bahl Street
Cupertino, California 95014
Office of International Geology
Menlo Park — **(415) 323-8111 x2111**
U. of Oregon-BS, MS
California phosphorite project
Sedimentary mineral resources

ROBERTS, CAROL E.
Administrative Clerk
118 W. Church Road
Sterling, Virginia 22170
Office of Geochemistry & Geophysics
Reston — **(703) 860-6585**
U. of Maryland-BS

ROBERTS, CARTER W.
PST — (415) 524-2146
5 Highgate Court
Berkeley, California 94707
Br. of Regional Geophysics
Menlo Park — **(415) 323-8111 x2638**
U. of California-BA; San Jose State U.
Isostatic residual gravity map of California
High precision gravity measurements

ROBERTS, JOHN K., III
Librarian — **(703) 860-4711**
11882 St. Trinians Court
Reston, Virginia 22091
Library
Reston — **(703) 860-6613**
Presbyterian Coll.-BA; U. of North Carolina (Chapel Hill)

ROBERTS, RALPH J. — **Arleda**
Geologist — **(415) 326-8037**
544 Forest
Palo Alto, California 94301
Br. of Western Mineral Resources
Menlo Park — **(415) 323-8111**
Wash. State U.; U. of Wash.-BS, MS; Yale-PhD
Mineral resources of Nevada
Massive sulfide deposits; carlin-type gold deposits

ROBERTSON, EUGENE C. — **Sheila**
Geophysicist — **(202) 362-3043**
3917 McKinley Street, NW
Washington, D.C. 20015
Br. of Engineering Geology
Reston — **(703) 860-7404**
U. of Illinois-BS; Harvard-MA, PhD
Rock deformation
Experimental geology; structural geology; geophysics

ROBERTSON, JACQUES F. — **Janet**
Geologist — **(303) 936-1903**
1025 S. Upham Street
Lakewood, Colorado 80226
Br. of Uranium-Thorium Resources
Denver — **(303) 234-2826**
U. of Washington-BS
Crownpoint uranium studies, New Mexico
Uranium deposits; stratigraphy; areal geology

ROBIE, RICHARD A. — **Carol**
Geophysicist
1114 Dale Drive
Silver Spring, Maryland 20910
Br. of Exper. Geochem. & Mineralogy
Reston — **(703) 860-7486**
Dartmouth-BA; U. of Chicago-PhD
Thermodynamic properties of minerals
Experimental thermochemistry

ROBINSON, ALLEN C.
Geologist
Br. of Isotope Geology
Menlo Park — **(415) 323-8111 x2675**
Stanford-BS, MS
Intrusive geochronology
Igneous geochronology & petrology

ROBINSON, GERSHON D. — **Valerie**
Geologist — **(415) 493-8399**
644 Maybell Avenue
Palo Alto, California 94306
Br. of Western Environmental Geology
Menlo Park — **(415) 323-8111 x2871**
Northwestern U.-BS; U. of California (Berkeley)-MA
Federal geothermal environmental advisory panel
Structural geology; Rocky Mtn. stratigraphy

ROBINSON, GILPIN R., JR.
Geologist
Br. of Exper. Geochem. & Mineralogy
Reston — **(703) 860-6911**
Tufts U.-BS; Harvard-PhD
Igneous & metamorphic petrology; geochemistry; structural geology

ROBINSON, KEITH — **Molly**
Geologist — **(303) 986-7452**
12310 W. Idaho Drive
Lakewood, Colorado 80228
Br. of Uranium-Thorium Resources
Denver — **(303) 234-3764**
Manchester U. (England)-BS; Northwestern U.; Virginia Polytech. Inst.-PhD
Geochemical patterns in uranium exploration
Crystal chemistry; exploration geochemistry; energy resource assessment

ROBINSON, STEPHEN W.
Physicist — (415) 328-2077
237 Marmona Drive
Menlo Park, California 94025
Br. of Isotope Geology
Menlo Park — **(415) 323-8111 x2858**
Indiana U.-PhD
Menlo Park radiocarbon lab
Radiocarbon geochemistry & geophysics

RODDY, DAVID J. — **Jeannie**
Geologist
Route 4, Suzette Lane
Flagstaff, Arizona 86001
Br. of Astrogeologic Studies
Flagstaff — **(602) 779-3311 x1355**
Miami U.-AB, MS; Cal. Tech-PhD
Impact & explosion cratering mechanics
Very high pressure deformation; shock metamorphism

RODGERS, JOHN
Geologist
Department of Geology & Geophysics
Yale U., New Haven, Connecticut 06520
Br. of Eastern Environmental Geology
New Haven, Connecticut — (203) 436-0616
Cornell U.-AB, MS; Yale-PhD
Geologic map of Connecticut (bedrock)
Structural & regional geology; stratigraphy

RODRIGUEZ, EDUARDO A. — **Thelma**
PST
44 Renato Court, Apt. 20
Redwood City, California 94061
Br. of Western Environmental Geology
Menlo Park — (415) 323-8111 x2504
U. of California (Santa Cruz)

RODRIGUEZ, THELMA R. — **Eduardo**
Secretary
44 Renato Court, Apt. 20
Redwood City, California 94061
Br. of Tectonophysics
Menlo Park — **(415) 323-8111 x2894**
U. of Cal. (Santa Cruz)

ROEDDER, EDWIN W. — **Kathleen**
Geologist — **(301) 530-4757**
8405 Rayburn Road
Bethesda, Maryland 20034
Br. of Exper. Geochem. & Mineralogy
Reston — **(703) 860-6630**
Lehigh U.-BA; Columbia U.-MA, PhD
Fluid & magmatic inclusions in minerals
Atomic waste disposal; microchemical analysis; silicate melts

ROEHLER, HENRY W. Salley
Geologist (303) 986-6368
1874 S. Van Gordon Street
Lakewood, Colorado 80228
Br. of Coal Resources
Denver (303) 234-3558
U. of Wyoming-BS, BA, MS
Energy resource stratigraphy, sedimentation and economics

ROEMING, SUSAN S.
Cartographer (602) 779-3084
Route 6, Box 170
Flagstaff, Arizona 86001
Br. of Astrogeologic Studies
Flagstaff (602) 779-3311 x1515
Northern Arizona U.-BS

ROEN, JEANNE H. John
Administrative Clerk (301) 430-7596
Route 2, Box 469, Lakeland
Severna Park, Maryland 21146
Office of International Geology
Reston (703) 860-6348

ROEN, JOHN B. Jeanne
Geologist (301) 647-4841
Br. of Oil & Gas Resources
Reston (703) 860-6595
U. of Arizona; UCLA-BA, MA
Fuel resources, Appalachian basin

ROGERS, AL M. Susan
Geophysicist (303) 279-0338
17214 Rimrock Drive
Golden, Colorado 80401
Br. of Ground Motion & Faulting
Denver (303) 234-2869
St. Louis U.-BS, PhD
Seismic zonation studies in Los Angeles Basin
Seismic ground response; natural and induced seismicity

ROGERS, BRUCE W.
PST (415) 858-0535
889 Lobrado Avenue
Palo Alto, California 94303
Br. of Western Environmental Geology
Menlo Park (415) 323-8111 x2958
San Jose State U.; UCLA; U. of Santa Clara
Mineralogy; sedimentary petrology; speleology

ROGERS, JOHN A.
Electronics Engineer (415) 494-6192
921 Maddux Drive
Palo Alto, California 94303
Br. of Ground Motion & Faulting
Menlo Park (415) 323-8111 x2514
Montana State U.-MSEE
Alaska seismic studies
Digital processing; programming; system analysis

ROGGE, BETTY M. Franklin
Library Technician
7505 W. Calahan Avenue
Lakewood, Colorado 80226
Library
Denver (303) 234-4133

ROHRET, DONALD H. Irene
Electronics Technician (303) 421-2815
7041 Saulsbury Street
Arvada, Colorado 80003
Br. of Electromag. & Geomagnetism
Denver (303) 234-2589

ROONEY, LAWRENCE F. Rosalia
Geologist
USGS, c/o American Embassy
APO New York 09697
Br. of Latin Am. & African Geology
Jiddah, Saudi Arabia 674188
U. of Montana-BA, MA; U. of Indiana-PhD
Industrial minerals

ROOT, DAVID H. Cherie
Mathematician (703) 860-0989
2410 Ansdel Court
Reston, Virginia 22091
Office of Resource Analysis
Reston (703) 860-6455
MIT-BS; U. of Washington-PhD
Petroleum resource estimates

ROSARIO, HENRY Anne
Procurement & Transportation Ass't.
1213 Clagett Drive
Rockville, Maryland 20851
Office of International Geology
Reston (703) 860-6531
Fall River Business Institute

ROSE, HARRY J. Tina
Supervisory Chemist (301) 460-5015
14413 Ash Court
Rockville, Maryland 20853
Br. of Analytical Laboratories
Reston (703) 860-7543
St. Francis Coll.-BS; U. of Maryland-MS
X-ray spectroscopy
X-ray fluorescence; geochemistry

ROSEBOOM, EUGENE H., JR. Joan
Geologist (301) 530-1059
5502 Beech Avenue
Bethesda, Maryland 20014
Br. of Exper. Geochem. & Mineralogy
Reston (703) 860-7428
Ohio State U.-BS, MS; Harvard-PhD
Nuclear waste disposal
Experimental petrology; phase equilibria

ROSENBAUER, ROBERT J. Terri
Geologist (415) 726-7305
132 Jib Court
Half Moon Bay, California 94019
Br. of Pacific-Arctic Geology
Menlo Park (415) 856-7164
Holy Cross-BS; USC-MS
Marine minerals
Geochemistry; hydrothermal; uranium series dating

ROSENBAUM, JOSEPH G. Joann
Geophysicist (303) 494-0139
701 75th Street
Boulder, Colorado 80303
Br. of Special Projects
Denver (303) 234-2365
Swarthmore Coll.-BA; U. of Colorado-PhD
Rock & paleomagnetism

ROSENBLUM, LENORE Sam
Secretary (303) 986-5661
12165 W. Ohio Place
Lakewood, Colorado 80228
Office of International Geology
Denver (303) 234-3708
International geology-participants programs

ROSENBLUM, SAM Lenore
Geologist (303) 986-5661
12165 W. Ohio Place
Lakewood, Colorado 80228
Br. of Exploration Research
Denver (303) 234-2816
CCNY-BS; Stanford U.-MS
Mineralogical research
Economic mineralogy; rare-earth element resources; geochemical exploration research

ROSHOLT, JOHN N.
Chemist (303) 238-1570
125 Dudley Street
Lakewood, Colorado 80226
Br. of Isotope Geology
Denver (303) 234-4201
U. of Colorado-BS; U. of Miami-MS, PhD
Applied uranium geochemistry
Geochemistry; geochronology

ROSS, DONALD C. Lou
Geologist (415) 493-7806
301 Barclay Court
Palo Alto, California 94306
Br. of Earthquake Tectonics & Risk
Menlo Park (415) 323-8111 x2341
U. of Iowa-BA, MS; UCLA-PhD
Basement tectonic framework studies, southern California
Igneous & metamorphic petrology; tectonics

ROSS, MALCOLM
Mineralogist
Br. of Exper. Geochem. & Mineralogy
Reston (703) 860-6607
Utah State U.-BS; U. of Maryland-MS; Harvard-PhD
Asbestiform minerals
Mineralogy; petrology; crystallography

ROSS, REUBEN JAMES, JR. Jill
Geologist (303) 794-1362
5255 Ridge Trail, Bow Mar
Littleton, Colorado 80123
Br. of Paleontology & Stratigraphy
Denver (303) 234-5859
Princeton-BA; Yale-PhD
Ordovician correlations in the USA
Regional Ordovician stratigraphy & paleotectonics; paleontology of trilobites, brachiopods, graptolites

ROSS, WILLIAM O.
PST (703) 670-4364
3812 N. Fortuna Court
Woodbridge, Virginia 22193
Br. of Paleontology & Stratigraphy
Washington, D.C. (202) 343-8098
U. of Tennessee-BS

ROTH, EDWARD F. Pat
PST
Br. of Ground Motion & Faulting
Menlo Park (415) 323-8111 x2779
San Jose State U.
Seismic velocity studies
Shear wave studies

ROUBIQUE, CHARLES J. Eleanor
Electronics Engineer (303) 424-2857
9944 W. 85th Place
Arvada, Colorado 80005
Br. of Electromag. & Geomagnetism
Denver (303) 234-5398
Louisiana State U.-BS, MS
Induction resistivity probe development

ROUSE, MARY A.
Administative Operations Assistant
Br. of Paleontology & Stratigraphy
Menlo Park (415) 323-8111 x2150

ROWAN, LAWRENCE C. Frances
Geologist (703) 860-5234
2024 Mock Orange Court
Reston, Virginia 22091
Br. of Petrophysics & Remote Sens.
Reston (703) 860-7461
U. of Virginia-BA; U. of Cincinnati-PhD
Remote sensing; research & applications in mineral resources
Remote sensing; structural geology

ROWLAND, ROBERT W. Linda
Geologist (703) 860-5895
2400 Southgate Square
Reston, Virginia 22090
Office of Marine Geology
Reston (703) 860-7241
San Diego State U.-BS; U. of Cal. (Davis)-MS, PhD
Environmental studies, Alaska
Marine geology; paleoecology

ROWLEY, PETER D.
Geologist (303) 934-5242
6618 West Mississippi Way
Lakewood, Colorado 80226
Br. of Central Environmental Geology
Denver (303) 234-3940
Carleton Coll.-BA; U. of Texas (Austin)-PhD
Richfield 2°, Utah
Structural geology; igneous geology; field geology

RUANE, PAUL J. Eileen
Geologist (703) 451-7932
6902 Dawley Court
Springfield, Virginia 22152
Br. of Eastern Environmental Geology
Reston (703) 860-6421
Syracuse U.-AB
Safe mine waste disposal; landslides
Engineering geology

RUBIN, DAVID M. Michelle
Geologist (415) 327-3144
1102 Emerson Street
Palo Alto, California 94301
Br. of Pacific-Arctic Geology
Menlo Park (415) 856-7016
U. of Rochester-BA, MS; Rensselaer Polytech. Inst.-PhD
Dynamics of sediment bedforms
Sediment bedforms; faulting offshore northern California; carbonate sediments

RUBIN, JASON S.
PST (415) 326-7862
859 Lytton Avenue
Palo Alto, California 94301
Br. of Pacific-Arctic Geology
Menlo Park (415) 856-7128
Middlebury College-BA
Submarine ore deposition—North Pacific manganese nodules
Ophiolitic terranes & sulphide deposits; metallogenesis

RUBIN, MEYER Mary
Geologist (703) 560-4541
8215 Cottage Street
Vienna, Virginia 22180
Br. of Isotope Geology
Reston (703) 860-6114
U. of Chicago-BS, MS, PhD
Carbon 14 dating
Volcanic & earthquake hazards; paleoclimate

RUDER, VIC C.
Administrative Technician
Br. of Pacific-Arctic Geology
Menlo Park (415) 467-7136
San Jose City College-AA
Accounting, administrative office

RUPPEL, EDWARD T. Phyllis
Geologist
Br. of Central Environmental Geology
Denver (303) 234-4954
U. of Montana-BA; U. of Wyoming-MA; Yale-PhD
Structural & economic geology

RUSLING, DONALD H. Evelyn
Nuclear Engineer (303) 988-1231
722 S. Kline Court
Lakewood, Colorado 80226
Office of the Chief Geologist
Denver (303) 234-2608
Texas A&I-BS; Southern Methodist U.-MSE

RUSMORE, MARGARET E.
PST
Br. of Western Environmental Geology
Menlo Park **(415) 323-8111 x2929**
U. of Cal. (Santa Cruz)-BS; U. of Washington
Foothills fault systems, Sierra Nevada, California
Structural geology & regional tectonics

RUSS, DAVID P.
Geologist **(303) 425-1658**
8773 Kendall Court
Arvada, Colorado 80003
Br. of Earthquake Tectonics & Risk
Denver **(303) 234-5065**
Penn State-BS, PhD; West Virginia U.-MS
Mississippi embayment seismotectonics
Geomorphology; Quaternary geology; structural geology

RUSSELL, DALE M.
Museum Specialist
275 Hawthorne, #138
Palo Alto, California 94301
Br. of Paleontology & Stratigraphy
Menlo Park **(415) 323-8111 x2540**
Humboldt State U.-BA; San Diego State U.
Tertiary Alaska megafossils

RUSSELL, JERRY M.
Cartographic Technician **(703) 430-0853**
124 W. Church Road
Sterling, Virginia 22170
Office of Scientific Publications
Reston **(703) 860-6492**
U. of Maryland
Geologic map editor
Piedmont geology; Virginia Triassic rocks

RUSSELL, PAUL C.
PST
P.O. Box 2304
Stanford, California
Br. of Western Environmental Geology
Menlo Park **(415) 323-8111 x2658**
San Francisco State U.
Tephrochronology
Volcanic rocks

RUSSELL-ROBINSON, SUSAN L. **Rob**
Geologist **(703) 435-0391**
1560 Scandia Circle
Reston, Virginia 22090
Br. of Field Geochemistry & Petrology
Reston **(703) 860-7451**
Bates Coll.-BS; George Washington U.-MAT
Regional volcanology
Field petrology; geochemistry of silicic volcanic systems

RYDER, ROBERT T. **Mimi**
Geologist
Br. of Oil & Gas Resources
Denver **(303) 234-5008**
Michigan State U.-BS; Penn State-PhD
Seismic detection of stratigraphic traps
Stratigraphy; petroleum geology; sedimentation & tectonics

RYE, ROBERT O. **Frances**
Geologist **(303) 238-2716**
11863 W. 27th Drive
Lakewood, Colorado 80215
Br. of Isotope Geology
Denver **(303) 234-3876**
Occidental Coll.-BA; Princeton-PhD
Stable isotope geochemistry; ore deposits

RYMER, MICHAEL J.
Geologist **(415) 327-6625**
2710 Emerson Street
Menlo Park, California 94306
Br. of Earthquake Tectonics & Risk
Menlo Park **(415) 323-8111 x2081**
San Jose State U.-BS, MS
Quaternary tectonics of San Andreas fault
Sedimentation stratigraphy

SABLE, EDWARD G. **Vera H.**
Geologist
USGS c/o American Embassy
APO New York 09697
Br. of Latin Am. & African Geology
Jiddah, Saudi Arabia **674188**
U. of Minnesota-BS; U. of Michigan-MS, PhD
Mapping Precambrian shield, Saudi Arabia
Regional stratigraphic analysis; petroleum geology

SABLE, VERA H. **Edward**
Geologist
680 Range View Trail
Golden, Colorado 80401
Office of Scientific Publications
Denver
U. of Michigan-BS

SAKAMOTO, KENJI **Noriko**
Scientific & Tech. Photographer **(415) 493-5551**
768 Charleston Road
Palo Alto, California 94303
Br. of Paleontology & Stratigraphy
Menlo Park **(415) 323-8111 x2495**
Kumamoto U.; Coll. of San Mateo-AA; San Jose State
Fossil photography
Scientific photography

SAKO, MAURICE K. **Ehukai**
PST
Br. of Field Geochemistry & Petrology
Hawaiian Volcano Observ. **(808) 967-7328**
Deformation studies; geoelectrical studies

SAKSS, YULA E. **Uldis**
Computer Aid **(703) 860-1875**
11442 Tanbark Drive
Reston, Virginia 22091
Office of Scientific Publications
Reston **(703) 860-7297**
U. of Bridgeport

SALDUKAS, R. BIRUTE
Geologist
Br. of Paleontology & Stratigraphy
Reston **(703) 860-6179**
Kaunas & Vilnius (Lithuania); Greifswald & Kiel (Germany); Ohio State-MS
Carboniferous paleogeography of the Northern hemisphere
Paleogeography

SALEM, BRUCE B.
Geologist **(415) 326-1442**
1325 Laurel Street, #2
Menlo Park, California 94025
Office of Resource Analysis
Menlo Park **(415) 323-8111 x2906**
Cal. State U.-BS; Stanford-MS
Metal deposits data base
Computer applications in geology

SALLENGER, ASBURY H. **Delores**
Oceanographer
Br. of Pacific-Arctic Geology
Menlo Park **(415) 856-7075**
U. of Virginia-BA, PhD
Coastal processes project
Nearshore sedimentary processes

SALTUS, RICHARD W.
Geophysicist **(415) 493-3032**
451 Wilton Avenue
Palo Alto, California 94306
Br. of Regional Geophysics
Menlo Park **(415) 323-8111 x2784**
Stanford U.-BS
Nuclear waste isolation
Gravity exploration; computer methods

SAMPSON, JAY A. **Susan**
PST
8357 Benton Way
Arvada, Colorado 80003
Br. of Petrophysics & Remote Sens.
Denver **(303) 234-5390**
Fort Lewis Coll.-BA; U. of Colorado-BA; Colorado Schl. of Mines-MS
Coal seismic system
Feasibility studies; project evaluation; coal gasification

SANDERS, REX
Computer Programmer
Br. of Pacific-Arctic Geology
Menlo Park **(415) 323-8111 x2067**
U. of California (Riverside)-BS

SANDERS, WILLIAM
Librarian
Library
Menlo Park **(415) 323-8111 x2208**
U. of Denver-BA; Sorbonne (Paris); U. of California-MLS

SANDO, WILLIAM J.
Geologist **(703) 370-3360**
5348 Thayer Avenue
Alexandria, Virginia 22304
Br. of Paleontology & Stratigraphy
Washington, D.C. **(202) 343-3510**
Johns Hopkins U.-BA, MA, PhD
Mississippian stratigraphy & coral faunas, western U.S.
Biostratigraphy; paleobiology

SANDOVAL, NANCY C. **Michael**
Secretary
Br. of Tectonophysics
Menlo Park **(415) 323-8111 x2646**

SANGREE, ANNE C.
Geologist
3210 Wisconsin Avenue, NW
Washington, D.C. 20016
Office of Scientific Publications
Reston **(703) 860-6493**
Wellesley-BA; McGill U.-MS; Drexel U.-MLS
Editing

SANTOS, ELMER S. **Katherine**
Geologist
566 S. Flower Street
Lakewood, Colorado 80226
Br. of Uranium-Thorium Resources
Denver **(303) 234-5254**
Western Reserve-BS; U. of Michigan-MS
Powder River Basin—uranium

SANZOLONE, FRANK V. **Lucille**
Production Control Supr. **(303) 421-0709**
4160 Yarrow Court
Wheatridge, Colorado 80033
Br. of Exploration Research
Denver **(303) 234-6184**

SANZOLONE, RICHARD F.
Chemist **(303) 426-4528**
5620 W. 71st Circle
Arvada, Colorado 80003
Br. of Exploration Research
Denver **(303) 234-6177**
U. of Colorado-BA
Research in methods of chemical analysis for geochemical exploration
Atomic absorption, methods development

SARGENT, KENNETH A. **Ruth**
Geologist **(303) 238-0219**
Br. of Central Environmental Geology
Denver **(303) 234-4118**
Bates Coll.-BS; U. of Iowa-MS, PhD
Environmental geologic studies of energy lands, Cedar City 2° sheet, Utah
Colorado Plateau geology; volcanic rocks of the Basin and Range

SARNA-WOJCICKI, ANDREI **Deborah Harden**
Geologist **(415) 325-4140**
868 14th Avenue
Menlo Park, California 94025
Br. of Western Environmental Geology
Menlo Park **(415) 323-8111 x2745**
Columbia Coll.-BA; U. of Cal. (Berkeley)-PhD
Tephrochronology; earthquake hazards of Western Transverse Ranges
Volcanic ash chronology; Quaternary stratigraphy & tectonics

SASS, JOHN H. **Maureen**
Geophysicist **(415) 854-3715**
20 Barney Court
Menlo Park, California 94025
Br. of Tectonophysics
Menlo Park **(415) 323-8111 x2273**
U. of Western Ontario-BSc, MSc; Australian National U.-PhD
Regional & site geothermal energy reconnaissance
Heat flow, earthquake research

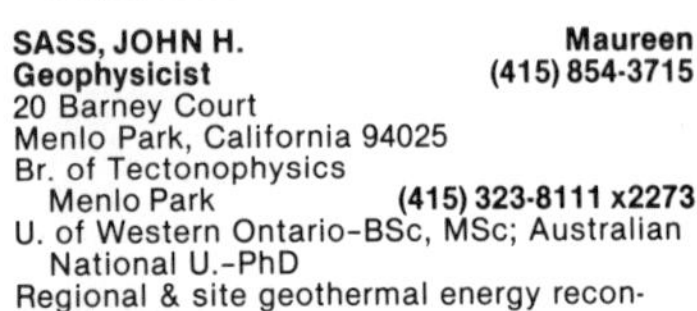

SASSCER, RICHARD S.
Librarian **(301) 559-0191**
5606 Chillum Heights Drive, Apt. 303
Hyattsville, Maryland 20782
Library
Reston **(703) 860-6612**
Belmont Abbey Coll.-BA; Catholic U.-MSLS

SATO, MOTOAKI **Ann**
Geologist **(703) 356-3805**
1312 Darnall Drive
McLean, Virginia 22101
Br. of Experimental Geochem. & Mineralogy
Reston **(703) 860-6600**
U. of Tokyo-BS, MS; U. of Minnesota-PhD
Geochemistry of gas-forming elements
Oxidation-reduction reactions; gas fugacities; volcanic gas monitoring

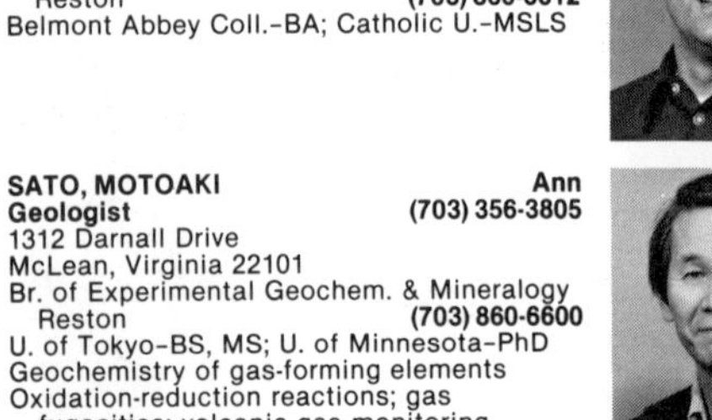

SAUNDERS, HAROLD I. **Louise**
SPST **(703) 370-3167**
727 N. Van Dorn Street
Alexandria, Virginia 22304
Br. of Paleontology & Stratigraphy
Washington, D.C. **(202) 343-8620**
West Virginia Inst. of Tech.; West Virginia U.-BS
Fossil collections

SAUTER, EDWARD A. **Elizabeth**
Electronics Technician **(907) 479-8151**
College Observ.—Yukon Dr. On. W. Ridge
Fairbanks, Alaska 99701
Br. of Electromag. & Geomagnetism
College, Alaska **(907) 479-6146**
College Observatory

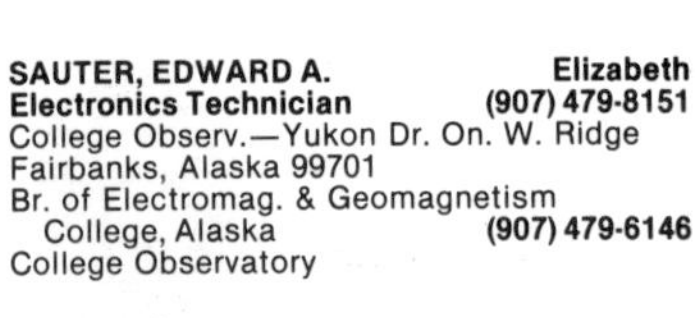

SAVAGE, JAMES C.
Geophysicist **(415) 322-8557**
451 Menlo Oaks Drive
Menlo Park, California 94025
Br. of Tectonophysics
Menlo Park **(415) 323-8111 x2633**
U. of Arizona-BS; Cal Tech.-PhD
Crustal strain
Crustal deformation; earthquake prediction

SAVAGE, WILLIAM Z. **Marjory**
Geologist
Br. of Engineering Geology
Denver **(303) 234-3138**
Lawrence U.-BA; Syracuse-MS; Texas A&M-PhD
Computer modeling research for engineering geology
Geomechanics & computer applications

SAWATZKY, DON L.
Geologist
Br. of Petrophysics & Remote Sens.
Denver **(303) 234-5475**
Colorado Schl. of Mines-DSc
Digital image processing
Remote sensing, computer applications

SCANLON, KATHRYN M.
Geologist **(617) 540-2726**
P.O. Box 543
W. Falmouth, Massachusetts 02574
Br. of Atlantic-Gulf of Mex. Geology
Woods Hole **(617) 548-8700**
Cornell U.-BA; SUNY (Albany)-MS
Geologic hazards on Georges Bank
Marine geology

SCAUN, ANATOLE
Librarian **(703) 860-0458**
2239 Castle Rock Square
Reston, Virginia 22091
Library
Reston **(703) 860-6671**
Columbia U.-BA, MA; Pratt Inst.-MLS

SCHABER, GERALD G. **Sandra**
Geologist **(602) 526-1630**
3312 N. Patterson
Flagstaff, Arizona 86001
Br. of Astrogeologic Studies
Flagstaff **(602) 779-3311 x1455**
U. of Kentucky-BS; U. of Cincinnati-MS, PhD
Geologic mapping of Io; Pioneer-Venus radar data of Venus; radar backscatter research
Terrestrial & planetary radar; planetary geology, remote sensing

SCHACK, ARTIS M. **Richard**
Editorial Assistant **(303) 428-4170**
6428 Xavier Street
Arvada, Colorado 80003
Br. of Sedimentary Mineral Resources
Denver **(303) 234-3786**
Kearney State U.

SCHAFER, CONSTANCE M. **Ken**
Scientific Data Assistant **(703) 860-8038**
11404 Tanbark Drive
Reston, Virginia 22091
Br. of Exper. Geochem. & Mineralogy
Reston **(703) 860-6911**
Georgia State U.; Northern Virginia Community Coll.-AS

SCHAFER, FRANCIS J.
Physical Scientist
Br. of Astrogeologic Studies
Flagstaff **(602) 261-1455**
Utica Coll.-BS
Mars mapping
Photogrammetic applications to space imaging systems

SCHAFER, JOHN PHILLIP
Geologist **(703) 437-9187**
11629 Charter Oak Court, #101
Reston, Virginia 22090
Br. of Eastern Environmental Geology
Reston **(703) 860-6595**
Harvard-BS, AM
Glacial geology, Connecticut
Glacial geology; geomorphology

SCHENK, CHRISTOPHER J.
Geologist **(303) 238-8821**
1059 Ammons Street, #8
Lakewood, Colorado 80215
Br. of Oil & Gas Resources
Denver **(303) 234-2949**
Salem State U.-BS; U. of Michigan-MS
Experimental formation of eolian deposits
Eolian deposits; sediment transport mechanics

SCHLEE, JOHN S. **Susan**
Geologist
147 Sippewissett Road
Falmouth, Massachusetts 02540
Br. of Atlantic-Gulf of Mex. Geology
Woods Hole **(617) 548-8700**
U. of Michigan-BS; UCLA-MA; Johns Hopkins-PhD
Stratigraphy & resource assessment of Georges Bank
Seismic stratigraphy; structure & stratigraphy

SCHLOCKER, JULIUS **Lois**
Geologist **(415) 592-1595**
197 Vine Street
San Carlos, California 94070
Br. of Western Environmental Geology
Menlo Park **(415) 323-8111 x2383**
U. of California (Berkeley)-BA, MA; Colorado Schl. of Mines
Engineering geology; mineralogy (including clays); petrology

SCHMIDT, ARLENE C.
Clerk-Typist
Br. of Central Mineral Resources
Denver **(303) 234-3836**

SCHMIDT, DWIGHT L. **Carol**
Geologist
13560 W. Dakota Place
Lakewood, Colorado 80228
Br. of Central Environmental Geology
Denver **(303) 234-4111**
U. of Washington-PhD
Regional geology; economic geology; geomorphology

SCHMIDT, PAUL W. **Deanna**
Geologist **(303) 986-6141**
10475 W. Warren Avenue
Lakewood, Colorado 80227
Office of Scientific Publications
Denver **(303) 234-3283**
U. of Colorado-BA
Editor
Mountain soils; Front Range urban corridor

SCHMIDT, ROBERT GEORGE **Jessica**
Geologist **(301) 949-2351**
9623 Culver Street
Kensington, Maryland 20795
Br. of Central Environmental Geology
Reston **(703) 860-6503**
U. of Colorado-AB; Harvard-MA, PhD
Butte 1° x 2° quadrangle (CUSMAP)
Geologic mapping; volcanology; earthquake hazards

SCHMIDT, ROBERT GORDON **Alice**
Geologist
3732 N. Nelson Street
Arlington, Virginia 22207
Br. of Eastern Mineral Resources
Reston **(703) 860-7358**
U. of Wisconsin-BS, MS
Porphyry Cu-Mo deposits, Eastern U.S.
Metallic mineral deposits; marine placers; satellite remote sensing

SCHMIEDER, WILLIAM H. **Mary J.**
Geophysicist
75 Zinnia Street
Denver, Colorado 80228
Br. of Global Seismology
Denver **(303) 234-3994**
Penn State-BS

SCHMOLL, HENRY R.
Geologist **(303) 965-8265**
2550 S. Meade Street
Denver, Colorado 80219
Br. of Engineering Geology
Denver **(303) 234-3290**
Columbia U.-AB, AM
Regional engineering geology, Cook Inlet, Alaska
Quaternary & engineering geology

SCHNABEL, DIANE C.
Technical Editor **(303) 422-0195**
5757 W. 35th Avenue
Denver, Colorado 80212
Office of Scientific Publications
Denver **(303) 234-2445**
Mount Holyoke Coll.-BA

SCHNEIDER, GARY B.
Geologist
Br. of Coal Resources
Denver **(303) 234-3578**
Knox Coll.-BA; Montana State U.-MS; South Dakota Schl. Mines-PhD

SCHNEIDER, JONATHAN N.
Museum Technician
Br. of Paleontology & Stratigraphy
Menlo Park **(415) 323-8111 x2261**
U. of California (Santa Cruz)-BS
Arctic benthic foraminifers

SCHNEIDER, RAY R. **Barbara**
Geologist **(703) 435-2392**
606 Austin Lane
Herndon, Virginia 22070
Br. of Eastern Environmental Geology
Reston **(703) 860-6421**
Edinboro State Coll.-BS
Charleston earthquake investigation
X-ray diffraction; sedimentary analysis

SCHNEPFE, MARIAN M. **William**
Chemist
Br. of Analytical Laboratories
Reston **(703) 860-6964**
George Washington U.-AA, BS, MS, PhD
Trace elements, sulfur speciation

SCHOENFELD, RICHARD E.
Cartographic Technician
Br. of Central Mineral Resources
Denver **(303) 234-5108**
Denver Community Coll.-AA

SCHOLLE, PETER A. **Mary**
Geologist **(303) 674-7503**
29254 Thimbleberry Lane
Evergreen, Colorado 80439
Br. of Oil & Gas Resources
Denver **(303) 234-2993**
Yale-BS; Princeton-MS, PhD
Reservoir studies of Niobrara Limestone
Chalks; carbon isotopes; sedimentology

SCHOLTEN, JOHN R. **Gail**
PST **(303) 973-9522**
8212 W. Quarto Drive
Littleton, Colorado 80123
Br. of Paleontology & Stratigraphy
Denver **(303) 234-5851**
Metro State Coll.-BS, BA
Palynology

SCHRAY, CAROLE A.
Administrative Officer
911 Elden Street
Herndon, Virginia 22070
Office of the Chief Geologist
Reston **(703) 860-6534**

SCHRAY, KAREN A. **Martin**
Computer Technician
Office of Scientific Publications
Reston **(703) 860-6738**
Geomap index

SCHRUBEN, PAUL G.
Geologist
11011 Becontree Lake Drive, Apt. 305
Reston, Virginia 22090
Office of Resource Analysis
Reston **(703) 860-6446**
U. of Maryland-BS

SCHULTZ, DAVID M. **Barbara**
Physical Scientist
633 Nash Street
Herndon, Virginia 22070
Br. of Atlantic-Gulf of Mex. Geology
Reston **(703) 860-6965**
Hartwick Coll.-BA; U. of Rhode Island-PhD
Marine organic geochemistry; resource & environmental assessment
Humic substances; organic geochemistry

SCHULTZ, LEONARD G. **Hildreth**
Geologist **(303) 233-6276**
500 Garland Street
Lakewood, Colorado 80226
Br. of Regional Geochemistry
Denver **(303) 234-3675**
Cal Tech.-BS; U. of Illinois-PhD
Geochemistry; Tippecanoe Sequence
Clays; mineralogy of sedimentary rocks

SCHULZ, SANDRA S.
PST
Br. of Tectonophysics
Menlo Park **(415) 323-8111 x2763**
U. of California (Santa Cruz)-BS
Fault zone tectonics

SCHUSTER, ROBERT L. **Patricia**
Geologist **(303) 279-9849**
1941 Golden Vue Drive
Golden, Colorado 80401
Br. of Engineering Geology
Denver **(303) 234-4697**
Washington State Coll.-BS; Ohio State-MS; Purdue U.-MS, PhD; U. of London-Dipl.
Ground failure hazards in Columbia River valley, Oregon & Washington
Eng. geology; geotechnical engineering

SCHWAB, CARL E.
PST **(415) 493-8242**
891 San Jude Avenue
Palo Alto, California 94306
Br. of Alaskan Geology
Menlo Park **(415) 323-8111 x2248**
U. of California (Santa Cruz)-BS
AMRAP report, Medfra quadrangle, Alaska
Regional mapping

SCHWARZ, LOUIS J. **Doris**
Chemist
53 Orchard Way North
Rockville, Maryland
Br. of Analytical Laboratories
Reston **(703) 860-6852**
Gallaudet Coll.-BA
Radiochemistry & radioactiviation project
Instrumental neutron activation analysis

SCHWEINFURTH, STANLEY P. **Kitty**
Geologist
Br. of Coal Resources
Reston **(703) 860-7570**
Muhlenberg Coll.; U. of Cincinnati-BS, MS
Alabama coal resources
Regional geology; organic fuel deposits

SCOTT, DAVID H. **Lilia**
Geologist **(602) 774-6195**
1660 Appalachian Way
Flagstaff, Arizona 86001
Br. of Astrogeologic Studies
Flagstaff **(602) 779-3311 x1381**
Cal Tech-BS; UCLA-PhD
Mars geology
Planetary mapping; volcanology; tectonics

SCOTT, DOROTHY L. **Clarence**
Administrative Officer **(703) 938-9653**
415 Courthouse Road, SW
Vienna, Virginia 22080
Office of Geochemistry & Geophysics
Reston **(703) 860-6585**

SCOTT, EDWARD W. **Winifred**
Geologist **(714) 970-7889**
5212 Stone Canyon
Yorba Linda, California 92686
Br. of Oil & Gas Resources
Laguna Niguel, CA **(714) 831-4232**
UCLA-BA
Oil & gas resource appraisal
Basin evaluation; petroleum exploration

SCOTT, GLENN R. **Juanita**
Geologist **(303) 233-4568**
60 Estes Street
Lakewood, Colorado 80226
Br. of Central Environmental Geology
Denver **(303) 234-3545**
Colorado U.-BA
Southern Raton basin & San Juan basin
Environmental & surficial geology; stratigraphy

SCOTT, JAMES H.
Geophysicist
12372 W. Louisiana Avenue
Lakewood, Colorado 80228
Br. of Petrophysics & Remote Sens.
Denver **(303) 234-3298**
Union Coll.-BS
Borehole geophysics
Borehole geophysics; seismic refraction

SCOTT, WILLIAM A. **Sue**
Computer Programmer
Office of Resource Analysis
Reston **(703) 860-6452**

SCOTT, WILLIAM E. **Mary Pollock**
Geologist **(303) 674-3296**
33913 Upper Bear Creek Road
Evergreen, Colorado 80439
Br. of Central Environmental Geology
Denver **(303) 234-5215**
St. Lawrence U.-BS; U. of Washington-MS, PhD
Quaternary stratigraphy, Wasatch front
Quaternary stratigraphy; neotectonics; geomorphology

SEELAND, DAVID A. **Bette**
Geologist **(303) 238-3306**
47 Flower Street
Lakewood, Colorado 80226
Br. of Uranium-Thorium Resources
Denver **(303) 234-5668**
U. of Minnesota-BA, MS; U. of Utah-PhD
Tertiary sedimentology of Wyoming basins
Sedimentology; ore deposits; geologic mapping

SEELEY, JAMES L. **Sharon**
Chemist **(303) 674-2052**
25693 Independence Trail
Evergreen, Colorado 80439
Br. of Analytical Laboratories
Denver **(303) 234-2521**
Hamilton Coll.-BA; Colorado State U.-PhD
Emission spectroscopy
Atomic spectroscopy, analytical chemistry

SEELEY, ROBERT L.
Electronics Engineer
4560 W. Evans Avenue
Denver, Colorado 80219
Br. of Petrophysics & Remote Sens.
Denver **(303) 234-5111**
U. of California (Berkeley)-BS; U. of Washington-MS
Digital IR scanner system
High speed digital processing; sensor development

SEGAL, DONALD B. **Sharon**
PST **(703) 860-8222**
11921 Winterthur Lane, #109
Reston, Virginia 22091
Br. of Petrophysics & Remote Sens.
Reston **(703) 860-6994**
Franklin & Marshall-BS
Remote sensing; mineral deposits
Structural geology; photo interpretation

SEGERSTROM, KENNETH **Mildred**
Geologist **(303) 233-3202**
41 Morningside Drive
Denver, Colorado 80215
Br. of Central Mineral Resources
Denver **(303) 234-2890**
U. of Denver-BA; Pomona Coll.; Harvard-MS
Stillwater complex, Montana
Geologic mapping; metallic ore deposits; geomorphology

SEGINAK, EMIL P. **Leona**
Cartographic Technician **(703) 435-9568**
11619 Charter Oak Court, Apt. 101
Reston, Virginia 22090
Br. of Regional Geophysics
Reston **(703) 860-6507**
St. Procopius Acad.; American U.

SEIDERS, VICTOR M. **Wanda**
Geologist **(408) 733-7673**
1198 Sesame Drive
Sunnyvale, California 94087
Br. of Western Environmental Geology
Menlo Park **(415) 323-8111 x2919**
Franklin & Marshall-BS; Princeton-MS, PhD
Tectonics of the Nacimiento Block, California
Regional geology & petrology

SEIDERS, WANDA H. **Victor**
Program Assistant **(408) 733-7673**
1198 Sesame Drive
Sunnyvale, California 94087
Office of Earthquake Studies
Menlo Park **(415) 323-8111 x2565**

SEITSINGER, KATHERYN M.
Administrative Assistant
Office of Geochemistry & Geophysics
Denver **(303) 234-5461**
Denver Community Coll.

SEITZ, JAMES F.
Geologist **(415) 665-7540**
1119 Stanyan Street
San Francisco, California 94117
Br. of Western Mineral Resources
Menlo Park **(415) 323-8111 x2472**
U. of Iowa-BA, BS; U. of Washington-MS
Mapping roof pendant of Sierra Nevada Mts.

SEKULICH, MICHAEL J.
PST **(303) 935-6801**
1870 W. Mosier Place
Denver, Colorado 80223
Br. of Central Mineral Resources
Denver **(303) 234-3830**
U. of Colorado (Denver)
Heavy minerals laboratory
Mineral separations; geophysics

SELANDER, LAURIE L. Paul
Clerk-Typist
Office of Scientific Publications
Denver **(303) 234-2445**
U. of Wyoming-BA

SELLERS, GEORGE A. Louise
Geologist **(703) 281-5576**
9629 Podium Drive
Vienna, Virginia 22180
Br. of Analytical Laboratories
Reston **(703) 860-7543**
Penn State-BS; Cal Tech-MS, PhD
X-ray spectroscopy trace elements
Sedimentary petrology; organic geochemistry

SELLIN, JON B. Linda
Librarian **(301) 345-6724**
9108 Edmonston Road
Greenbelt, Maryland 20770
Library
Reston **(703) 860-6679**
Texas Christian U.-BA; Catholic U.-MLS

SENECAL, JOSEPH M. Dora
Warehouseman **(505) 898-9128**
209 Santa Elena Road, SE
Rio Rancho, New Mexico 89124
Br. of Global Seismology
Albuquerque **(505) 844-4637**

SENFTLE, FRANK E. Anne
Physicist **(301) 933-4729**
3619 Glenmoor Drive
Chevy Chase, Maryland 20015
Br. of Isotope Geology
Reston **(703) 860-7662**
U. of Toronto-PhD
Borehole neutron activation for mineral exploration
Physics; geophysics; geology

SENTERFIT, ROBERT M. Marie-Louise
Geologist **(303) 449-5350**
540 University
Boulder, Colorado 80302
Br. of Electromag. & Geomagnetism
Denver **(303) 234-5157**
U. of Missouri-BS; U. of Innsbruck (Austria)
Geothermal studies
Audiomagnetotellurics; gravity; geology

SEVERSON, RONALD C.
Soil Scientist
Br. of Regional Geochemistry
Denver **(303) 234-5242**
U. of Minnesota-PhD
Element availability in soils
Trace elements in soils; spatial variability of soils; soil-plant element relations

SEYLER, DAVID A. Dorothy
Program Officer **(703) 860-8018**
2076 Amberjack Court
Reston, Virginia 22090
Office of the Chief Geologist
Reston **(703) 860-6544**
Ohio State-BS, MA; SUNY-Albany
Program planning; program evaluation; management systems

SHAFFER, GLENN L.
Geologist **(703) 435-1248**
525 Florida Avenue, T4
Herndon, Virginia 22070
Office of Resource Analysis
Reston **(703) 860-6451**
Montgomery Coll.-AA; Old Dominion U.-BS; George Washington U.
Metallogenic map of North America
Resource appraisal/analysis; data bank/file management

SHALER, SAM
Engineering Technician **(415) 965-4107**
312 Camille Court
Mountain View, California 94040
Br. of Engineering Geology
Menlo Park **(415) 856-7114**
U. of Arizona
Soil engineering research
Drill rig operator

SHARKEY, BEATRICE D. John
Administrative Clerk **(303) 697-4019**
P.O. Box 310
Evergreen, Colorado 80439
Office of the Chief Geologist
Denver **(303) 234-3622**
Denver Community Coll., Red Rocks Coll., Metropolitan State Coll.

SHARP, WILLIAM N. Grace
Geologist **(303) 237-1244**
3275 Miller Street
Wheatridge, Colorado 80033
Br. of Central Mineral Resources
Denver **(303) 234-2818**
UCLA-BA
Silver City, New Mexico 2° quadrangle
Intrusive rocks; accessory assemblages; petrology & mineralogy

SHARPS, JOSEPH A. Patricia
Geologist
5415 Tenino Avenue
Boulder, Colorado 80303
Br. of Central Environmental Geology
Denver **(303) 234-5375**
U. of New Hampshire-BA
Energy resources of New Mexico
Quaternary geology; stratigraphy; fuels

SHAW, EFFIE G. Douglas
PST **(703) 450-4246**
108 N. Alder Avenue
Sterling, Virginia 22170
Br. of Paleontology & Stratigraphy
Reston **(703) 860-7445**
Old Dominion U.-BS

SHAW, HERBERT R.
Geologist
38623 Cherry Lane, #214
Fremont, California 94536
Br. of Exper. Geochem. & Mineralogy
Menlo Park **(415) 323-8111**
U. of California (Berkeley)-PhD
Geology of radioactive waste disposal
Volcanology; global tectonics; tidal energy

SHAW, VAN E.
Chemist **(303) 986-0914**
9893 West Maryland Drive
Lakewood, Colorado 80226
Br. of Analytical Laboratories
Denver **(303) 234-6406**
Colorado State U.-BS; U. of Nevada
Analytical services and research
Oil shale; rare earths; analytical development

SHAWE, DANIEL R. Helen
Geologist **(303) 237-7195**
8920 W. 2nd Avenue
Lakewood, Colorado 80226
Br. of Central Mineral Resources
Denver **(303) 234-4839**
Stanford-BS, MS, PhD
Geology & mineral resources, Round Mtn & Tonopah quadrangles, Nevada
Mineral deposits; Great Basin geology; Colorado Plateau geology

SHAWE, FRED R. Janet
Geologist **(303) 986-5359**
942 South Quail Way
Lakewood, Colorado 80226
Br. of Central Environmental Geology
Denver **(303) 234-5116**
U. of Nevada (Reno)-BS
S.W. Utah coal
Sedimentation; stratigraphy; coal geology

SHEDLOCK, KAYE M.
Mathematician
Br. of Earthquake Tectonics & Risk
Denver **(303) 234-5605**
U. of Maryland-BS; Johns Hopkins-MS
Seismic risk

SHELDON, RICHARD P. **Claude**
Geologist
Br. of Sedimentary Mineral Resources
Reston **(703) 860-6431**
Yale-BS; Stanford-PhD
Phosphate resource specialist
Phosphorite petrology & resources

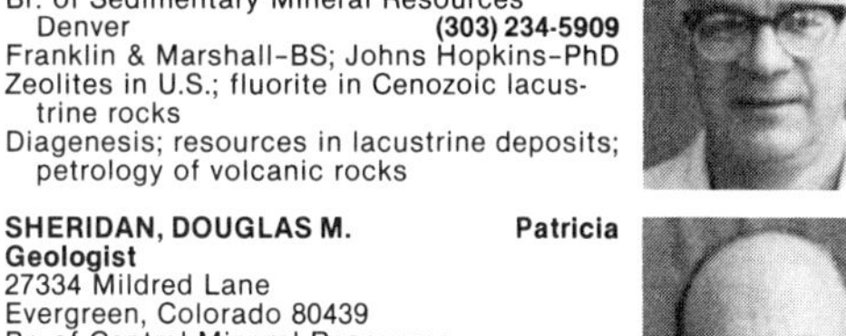

SHEPPARD, RICHARD A. **Evenne**
Geologist **(303) 422-4492**
11647 W. 37th Place
Wheatridge, Colorado 80033
Br. of Sedimentary Mineral Resources
Denver **(303) 234-5909**
Franklin & Marshall-BS; Johns Hopkins-PhD
Zeolites in U.S.; fluorite in Cenozoic lacustrine rocks
Diagenesis; resources in lacustrine deposits; petrology of volcanic rocks

SHERIDAN, DOUGLAS M. **Patricia**
Geologist
27334 Mildred Lane
Evergreen, Colorado 80439
Br. of Central Mineral Resources
Denver **(303) 234-3466**
Carleton Coll.-BA; U. of Minnesota-MS
Precambrian sulfide deposits of Colorado
Mineral deposits in hard-rock terrane

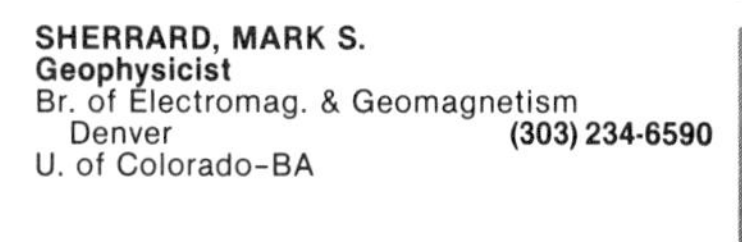

SHERRARD, MARK S.
Geophysicist
Br. of Electromag. & Geomagnetism
Denver **(303) 234-6590**
U. of Colorado-BA

SHERRILL, NATHANIEL D. **Jan**
Electronics Engineer **(415) 941-6023**
P.O. Box 4155
Woodside, California 94062
Br. of Isotope Geology
Menlo Park **(415) 323-8111 x2023**
Northeastern U.
Electronics development
Data acquisition systems; mass spectrometers; microprocessors

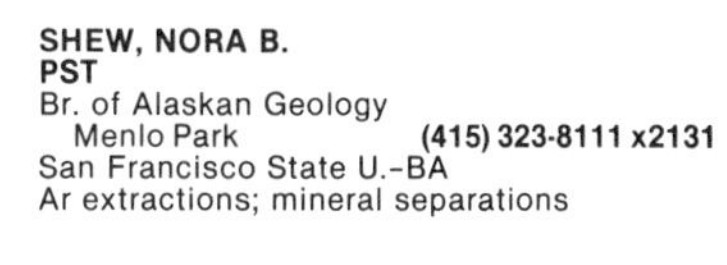

SHEW, NORA B.
PST
Br. of Alaskan Geology
Menlo Park **(415) 323-8111 x2131**
San Francisco State U.-BA
Ar extractions; mineral separations

SHIDELER, GERALD L. **Marie Jeannette**
Geologist
Br. of Atlantic-Gulf of Mex. Geology
Corpus Christi **(512) 888-3294**
Michigan State U.-BS; U. of Illinois-MS; U. of Wisconsin-PhD
Sediment studies in marine & coastal zone environments
Sedimentology; marine geology; oceanography

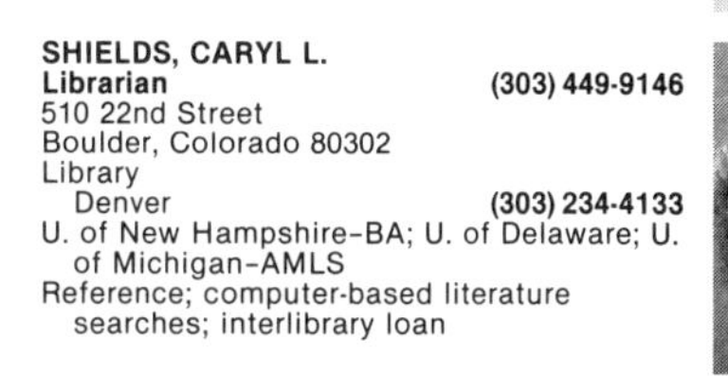

SHIELDS, CARYL L.
Librarian **(303) 449-9146**
510 22nd Street
Boulder, Colorado 80302
Library
Denver **(303) 234-4133**
U. of New Hampshire-BA; U. of Delaware; U. of Michigan-AMLS
Reference; computer-based literature searches; interlibrary loan

SHIFFLETT, CAROL M.
Geologist
11635 Charter Oak
Reston, Virginia 22090
Br. of Engineering Geology
Reston **(703) 860-6595**
U. of Maryland-BS
Coal-mine subsidence, Appalachian region; engineering hazards, Big Horn Basin, Wyo.

SHINN, EUGENE A. **Pat**
Geologist **(305) 253-3230**
8045 S. 133 Street
Miami, Florida 33156
Br. of Oil & Gas Resources
Miami Beach **(305) 672-1784**
U. of Miami-BS
Diagenesis of modern & ancient limestones
Diagenesis; sedimentation; burrowing

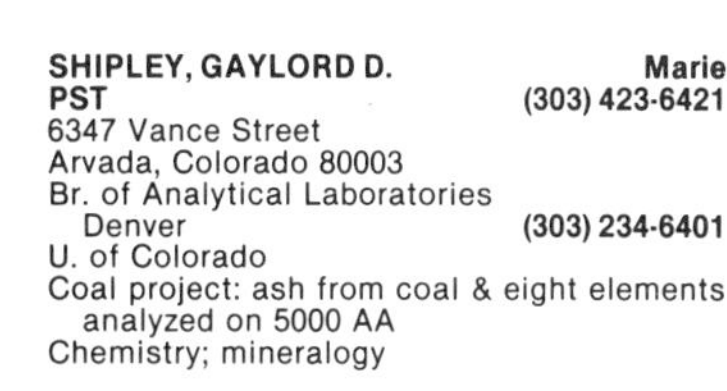

SHIPLEY, GAYLORD D. **Marie**
PST **(303) 423-6421**
6347 Vance Street
Arvada, Colorado 80003
Br. of Analytical Laboratories
Denver **(303) 234-6401**
U. of Colorado
Coal project: ash from coal & eight elements analyzed on 5000 AA
Chemistry; mineralogy

SHOCK, EVERETT L.
PST
Br. of Western Mineral Resources
Menlo Park **(415) 323-8111 x2549**
U. of California (Santa Cruz)-BS
Lab supervisor
Contoured geochemical mapping; mineral separation

SHOEMAKER, EUGENE M. **Carolyn**
Geologist **(602) 774-4350**
Box 984
Flagstaff, Arizona 86002
Br. of Petrophysics & Remote Sens.
Flagstaff **(602) 779-3311 x1544**
Cal Tech-BS; Princeton-PhD
Origin of the Earth
Collision processes in the solar system; Southwestern U.S. geology & paleomagnetism

SHOUKIMAS, MARY E.
Computer Specialist
Br. of Atlantic-Gulf of Mex. Geology
Woods Hole **(617) 548-8700**
Vassar Coll.-BA

SHRIDE, ANDREW F. **Stella**
Geologist **(303) 237-2594**
2055 Miller Court
Lakewood, Colorado 80215
Br. of Eastern Environmental Geology
Denver **(303) 234-5151**
U. of Washington-BS; U. of Arizona-PhD
Fault definition, N.E. Massachusetts
Regional geology; mineral deposits

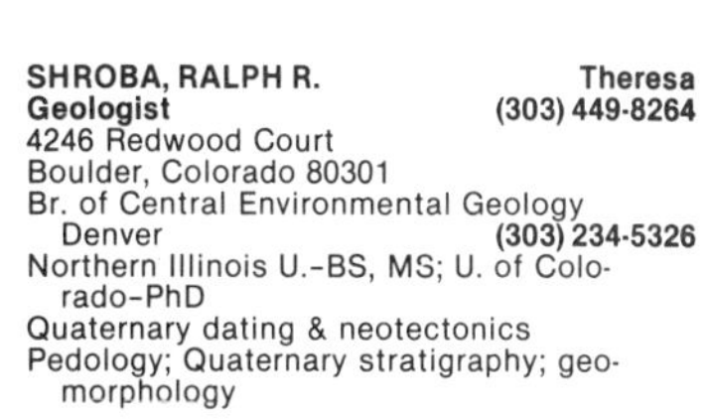

SHROBA, RALPH R. **Theresa**
Geologist **(303) 449-8264**
4246 Redwood Court
Boulder, Colorado 80301
Br. of Central Environmental Geology
Denver **(303) 234-5326**
Northern Illinois U.-BS, MS; U. of Colorado-PhD
Quaternary dating & neotectonics
Pedology; Quaternary stratigraphy; geomorphology

SHURR, GEORGE W. **Margaret**
Geologist **(612) 253-7810**
RRt 5, Valley View
St. Cloud, Minnesota 56301
Br. of Oil & Gas Resources
Denver **(303) 243-4750**
U. of S. Dakota-BA; Northwestern U.-MS; U. of Montana-PhD
Northern Great Plains tight gas reservoirs
Stratigraphy; tectonics

SHURTLEFF, JOAN L.
Clerk-Typist
RFD #1
Buzzards Bay, Massachusetts 02532
Br. of Atlantic-Gulf of Mex. Geology
Woods Hole **(617) 548-8700**

SICARD, DOROTHY M.
Secretary
501 Forest Avenue, #707
Palo Alto, California 94301
Br. of Pacific-Arctic Geology
Menlo Park **(415) 856-7090**
Southwestern Coll.; U. of San Diego

SIEMS, DAVID F.
Chemist
Br. of Exploration Research
Denver **(303) 234-6190**
Regis College-BS
CUSMAP: Charlotte, North Carolina; Dillon, Montana
Emission spectroscopy

SIGLEO, WAYNE R. **Susan**
Geologist **(703) 354-4334**
7319 Hogarth Street
Springfield, Virginia 22151
Br. of Coal Resources
Reston **(703) 860-7734**
U. of New Mexico-BA; U. of Arizona-MA; U. of Tasmania (Australia)-PhD
Coal resources of the eastern U.S.
Quaternary geology, palynology, geomorphology; Carboniferous stratigraphy, coal

SIKKINK, PAMELA G. **Arthur**
Geologist
Box 638
Golden, Colorado 80401
Br. of Uranium-Thorium Resources
Denver **(303) 234-5664**
Bemidji State U.-BS
Tertiary rocks of San Juan Basin

SIKORA, ROBERT F. **Pat**
PST **(415) 366-9177**
165 Atherwood Avenue
Redwood City, California 94061
Br. of Regional Geophysics
Menlo Park **(415) 323-8111 x2638**
San Francisco State Coll.-BA; San Jose State Coll.-BA
Cores of Newberry Crater, Oregon
2-D gravity & magnetics modeling; gravity measurements & reductions

SILBERLING, NORMAN J.
Geologist
Br. of Paleontology & Stratigraphy
Denver **(303) 234-5933**
Stanford-BS, MS, PhD
Accretionary pre-Tertiary terranes; Alaskan fragments
Lower Mesozoic biostratigraphy; pre-Tertiary tectonic history of western North America; carbonate petrology

SILBERMAN, MILES L.
Geologist
Br. of Alaskan Geology
Menlo Park **(415) 323-8111 x2267**
U. of Rochester-MS, PhD
Geochemistry & geochronology of Alaskan ore deposits
Economic geology; isotope geology

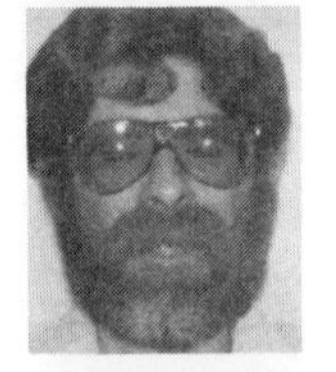

SILK, ELEANA S.
PST
Br. of Analytical Laboratories
Reston **(703) 860-6689**
Michigan State U.-BS; George Washington U.-BS
Geology; zoology; computers

SILVERMAN, STAN A.
Computer Programmer
Br. of Tectonophysics
Menlo Park **(415) 323-8111 x2933**
Purdue U.-BS; Ohio State-MS
Earthquake prediction
Computer applications

SILVERSTEIN, BARRY L.
PST
Br. of Seismic Engineering
Menlo Park **(415) 323-8111 x2881**
SUNY, College at New Paltz-BS
Geologic site description of strong-motion accelerograph stations

SIMIRENKO, MARIE J.
Research Technician
Office of Earthquake Studies
Menlo Park **(415) 323-8111 x2632**
U. of Minnesota-BS, MS
Seismicity in Sierra Foothills
Sierra Foothills seismic data & fault plane analysis

SIMMONS, FLORENCE M. **Hal**
Administrative Officer
1774 Robb Street
Lakewood, Colorado 80215
Br. of Sedimentary Mineral Resources
Denver **(303) 234-3800**

SIMMONS, GEORGE C. **Virginia**
Geologist
Br. of Central Mineral Resources
Denver **(303) 234-6296**
Tulane U.-BS; Washington State U.-MS
Hells Canyon wilderness
Mineral evaluation; geologic mapping; geochemical reconnaissance

SIMMONS, KATHLEEN R. **Craig**
Geologist
Br. of Uranium-Thorium Resources
Denver **(303) 234-4201**
SUNY (Stony Brook)-BS, MS
Isotope geochemistry

SIMONS, FRANK S. **Mary**
Geologist **(303) 238-2338**
9005 W. 2nd Avenue
Lakewood, Colorado 80226
Br. of Central Mineral Resources
Denver **(303) 234-5015**
UCLA-BA; Stanford-PhD
Madison-Gallatin study area, Montana
Areal mineral evaluation; geologic mapping

SIMPSON, HOWARD E. **Elizabeth**
Geologist
2020 Washington Avenue
Golden, Colorado 80601
Br. of Engineering Geology
Denver **(303) 234-3426**
U. of N. Dakota-BS; U. of Illinois-MS; Yale-PhD
Shale evaluation for nuclear waste repository
Eng., urban, & environmental geology

SIMPSON, SHIRLEY L.
Geologist
140 Flower Street
Lakewood, Colorado 80226
Br. of Petrophysics & Remote Sens.
Denver **(303) 234-5493**
Michigan State U.-BS
Remote sensing geophysics; Landsat

SIMS, JOHN D. **Christina**
Geologist **(415) 326-7175**
2784 South Court
Palo Alto, California 94306
Br. of Earthquake Tectonics & Risk
Menlo Park **(415) 323-8111 x2252**
U. of Illinois-BS; U. of Cincinnati-MS; Northwestern U.-PhD
Paleoseismic indicators in sediments & Quaternary reference core
Quaternary geology, sedimentology

SIMS, PAUL K. **Dolores**
Geologist
1315 Overhill Road
Golden, Colorado 80401
Br. of Central Mineral Resources
Denver **(303) 234-2105**
U. of Illinois-BA, MS; Princeton-PhD
Precambrian geology & mineral deposits

SINGER, DONALD A. **Jan**
Geologist **(408) 255-2687**
10191 N. Blaney Avenue
Cupertino, California 95014
Office of Resource Analysis
Menlo Park **(415) 323-8111**
San Francisco State U.-BA; Penn State-MS, PhD
Methods of mineral resource assessment
Statistical & economic analysis of mineral deposits

SINNOCK, SHANNON K.
Administrative Technician
Br. of Pacific-Arctic Geology
Menlo Park **(415) 856-7136**

SINNOTT, TRUDY M. **Allen**
Librarian
2407 Paddock Lane
Reston, Virginia 22091
Library
Reston **(703) 860-6671**
Radcliffe-BA; Rutgers U.-MLS

SIPKIN, STUART A.
Geophysicist
Br. of Global Seismology
Denver **(303) 234-3994**
Cal Tech-BS; Princeton-MS; U. of California (San Diego)-PhD
Seismic source parameter determination
Seismic attenuation; lateral heterogeneity in the mantle

SKEEN, CAROL J. **David**
Physical Scientist
Br. of Analytical Laboratories
Reston **(703) 860-7652**
Emporia State U.-AA; U. of Maryland-BS
Optical emission spectroscopy
Analytical chemistry; geochemistry

SKIPP, BETTY A.
Geologist **(303) 442-0463**
2035 Grape Avenue
Boulder, Colorado 80302
Br. of Central Environmental Geology
Denver **(303) 234-2885**
Northwestern U.-BS; U. of Colorado-MS
Geology of Dubois 1° x 2° quad., Idaho & Montana
Geologic mapping; stratigraphy; Mississippian calcareous foraminifers

SLACK, JOHN F.
Geologist **(703) 534-1763**
117 W. Marshall Street
Falls Church, Virginia 22046
Br. of Eastern Mineral Resources
Reston **(703) 860-7356**
West Virginia U.-BS; Miami U.-MS; Stanford-PhD
New England massive sulfide deposits
Economic geology; petrology; geochemistry

SLITER, WILLIAM V. **Edna**
Geologist **(703) 281-9432**
1902 Toyon Way
Vienna, Virginia 22180
Br. of Paleontology & Stratigraphy
Reston **(703) 860-6051**
UCLA-BA, PhD
Paleoecology, biostratigraphy, taxonomy, & zoogeography of Mesozoic foraminifers

SLOAN, BARBARA J.
Secretary **(303) 499-3853**
1195 Edinboro Drive
Boulder, Colorado 80303
Br. of Global Seismology
Denver **(303) 234-5924**

SMEDLEY, OCIE V.
Budget & Accounting Assistant
Office of International Geology
Reston **(703) 860-6521**

SMITH, BRIAN L. **Cheryl**
Electronics Technician **(703) 361-1539**
Br. of Analytical Laboratories
Reston **(703) 860-7441**
Memphis State U.; Northern Virginia Comm. Coll.-AA
Automation & maintenance of lab instrumentation
Data acquisition & processing; microprocessor applications; fiber optic communication

SMITH, CHARLES C. **Harriet**
Geologist **(703) 978-5571**
5520 Peppercorn Drive
Burke, Virginia 22105
Br. of Paleontology & Stratigraphy
Washington, D.C. **(202) 343-2875**
U. of Texas (Austin)-BS; U. of Houston-MS; U. of Texas (Dallas)-PhD
Geology of Tennessee-Tombigbee Waterway
Mesozoic calcareous plankton biostratigraphy; Gulf Coastal Plain stratigraphy

SMITH, DAVID B.
Chemist **(303) 425-6979**
7520 Brawn Court
Arvada, Colorado 80005
Br. of Exploration Research
Denver **(303) 234-4201**
Vanderbilt U.-BA, MS; Colorado Schl. of Mines-PhD
Exploration geochemistry; leachability of trace elements from volcanic rocks

SMITH, EUGENE P.
PST
Br. of Tectonophysics
Menlo Park **(415) 323-8111**

SMITH, GEORGE I. **Teruko**
Geologist **(415) 854-5270**
15 Siesta Court
Menlo Park, California 94025
Br. of Sedimentary Mineral Resources
Menlo Park **(415) 323-8111 x2286**
Colby College-AB; Cal Tech-MS, PhD
Pluvial history of Great Basin lakes
Climatic history of Great Basin; geochem. of saline lakes; history of Garlock fault

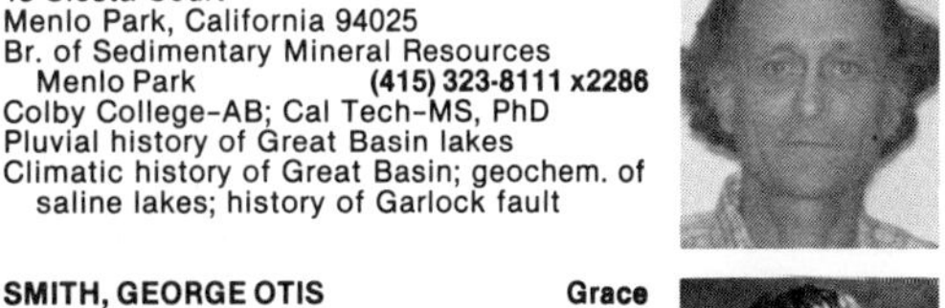

SMITH, GEORGE OTIS **Grace**
Geologist
Office of the Director
Reston
Colby College-BA, MA; Johns Hopkins-PhD
Igneous & metamorphic petrography; land classification studies; water power inventory
Scientist, administrator, lecturer

SMITH, HEZEKIAH LILLIAN
Chemist (202) 882-8746
6207-7th Street, NW
Washington, D.C. 20011
Br. of Analytical Laboratories
Reston (703) 860-6824
Allen U.-BS; Howard U.; Depart. of Agric. Grad School

SMITH, J. FRED, JR. Marjorie
Geologist
529 S. Xenon Court
Lakewood, Colorado 80228
Br. of Central Mineral Resources
Denver (303) 234-2334
Southern Methodist U.-BS; Harvard-MA, PhD
Snake Mountains, Elko Co., Nevada
Stratigraphy & structure, NE Nevada

SMITH, KATHLEEN S.
PST (303) 377-1186
1340 Race Street, Apt. 1
Denver, Colorado 80206
Br. of Regional Geochemistry
Denver (303) 234-5628
Beloit Coll.-BS

SMITH, MARJORIE C. Fred
Geologist
529 S. Xenon Court
Lakewood, Colorado 80228
Br. of Sedimentary Mineral Resources
Denver (303) 234-5120
U. of Missouri-AB
Oil Shale

SMITH, PAULA K. James
PST (303) 232-8759
10646 W. 7th Avenue
Lakewood, Colorado 80215
Br. of Global Seismology
Denver (303) 234-3994
U.S. earthquakes

SMITH, PEGGY A. Tim
PST (415) 257-9466
1079 Windsor Street
San Jose, California 95129'
Br. of Alaskan Geology
Menlo Park (415) 323-8111 x2659
San Jose State U.-BS
Alaskan Quaternary geology
Quaternary geology

SMITH, RICKE J.
Computer Specialist
Br. of Coal Resources
Denver (303) 234-3560

SMITH, ROBERT L. Barbara
Geologist (301) 424-5183
10116 Lloyd Road
Potomac, Maryland 20854
Br. of Field Geochemistry & Petrology
Reston (703) 860-7451
U. of Nevada-BS; Columbia U.; U. of California (Berkeley)
Geochemistry of silicic volcanic rocks
Igneous related resources; volcanoevolution; volcano tectonics

SMITH, WILLIAM K.
Geologist
6137 W. 65th Avenue
Arvada, Colorado 80003
Br. of Engineering Geology
Denver (303) 234-3138
Colorado Schl. of Mines-GE; U. of Arizona-MSc
Highwall stability, Powder River Basin
Engineering geology; slope stability; computer modeling

SMOOT, NANCY M. Jim
Secretary
Office of the Chief Geologist
Menlo Park (415) 323-8111 x2214

SNAVELY, PARKE D., JR. Anne
Geologist (415) 967-3079
1210 Larnel Place
Los Altos, California 94022
Br. of Pacific-Arctic Geology
Menlo Park (415) 856-7049
UCLA-AB, MA
Tertiary geology; Oregon & Wash. continental margin
Structure; stratigraphy; mineral fuels

SNEDDON, RICHARD A.
Electronics Technician (303) 233-8172
Br. of Electromag. & Geomagnetism
Denver (303) 234-2589
Magnetotellurics

SNYDER, DAVID B.
Geophysicist (415) 493-3032
451 Wilton Avenue
Palo Alto, California 94306
Br. of Regional Geophysics
Menlo Park (415) 323-8111
U. of Massachusetts; Stanford-BS, MS
Nuclear waste isolation
Gravity exploration

SNYDER, GEORGE L. Margaret
Geologist (303) 985-5670
2121 South Allison Court
Lakewood, Colorado 80227
Br. of Central Environmental Geology
Denver (303) 234-3593
Dartmouth Coll.-AB; U. of Chicago-MS
Central Wyoming Precambrian
Field geology; igneous & metamorphic petrology

SNYDER, RICHARD P. Donna
Geologist (303) 794-0952
Br. of Special Projects
Denver (303) 234-2261
Ohio State U.-BS
Radioactive waste disposal in geologic media

SNYDER, STEPHEN L.
PST (703) 860-4934
11701 Briary Branch Court
Reston, Virginia 22091
Br. of Regional Geophysics
Reston (703) 860-7233
U. of Maryland-BS
National magnetic anomaly map

SODERBERG, NANCY K.
Clerk-Typist
Br. of Atlantic-Gulf of Mex. Geology
Woods Hole (617) 837-4155
Smith College; NE School of Law

SODERBLOM, LAURENCE A. Barbara
Physical Scientist (602) 779-0030
3940 North Paradise Road
Flagstaff, Arizona 86001
Br. of Astrogeologic Studies
Flagstaff (602) 779-3311 x1455
Cal Tech-PhD
Voyager mission to Jupiter & Saturn
Planetary remote sensing & crustal histories

SOHL, NORMAN F. **Dorothy**
Geologist **(703) 938-5256**
10629 Marbury Road
Oakton, Virginia 22124
Br. of Paleontology & Stratigraphy
Washington, D.C. **(202) 343-8098**
U. of Illinois-BS, MS, PhD
Cretaceous stratigraphy; Mesozoic Gastropoda & Bivalvia

SOHN, ISRAEL GREGORY **Bonnie**
Geologist **(202) 244-5559**
3930 Livingston Street, NW
Washington, D.C. 20015
Br. of Paleontology & Stratigraphy
Washington, D.C. **(202) 343-2885**
CCNY; Columbia U; Hebrew U. of Jerusalem-PhD
Upper Paleozoic & younger marine & nonmarine Ostracoda; micropaleontology

SOLT, MERLYN W. **Violette**
Chemist
13855 Braun Road
Golden, Colorado 80401
Br. of Analytical Laboratories
Denver **(303) 234-4201**
Hastings Coll.-BA; U. of Colorado; Iowa State
Radioactivation analysis
Analytical chemistry; INAA, X-ray fluorescence analysis

SOMERA, ROBERT G.
Computer Technician
920 W. Remington Drive, 10A
Sunnyvale, California 94087
Br. of Network Operations
Menlo Park **(415) 323-8111**

SONNEVIL, RONALD A.
Geologist
Br. of Alaskan Geology
Menlo Park **(415) 323-8111 x2637**
Stanford-BS, MS
Petersburg
Igneous & metamorphic petrology; ore deposits; skarns

SORENSEN, MARTIN L. **Lauralee**
Geologist **(408) 732-9536**
10284 Creston Drive
Cupertino, California 95014
Br. of Western Mineral Resources
Menlo Park **(415) 467-2377**
U. of California-BS; San Jose State U-MS
Massive sulfides of Nevada
Structural & stratigraphic studies, Basin & Range

SORENSEN, SCOTT B.
GFA
Br. of Regional Geophysics
Denver **(303) 234-4938**
Fort Lewis Coll.-BS

SORG, DENNIS H. **Marilyn**
PST
10791 East Estates Drive
Cupertino, California 95014
Br. of Western Environmental Geology
Menlo Park **(415) 323-8111 x2725**
Coll. of San Mateo-AA; San Jose State U.

SPALL, HENRY **Juliet**
Geophysicist **(703) 860-1356**
11319 French Horn Lane
Reston, Virginia 22091
Office of Scientific Publications
Reston **(703) 860-6575**
U.of London-BSc; Southern Methodist U.-MS; U. of London-PhD

SPARKS, DELORES M. **Charles**
Geologic Inquiries Asst. **(703) 754-4862**
3307 Aldie Road
Catharpin, Virginia 22018
Office of Scientific Publications
Reston **(703) 860-6517**

SPENCE, WILLIAM J.
Geophysicist **(303) 399-5188**
2118 Race Street
Denver, Colorado 80205
Br. of Global Seismology
Denver **(303) 234-4041**
SUNY (Albany)-BS, MS; Penn State-PhD
Seismicity & tectonics
Plate tectonics; crustal & upper mantle structure; earthquake prediction

SPENCER, CHARLES W.
Geologist
Br. of Oil & Gas Resources
Denver **(303) 234-3893**
U. of Wyoming; Colby Coll.-AB; U. of Illinois-MS
Western tight gas reservoirs; land classification-oil & gas
Stratigraphy; unconventional reservoirs; hydrodynamics

SPENGLER, RICHARD W. **Roberta**
Geologist **(303) 989-3181**
1521 S. Owens Street, #20
Lakewood, Colorado 80226
Br. of Special Projects
Denver **(303) 234-2146**
Humboldt State U.-BA; U. of Nevada (Las Vegas); Colorado State U.
Volcanic studies; Nevada nuclear waste isolation project
Volcanology; granite; engineering geology

SPIETH, MARY ANN
Geophysicist
Br. of Seismology
Menlo Park **(415) 323-8111 x2823**
Miami U.-AB, MA
Sierran crustal structure
Seismic refraction

SPIKE, BARBARA P.
Secretary
2039 Chadds Ford Drive
Reston, Virginia 22091
Office of International Geology
Reston **(703) 860-6410**
Miami U.

SPIKER, ELLIOTT C.
Geologist
Br. of Isotope Geology
Reston **(703) 860-6113**
George Washington U.-BS, MS
Radiocarbon geochronology
Carbon isotope geochemistry

SPIRAKIS, CHARLES S.
Geologist
Br. of Uranium-Thorium Resources
Denver **(303) 234-5040**
U. of Illinois-BS; Northwestern U.-MS; U. of Colorado
Mineral deposits

SPYDELL, D. RANDALL
Geophysicist
Br. of Regional Geophysics
Denver **(303) 234-6143**
Dartmouth-BA, MA
Geothermal resource assessment
Recent volcanism; plate tectonics; environmental geology

STAATZ, MORTIMER H. — **Rosemary**
Geologist — **(303) 279-6169**
13435 Braun Road
Golden, Colorado 80401
Br. of Uranium-Thorium Resources
Denver — **(303) 234-2986**
Cal Tech-BS; Northwestern U.-MS; Columbia U.-PhD
Thorium resources of the U.S.
Economic geology & mineralogy

STACEY, JOHN S.
Physicist
Br. of Isotope Geology
Denver — **(303) 234-5531**
U. of Durham-BSc; U. of British Columbia-MA PhD
U-Th-Pb isotope studies; mass spectrometry

STAGER, HAROLD K.
Geologist — **(415) 494-6228**
3375 Alma Street, Apt. 258
Palo Alto, California 94306
Br. of Western Mineral Resources
Menlo Park — **(415) 323-8111 x2522**
UCLA-BA
Tungsten resources of Nevada
Mineral deposits; mining & exploration

STAIRS, JOSEPH G.
Accounting Technician — **(303) 279-8598**
26 S. Holman Way
Golden, Colorado 80401
Br. of Central Mineral Resources
Denver — **(303) 234-4842**

STAMM, MARGARET E.
GFA — **(303) 499-8017**
1491 Brown Circle
Boulder, Colorado 80303
Br. of Coal Resources
Denver — **(303) 234-3536**
U. of South Alabama-BS

STANIN, S. ANTHONY — **Barbara**
Geologist
Office of International Geology
Jiddah, Saudi Arabia
U. of Utah-BS
Saudia Arabian program
Mineral exploration, economic geology

STANLEY, ROLFE S. — **Phyllis**
Geologist — **(802) 656-3396**
Dept. of Geology, U. of Vermont
Burlington, Vermont 05405
Br. of Eastern Environmental Geology
Reston — **(703) 860-6531**
Yale-PhD
Northern Vermont-ultramafic belt
Structural geology; tectonics

STARKEY, HARRY C. — **Ruth**
Geologist — **(303) 985-3613**
1636 South Yarrow Court
Lakewood, Colorado 80226
Br. of Central Mineral Resources
Denver — **(303) 234-2873**
West Virginia U.-BS
Sedimentary mineralogy
Clay mineralogy

STARKEY, RUTH W. — **Harry**
Administrative Technician
Br. of Special Projects
Denver — **(303) 234-2431**

STARRITT, BRUCE C.
Geologist
10471 W. 7th Place
Lakewood, Colorado
Office of Resource Analysis
Denver — **(303) 234-6377**
Northern Illinois U.-BS
Computerized resources information bank
Mineral resource appraisal

STEAD, FRANK W. — **Elinor**
Geologist — **(303) 233-3450**
8475 W. 4th Avenue
Lakewood, Colorado 80226
Br. of Special Projects
Denver — **(303) 234-2261**
CCNY-BS; Columbia U.-MS

STEARNS, CHARLES O. — **Lois**
Mathematician — **(303) 494-1959**
1415 Gillaspie Drive
Boulder, Colorado 80303
Br. of Electromag. & Geomagnetism
Denver — **(303) 234-5496**
U. of Northern Colorado-BA, MA; U. of Kansas; U. of Illinois
Electromagnetic computer modeling
Geophysical modeling; geomagnetic studies of Earth's interior; radio wave propagation

STEELE, WILLIAM C.
Geologist
Br. of Pacific-Arctic Geology
Menlo Park — **(415) 323-8111 x2011**
Indiana State U.-BS; Stanford-PhD
Geomorphology; remote sensing; computer programming

STEELE-MALLORY, BRENDA A. — **Carson**
Geologist — **(303) 278-8787**
17792 W. Lunnonhaus Drive, #5
Golden, Colorado 80401
Br. of Uranium-Thorium Resources
Denver — **(303) 234-5104**
U. of Wisconsin-BS; Colorado Sch. Mines-MS
Depositional systems studies, Permian Cutler Formation
Sedimentary petrology

STEHLE, JOHN P.
GFA
Br. of Petrophysics & Remote Sens.
Denver — **(303) 234-2588**
U. of Colorado-BA
Nuclear waste disposal

STELTING, CHARLES F. — **Linda**
PST — **(512) 992-3865**
1110 Southbay Drive
Corpus Christi, Texas 78412
Br. of Atlantic-Gulf of Mex. Geology
Corpus Christi — **(512) 888-3294**
U. of Minnesota; Texas A&I U.
Suspended sediment studies
Textural analysis

STEPHENS, CHRISTOPHER D.
Geophysicist
Br. of Ground Motion & Faulting
Menlo Park — **(415) 323-8111 x2578**
Cornell U.-BS, MS
Alaska seismic studies

STERN, THOMAS W. — **Lyn**
Geologist — **(202) 337-1778**
2400 Foxhall Road, NW
Washington, D.C. 20007
Br. of Isotope Geology
Reston — **(703) 860-6591**
U. of Chicago-AA, BS; U. of Texas-MA
U-Pb geochronology
Isotope geology; mineralogy; autoradiography

STEVEN, THOMAS A. **Grace**
Geologist
Br. of Central Mineral Resources
Denver **(303) 234-3567**
San Jose State U.-BA; UCLA-PhD
Richfield quadrangle, Utah
Volcanic rocks & associated ore deposits

STEVENSON, PETER R.
Mathematician
Br. of Network Operations
Menlo Park **(415) 323-8111 x2572**
U. of Colorado-BA; U. of Santa Clara-MS
Seismic data processing
Seismic data processing & plotting software

STEWART, DAVID B.
Geologist
10715 Midsummer Drive
Reston, Virginia 22091
Br. of Exper. Geochem. & Mineralogy
Reston **(703) 860-6691**
Harvard-BA, MA, PhD
Radioactive waste management
Petrology; mineralogy; physical chemistry

STEWART, JOHN H. **Sarah**
Geologist **(415) 851-7729**
30 Los Charros Lane
Portola Valley, California 94025
Br. of Western Mineral Resources
Menlo Park **(415) 323-8111 x2246**
U. of New Mexico-BS; Stanford-PhD
Walker Lake 2° sheet, CUSMAP
Nevada geology; regional stratigraphy & structure

STEWART, ROGER M. **Daphne**
Geophysicist **(703) 533-1071**
6049 N. 18th Street
Arlington, Virginia 22205
Office of Earthquake Studies
Reston **(703) 860-6473**
U. of California (Berkeley)-PhD
Rock physics; state, structure, composition of crust

STEWART, SAMUEL W. **Shirley**
Geophysicist
1111-Emerson Street
Palo Alto, California 94301
Br. of Seismology
Menlo Park **(415) 323-8111 x2525**
Princeton-AB; U. of Utah-MS; St. Louis U.-PhD
Earthquake data acquisition systems
Automated earthquake data processing; crustal structure studies

STOLTZ, MELISSA L. **Jim**
Computer Technician
Office of Resource Analysis
Reston **(703) 860-6455**
VA Commonwealth U.; Northern Virginia Community College
Computerized resources information bank

STONE, BYRON D. **Janet Radway**
Geologist
Br. of Eastern Environmental Geology
Reston **(703) 860-6503**
Ohio Wesleyan-BA; U. of Vermont-MS; Johns Hopkins-PhD
State surficial geologic map of Massachusetts
Glacial-Quaternary geology, clastic sedimentary geology

STONE, JANET R. **Byron**
Geologist **(703) 347-1866**
150 High Street
Warrenton, Virginia 22186
Br. of Eastern Environmental Geology
Reston **(703) 860-6503**
Birmingham-Southern Coll.; Wesleyan U.
Connecticut cooperative
Glacial geology

STONE, PAUL **Sean**
Geologist **(415) 364-9834**
225 Vera Avenue, #1
Redwood City, California 94061
Br. of Western Environmental Geology
Menlo Park **(415) 323-2282**
U. of California (Berkeley)-BA, MA

STONEHOCKER, LELAND K. **Doris**
Administrative Officer **(303) 232-3547**
930 Depew
Lakewood, Colorado 80214
Office of Geochemistry & Geophysics
Denver **(303) 234-2515**

STOVER, CARL W. **Pauline**
Geophysicist **(303) 499-4047**
5491 Omaha Place
Boulder, Colorado 80303
Br. of Global Seismology
Denver **(303) 234-3994**
Oklahoma City U.-BA; American U; George Washington U.
United States earthquakes
Seismicity; intensities; earthquake locations

STOVER, SARAH A.
Librarian **(703) 860-0113**
2424 Silver Fox Lane
Reston, Virginia 22091
Library
Reston **(703) 860-6613**
Penn State-BS; U. of Maryland-MLS

STRAVERS, JAY A. **Loreen**
PST **(303) 494-4072**
250 29th Street
Boulder, Colorado 80303
Br. of Central Environmental Geology
Denver **(303) 234-5371**
U. of Iowa-BS; U. of Colorado
Quaternary geologic map of the U.S.
Quaternary stratigraphy

STRICKER, GARY D.
Geologist
12338 W. Arizona Avenue
Lakewood, Colorado 80228
Br. of Coal Resources
Denver **(303) 234-3560**

STRICKER, MARY F. **Bob**
Secretary
Br. of Eastern Mineral Resources
Reston **(703) 860-6913**
Gardner School of Business

STROBELL, MARY E. **John**
Geologist **(602) 779-1590**
3824 Paradise Road
Flagstaff, Arizona 86001
Br. of Astrogeologic Studies
Flagstaff **(602) 779-3311 x1427**
Smith College-BA
Planetary nomenclature; planetary studies
Astrogeology; planetary nomenclature

STUART-ALEXANDER, DESIREE E.
Geologist **(415) 969-0269**
1595 Truman Avenue
Los Altos, California 94022
Br. of Western Environmental Geology
Menlo Park **(415) 323-8111 x2929**
U. of Richmond-BA; Stanford-MS, PhD
Foothills fault system, Sierra Nevada, Calif.
Metamorphic petrology, structural geology

STUCKLESS, JOHN Nan
Geologist (303) 674-7493
29948 Carriage Loop Drive
Evergreen, Colorado 80439
Br. of Uranium-Thorium Resources
Denver (303) 234-5531
Amherst Coll.-BA; Arizona State U.-MS; Stanford-PhD
Uranium source-rocks—granites
Igneous petrology; isotope geochemistry

SUESS, HANS E.
Geochemist
Chemistry Dept.-B-017, 5130 Mayer Hall
U. of Cal., San Diego (714) 452-3354
U. of Vienna-PhD
Natural variations in terrestrial C-14
The CO_2 problem; abundances of the elements; solar terrestrial correlations

SUITS, VIVIAN J.
PST
Br. of Uranium-Thorium Resources
Denver (303) 234-5990
Carleton College-BA
Statistics in data presentation; gravity & magnetics interpretations

SUMIDA, THEODORE T. Sumi
Administrative Officer (415) 326-7474
267 Santa Rita Avenue
Palo Alto, California 94301
Office of the Chief Geologist
Menlo Park (415) 323-8111 x2217
Georgetown U.-BS

SUTTON, ART L. Patricia
Chemist (303) 986-7941
11788 W. Jewell Drive
Lakewood, Colorado 80228
Br. of Analytical Laboratories
Denver (303) 234-6405
U. of Colorado-BA
Plasma direct reader
Laboratory automation

SUTTON, ROBERT L. Nona
Geologist (602) 525-1771
P.O. Box 924
Flagstaff, Arizona 86002
Br. of Central Environmental Geology
Flagstaff (602) 779-3311 x1389
Haverford Coll.-BA
Geomorphology; stratigraphy; environmental geology

SUTTON, ROBIN C. Jefferson
Cartographer (703) 435-5786
1813 Sycamore Valley Drive, Apt. 202
Reston, Virginia 22090
Br. of Eastern Environmental Geology
Reston (703) 860-6421
Mary Washington Coll.-BA
Safe mine waste disposal, Appalachians
Remote sensing; cartography

SVITEK, JOSEPH
PST (415) 364-4690
108 Murray Court
Redwood City, California 94061
Br. of Tectonophysics
Menlo Park (415) 323-8111 x2034
U. of Rochester-BS
In situ stress

SWADLEY, W. C. Katie
Geologist (303) 979-6928
9790 W. Frost Place
Littleton, Colorado 80123
Br. of Special Projects
Denver (303) 234-2365
U. of Texas-BA, MA
Tectonics, seismicity, volcanism & erosion of S. Great Basin
Geologic mapping; Quaternary geology

SWANN, GORDON A. Jody
Geologist (602) 779-3030
814 W. Murray Road
Flagstaff, Arizona 86001
Office of the Chief Geologist
Flagstaff (602) 261-1483
U. of Colorado-BA, PhD
Geology of the Apollo 15 landing site
Precambrian geology; structural geology; lunar geology

SWANN, INEZ L.
Library Technician
Library
Reston (703) 860-6613

SWANN, JODY Gordon
Tech. Information Specialist (602) 779-3030
814 West Murray Road
Flagstaff, Arizona 86001
Br. of Astrogeologic Studies
Flagstaff (602) 779-3311 x1505
Eastern New Mexico U.-BA

SWANSON, DONALD A. Barbara White
Geologist (408) 446-0170
10201 Firwood Drive
Cupertino, California 95014
Br. of Field Geochemistry & Petrology
Menlo Park (415) 323-8111 x2281
Washington State U.-BS; Johns Hopkins-PhD
Geologic map of Columbia Plateau
Volcanology of basaltic & andesitic rocks

SWANSON, JAMES R. Lynn
Geologist (408) 286-9406
136 S. 16th Street
San Jose, California 95112
Office of Resource Analysis
Menlo Park (415) 323-8111 x2906
William & Mary Coll.-BS
Geothermal computer file

SWANSON, ROGER W. Marie
Geologist (703) 378-6557
13208 Point Pleasant Drive
Fairfax, Virginia 22030
Office of Scientific Publications
Reston (703) 860-6511
U. of Minnesota-BA, MS; U. of Cincinnati; U. of California
Geologic Names staff
Areal mapping; mineral deposits; tectonics

SWEENEY, RONALD E.
Computer Scientist
2481 Quitman Street
Denver, Colorado 80212
Br. of Regional Geophysics
Denver (303) 234-5526
U. of Maryland-BA
Aeromagnetic & gravity analysis; computer support for multics & minicomputers

SWENSON, DORIS R.
Administrative Technician
Br. of Pacific-Arctic Geology
Menlo Park (415) 467-7136
U. of Virginia; DeAnza Junior College

SWIFT, B. ANN Stephen
Geophysicist
Br. of Atlantic-Gulf of Mex. Geology
Woods Hole (617) 548-8700
Stanford-BS, MS; Oregon State U.-MS
Subsidence of the Atlantic continental margin; lateral velocity variations on Blake Plateau
Marine geophysics

SWINT, THERESA R.
PST
Br. of Western Environmental Geology
Menlo Park **(415) 323-8111 x2504**
Humboldt State U.-BA

SWOLFS, HENRI S.
Geologist
Br. of Engineering Geology
Denver **(303) 234-3396**
City Coll. (CUNY)-BS; Texas A&M-MS, PhD
Geomechanics
Structural geology; rock mechanics; tectonophysics

SYLWESTER, RICHARD E. **Linda**
Physical Scientist **(617) 563-7212**
Br. of Atlantic-Gulf of Mex. Geology
Woods Hole **(617) 548-8700**
U. of Washington-BS, MS
High resolution marine geophysics

SZABO, BARNEY J.
Chemist
Br. of Isotope Geology
Denver **(303) 234-4201**
U. of Supron; U. of Miami-BS, MS; U. of Denver
Uranium-series dating
Geochronology; applied geochemistry; climate studies

TABOR, ROWLAND W. **Kajsa**
Geologist **(415) 851-7526**
108 Santa Maria Avenue
Portola Valley, California 94025
Br. of Western Environmental Geology
Menlo Park **(415) 323-8111 x2293**
Stanford-BS; U. of Washington-MS, PhD
Wenatchee 2° Project, Washington
Geology of northwestern Washington; structural geology & petrology; K-Ar dating

TAGGART, JAMES N. **Barbara**
Geophysicist **(303) 526-1362**
21899 Grandview
Golden, Colorado 80401
Br. of Global Seismology
Denver **(303) 234-5079**
S. Methodist U.-BS, MS; Harvard-PhD
National earthquake catalog
Earthquake location, size, & focal mechanisms;
accuracy of earthquake parameters

TAGGART, JOSEPH E., JR. **Beth**
Geochemist **(303) 979-9378**
6542 S. Yarrow Way
Littleton, Colorado 80123
Br. of Analytical Laboratories
Denver **(303) 234-6403**
Syracuse U.-BS; Miami U.-MS
X-Ray fluorescence spectroscopy
Methods development; mineralogy

TAMAMIAN, NANCY J.
Technical Publications Editor
592 Vista Avenue
Palo Alto, California 94303
Office of Scientific Publications
Menlo Park **(415) 323-8111 x2302**

TANG, ROGER W.
PST **(415) 739-7181**
1144 Susan Way
Sunnyvale, California 94087
Br. of Regional Geophysics
Menlo Park **(415) 323-8111 x2969**
Stanford U.-BA, BS
Radioactive waste isolation

TANNACI, NANCY E.
PST
40467 Seville Court
Fremont, California 94538
Br. of Engineering Geology
Menlo Park **(415) 856-7112**
U. of Cal. (Berkeley)-AB
Research on landslides caused by earthquakes
Post-earthquake field investigations

TANNER, ALLAN B.
Geophysicist **(703) 860-1033**
12125 Captiva Court
Reston, Virginia 22091
Br. of Isotope Geology
Reston **(703) 860-7662**
MIT-BS; U. of Utah
High-resolution borehole gamma spectrometry
Nuclear geophysics; radon migration

TANTI, DOROTHY V.
Secretary
917 Roble Avenue
Menlo Park, California 94025
Br. of Western Mineral Resources
Menlo Park **(415) 323-8111 x2521**

TAPIA-FRISKEN, MARIE L.
Secretary
1010 N. Ford
Golden, Colorado
Br. of Uranium-Thorium Resources
Denver **(303) 234-3624**

TARBERT, ANITA W. **Melvin**
Library Technician **(509) 747-4727**
Route 4, Box 701
Spokane, Washington 99204
Office of Scientific Publications
Spokane **(509) 456-4677**

TARR, ARTHUR C.
Geologist **(303) 499-2500**
1615 Gillaspie Drive
Boulder, Colorado 80303
Office of Scientific Publications
Denver **(303) 234-2445**
Carnegie-Mellon U.-BS; U. of Pittsburgh-PhD
Seismicity studies

TATLOCK, DONALD B. **Ruth**
Geologist **(022)82541**
Jl. Kapt. Tendean 65, Hegarmanah
Bandung, Indonesia
Br. of Middle Eastern & Asian Geology
Bandung, Indonesia
U. of Rochester-BA; Harvard U.
Indonesian Directorate of Mineral Resources
Petrology; economic geology

TATSUMOTO, MITSUNOBU **Kimiko**
Chemist **(303) 238-2602**
2512 Newcombe Street
Lakewood, Colorado 80215
Br. of Isotope Geology
Denver **(303) 234-3876**
Tokyo Bunrika U.-DSc
Isotopes in upper mantle
U-Th-Pb isotopic systematics in terrestrial & extraterrestrial materials

TAYLOR, ALFRED R. **Eugenia**
Geologist **(703) 369-2463**
9355 Birchwood Court
Manassas, Virginia 22110
Br. of Eastern Environmental Geology
Reston **(703) 860-6695**
U. of North Carolina-BS; Wisconsin Inst. of Tech.
Slope stability, Appalachians
Paleozoic stratigraphy; economic & field geology

TAYLOR, DAVID E. — **Jeanne**
PST — **(408) 252-4423**
22087 San Fernando Court
Cupertino, California 95014
Br. of Network Operations
Menlo Park — **(415) 323-8111 x2589**
U. of Denver

TAYLOR, DAVID J. — **Candace**
Geologist — **(303) 989-4952**
11351 W. Florida Avenue
Lakewood, Colorado 80226
Br. of Oil & Gas Resources
Denver — **(303) 234-5008**
Michigan State U.-BS
Seismic stratigraphic group
AOCS oil & gas potential; marine seismic data processing & interpretation

TAYLOR, FRED A.
Physical Scientist
697 Benvenue
Los Altos, California 94022
Br. of Western Environmental Geology
Menlo Park — **(415) 323-3214**
Eastern Oregon Coll.-BS; San Jose State U.
Seismic zonation, San Francisco Bay Region
Landslide economics

TAYLOR, JEANNE M. — **David**
Computer Operator — **(408) 252-4423**
22087 San Fernando Court
Cupertino, California 95014
Br. of Network Operations
Menlo Park — **(415) 323-8111 x2080**

TAYLOR, MARY E. — **John**
Secretary — **(703) 860-3510**
12339 Folkstone Drive
Herndon, Virginia 22070
Office of Marine Geology
Reston — **(703) 860-7243**

TAYLOR, RICHARD B. — **Phyllis**
Supervisory Geologist — **(303) 421-6748**
8485 W. 45th Avenue
Wheatridge, Colorado 80033
Br. of Central Mineral Resources
Denver — **(303) 234-3830**
Cal Tech-BS; U. of Minnesota-MS, PhD
Hardrock geology; ore deposits

TEASDALE, WARREN E. — **Dee**
Geologist — **(303) 985-2140**
1488 S. Ward Street
Lakewood, Colorado 80228
Br. of Coal Resources
Denver — **(303) 234-4230**
Hofstra U.-BA
OER drilling project
Drilling; sampling; operational interests

TEDDER, SANDI
Secretary — **(703) 968-7179**
4223 Mayport Lane
Fairfax, Virginia 22030
Office of the Chief Geologist
Reston — **(703) 860-6544**
U. of Maryland; Northern Virginia Community College

TELEKI, PAUL G.
Deputy
3403 Valewood Drive
Oakton, Virginia 22124
Office of Marine Geology
Reston — **(703) 860-7243**
U. of Florida-BS, MS; Louisiana State U.-PhD
Marine geology; physical oceanography; radar

TERMAN, MAURICE J. — **Gertrude**
Geologist — **(703) 532-9085**
616 Poplar Drive
Falls Church, Virginia 22046
Br. of Middle Eastern & Asian Geology
Reston — **(703) 860-6555**
Columbia Coll.; Columbia U.
Circum-Pacific map project
Tectonics

THADEN, ROBERT E. — **Mary**
Geologist — **(303) 279-5946**
16342 W. 55th Place
Golden, Colorado 80401
Br. of Uranium-Thorium Resources
Denver — **(303) 234-5002**
Michigan State U.-BS, MS
Window Rock—Fort Defiance area, Arizona
Geologic mapping; uranium deposits; gemstones resources

THATCHER, WAYNE — **Mary Ellen**
Geophysicist
409 Laurel
Menlo Park, California
Br. of Tectonophysics
Menlo Park — **(415) 323-8111 x2120**
McGill U.-BS; Cal Tech-PhD
Earthquakes & crustal deformation
Seismology

THAXTON, CHARLENE V.
Clerk-Typist — **(703) 437-9048**
1619 Becontree Lane, #2A
Reston, Virginia 22090
Br. of Analytical Laboratories
Reston — **(703) 860-6689**

THEEL, MARJORIE A.
Secretary — **(303) 526-1327**
26334 Centennial Trail
Golden, Colorado 80401
Br. of Regional Geophysics
Denver — **(303) 234-2623**

THEOBALD, PAUL K. — **Jean**
Geologist — **(303) 237-4972**
13400 W. 10th Avenue
Golden, Colorado 80401
Br. of Exploration Research
Denver — **(303) 234-5636**
Stanford-BS
Sonoran exploration technology; Ajo 2° sheet
Williams Fork RARE II
Exploration geochem.; economic geology

THEODORE, TED G. — **Dolores**
Br. of Western Mineral Resources
Menlo Park — **(415) 323-8111 x2655**
UCLA-AB, PhD
Molybdenum resources
Porphyry Mo-Cu deposits; fluid inclusion studies

THOMAS, JAMES A. — **Bessie**
PST — **(303) 429-1123**
3398 W. 80th Avenue
Westminster, Colorado 80030
Br. of Analytical Laboratories
Denver — **(303) 234-6401**
U. of Denver-BA
Mercury analysis

THOMAS, JULIA A.
Geologist
Office of Scientific Publications
Menlo Park — **(415) 323-8111**
San Francisco State U.; U. of Tennessee-BA

THOMAS, ROGER E.
PST
P.O. Box 603
Herndon, Virginia 22070
Br. of Eastern Environmental Geology
Reston **(703) 860-6998 x6595**
Northern Virginia Community Coll.; Virginia PolyTech. Inst.-BS
Massachusetts surficial resources potential
Carbonates & sedimentation

THOMAS, VIRGINIA L. **Richard**
ADP Prgm. Mngmt. Specialist **(703) 860-0514**
2203 Guildmore Road
Reston, Virginia 22091
Office of the Chief Geologist
Reston **(703) 860-6711**

THOMASSON, CAROL J.
Clerk-Stenographer
Br. of Earthquake Tectonics & Risk
Denver **(303) 234-5087**
U. of Denver-MALS; U. of Florida-MA

THOMPSON, CAROLYN L.
Geologist
Br. of Coal Resources
Reston **(703) 860-6789**
George Washington U.-BS

THOMPSON, CHARLES R. **Emma Jo**
Aerospace Engineer
Br. of Oil & Gas Resources
Flagstaff **(602) 261-1486**
U. of Texas-BS

THOMPSON, J. MICHAEL
Chemist
351 Nova Lane
Menlo Park, California 94025
Br. of Exper. Geochem. & Mineralogy
Menlo Park **(415) 323-8111 x2162**
UCD-BS; CSUS-MS
Analysis of geothermal brines

THOMPSON, JANET K.
Biologist
924 Sixth Avenue
Redwood City, California 94063
Br. of Pacific-Arctic Geology
Menlo Park **(415) 856-7194**
Lewis & Clark Coll.-BS; Calif. State U. (San Francisco)-MA
San Francisco Bay project
Benthic ecology; estuarine ecology

THOMPSON, JOHN B.
Electronics Technician
9200 Evangeline Avenue, NE
Albuquerque, New Mexico 87111
Br. of Global Seismology
Albuquerque **(505) 844-4637**

THOMPSON, MARIAN O.
Cartographic Technician
3511 Webster Street
Brentwood, Maryland 20722
Office of Scientific Publications
Reston **(703) 860-6495**

THORMAN, CHARLES H. **Barbara**
Geologist **(303) 234-5172**
1500 S. Zang Street
Lakewood, Colorado 80228
Br. of Central Mineral Resources
Denver **(303) 234-5172**
U. of Redlands-BS; U. of Washington-MS, PhD
Geology-Pinaleno Mountains, Arizona
Regional tectonics of Great Basins; tectonics of Oklahoma

THORNBER, CARL R. **Mary Love**
Geologist
11629 Stoneview Square
Reston, Virginia 22091
Br. of Exper. Geochem. & Mineralogy
Reston **(703) 860-6665**
U. of Massachusetts-BS; Queen's U. (Ontario)-MSc
Geological thermometry
Basaltic liquids; crystallization kinetics; diffusion

THROCKMORTON, CONSTANCE K.
PST
290 Laguna Drive
Milpitas, California
Br. of Western Environmental Geology
Menlo Park **(415) 323-8111 x2957**

THURNAU, ELLEN M.
Secretary **(703) 476-4373**
11820 Breton Court, 22-C
Reston, Virginia 22091
Br. of Exper. Geochem. & Mineralogy
Reston **(703) 860-6691**
U. of Virginia-BA; Duke-MA

TIBBETTS, REVA F.
Secretary **(303) 989-2453**
9350 W. Utah Avenue
Lakewood, Colorado 80226
Br. of Earthquake Tectonics & Risk
Denver **(303) 234-4022**

TIDBALL, RONALD R. **Jean**
Soil Scientist **(303) 988-6914**
9351 West Utah Place
Lakewood, Colorado 80226
Br. of Regional Geochemistry
Denver **(303) 234-4244**
Colorado State U.-BS; U. of Washington-MS; U. of California (Berkeley)-PhD
Soil geochemistry in western energy lands
Environmental geochemistry (soils)

TILBURY, CAROL A. **John**
Secretary **(415) 345-9659**
1344 Palos Verdes Drive
San Mateo, California 94403
Br. of Field Geochemistry & Petrology
Menlo Park **(415) 323-8111 x2058**
Central California Commerical College

TILLEY, SCOTT E. **Joanna**
Administrative Officer
Office of Environmental Geology
Reston **(703) 860-6417**
California State College, Sonoma-BA

TILLING, ROBERT I. **Susan**
Geologist **(703) 821-2267**
1042 Delf Drive
McLean, Virginia 22101
Office of Geochemistry & Geophysics
Reston **(703) 860-6584**
Pomona Coll.-BA; Yale-MS, PhD
Hawaiian volcanism; Boulder batholith; igneous petrology

TILTON, SANDRA P. **David**
Secretary
Br. of Electromag. & Geomagnetism
College, Alaska **(907) 479-6146**

TIMMINS, JANE D. **Larry**
Clerical Assistant **(703) 860-3807**
2932 Fort Lee Street
Herndon, Virginia 22070
Office of Scientific Publications
Reston **(703) 860-6491**
Middle Tennessee State U.

TINSLEY, JOHN C., III **Marilyn**
Geologist
1040 Oakland Avenue
Menlo Park, California 94025
Br. of Western Environmental Geology
Menlo Park **(415) 323-8111 x2037**
Colorado Coll.-BA; Stanford-MS, PhD
Quaternary framework for earthquake studies
Los Angeles, California
Quaternary geology; paleoclimatology; fluvial geomorphology

TIPPENS, CHARLES L. **Mary**
Electronics **(303) 237-0640**
638 Alkire Street
Golden, Colorado 80401
Br. of Electromag. & Geomagnetism
Denver **(303) 234-2589'**
Audio-magnetotellurics

TOBISCH, OTHMAR T. **Kathryn**
Geologist
Br. of Field Geochemistry & Petrology
Menlo Park **(408) 429-2777**
U. of California (Berkeley)-BA, MA; U. of London-PhD
Ritter Range & Mt. Goddard pendants, Cal.
Structural geology; quantitative strain; evolution of orogenic belts

TOLBERT, GENE E.
Geologist
1002 Moore Place
Vienna, Virginia 22180
Br. of Latin Am. & African Geology
Reston **(703) 860-6551**
Colorado Coll.; Harvard-BA, MA, PhD
Latin American mineral resources

TOMEY, JACK W.
Computer Operator
Br. of Network Operations
Menlo Park **(415) 323-8111 x2632**
San Jose State U.-BS

TOMPKINS, MARY L.
Computer Technician **(303) 985-0225**
2155 S. Depew
Denver, Colorado 80227
Br. of Regional Geology
Denver **(303) 234-2437**

TOOKER, EDWIN W. **Pauline**
Geologist
Br. of Western Mineral Resources
Menlo Park **(415) 323-8111 x2621**
Bates Coll.-BS; Lehigh U.-MS; U. of Illinois-PhD
Mineral resources of Utah
Economic geology; metallogenesis

TOULMIN, PRIESTLEY, III **Martha**
Geologist **(703) 549-2375**
418 Summers Drive
Alexandria, Virginia 22301
Br. of Exper. Geochem. & Mineralogy
Reston **(703) 860-6650**
Harvard-BA, PhD; U. of Colorado-MS
Mt. Princeton igneous complex, Colo.; planetary mineralogy
Igneous petrology; geochemistry of sulfide ore deposits; geochemistry of Mars

TOURTELOT, HARRY A. **Suzanne**
Geologist **(303) 989-0473**
12968 West Oregon Drive
Lakewood, Colorado 80228
Br. of Regional Geochemistry
Denver **(303) 234-3717**
U. of Nebraska-AB
Geochemistry & health
Geochemistry of sedimentary rocks

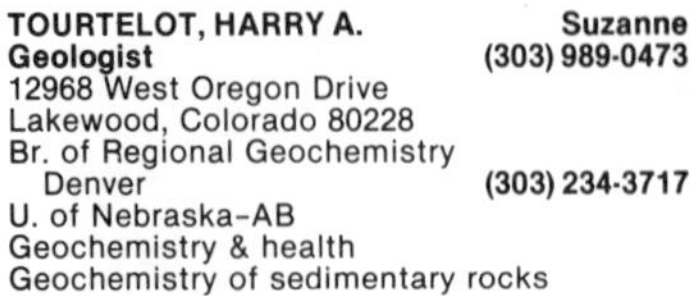

TOWLE, JAMES N.
Geophysicist
2920 Lafayette Drive
Boulder, Colorado 80303
Br. of Electromag. & Geomagnetism
Denver **(303) 234-5156**
Cornell U.-BS; U. of Wisconsin-MS, PhD
Geomagnetic deep sounding
Geomagnetic induction; geophysics

TOWNSHEND, JOHN B. **Frieda**
PST **(907) 479-2898**
College Observatory, Yukon Dr.-West Ridge
Fairbanks, Alaska 99701
Br. of Electromag. & Geomagnetism
College, Alaska **(907) 479-6146**
George Washington U.; U. of Wisconsin
Operation of College & Barrow Observatories
Geomagnetism; seismology; instrumentation

TRACEY, JOSHUA I., JR. **Frances**
Geologist **(703) 243-9479**
4023 N. 27th Road
Arlington, Virginia 22207
Office of Marine Geology
Reston **(703) 860-7538**
Yale U.-AB, MS, PhD
Resources of oceanic islands
Stratigraphy; ocean sediments; coral reefs

TRANTER, MARY C.
Secretary **(703) 476-9530**
2235 Castle Rock Square
Reston, Virginia 22091
Br. of Analytical Laboratories
Reston **(703) 860-7247**

TRASK, NEWELL J. **Esther**
Geologist **(703) 338-7546**
Route 1, Box H-20
Purcellville, Virginia 22132
Br. of Eastern Environmental Geology
Reston **(703) 860-7773**
MIT-BS; U. of Colorado-MA; Harvard-PhD
Radioactive waste
Structural geology; planetary geology; public risk

TRAVIS, ALLEN H. **Wanda**
Geophysicist **(703) 371-2564**
Br. of Electromag. & Geomagnetism
Corbin, Virginia **(703) 373-7601**
Oklahoma State U.-BS
Geomagnetism

TRENT, VIRGIL A. **Priscilla**
Geologist **(703) 684-8441**
712 25th Street S.
Arlington, Virginia 22202
Br. of Coal Resources
Reston **(703) 860-6697**
Michigan State Coll.-BS; Michigan State U.; U. of Minnesota
Coal geology
Economic geology

TREXLER, JAMES H., JR.
Geologist
Br. of Alaskan Geology
Menlo Park **(415) 323-8111 x2256**
U. of Maryland-BS; U. of Oklahoma-MS
Sedimentologist; stratigrapher

TRIMBLE, DEBORAH A.
PST **(408) 262-9120**
148 El Bosque
San Jose, California 95134
Br. of Isotope Geology
Menlo Park **(415) 323-8111 x2857**
U. of Cal. (Davis)-BS
Geochronology
Geology; radiocarbon dating

TRIPP, RICHARD B.
Geologist
Box 461
Golden, Colorado 80401
Br. of Exploration Research
Denver **(303) 234-5536**
Iowa State U.; U. of New Mexico
Geochemistry of California; Alaska Mineral Resource Assessment Program
Mineralogy of heavy-mineral concentrates of stream sediments

TRIPPET, ANITA L. **William**
Geologist **(512) 643-7400**
1807 Memorial Parkway
Portland, Texas 78374
Br. of Atlantic-Gulf of Mex. Geology
Corpus Christi **(512) 888-3294**
Baylor U.-BA, MA
Environmental geology of Pleistocene trend, NW Gulf of Mexico
Marine geology; seismic stratigraphy; environmental geology

TROLLMAN, WINIFRED M.
Editorial Clerk
836 Garland Drive
Palo Alto, California 94303
Br. of Alaskan Geology
Menlo Park **(415) 323-8111 x2231**
Oregon State U.

TROMBLEY, ROSE M. **Albert**
Editorial Asst. (Typing) **(415) 368-7650**
371 Rutherford Avenue
Redwood City, California 94061
Br. of Paleontology & Stratigraphy
Menlo Park **(415) 323-8111 x2171**
San Jose State Coll.-AB

TRUESDELL, ALFRED H.
Chemist **(415) 328-8099**
122 Santa Rita, #1
Palo Alto, California 94301
Br. of Exper. Geochem. & Mineralogy
Menlo Park **(415) 323-8111 x2164**
Oberlin-BA; Harvard-MA, PhD
Geochemical indicators in hydrothermal systems
Chemistry of geothermal systems; physical chemistry of aqueous solutions

TRUMBULL, JAMES V.A.
Geologist
Br. of Atlantic-Gulf of Mex. Geology
San Juan **(809) 722-3142**
Puerto Rico marine geology co-op
Marine geology

TSCHANZ, CHARLES M. **Virginia**
Geologist **(303) 986-2130**
876 S. Moore Street
Lakewood, Colorado 80226
Br. of Central Mineral Resources
Denver **(303) 234-5107**
U. of Idaho-BS; Stanford-MS
Boulder Mountains, Idaho
Geological mapping; mineral resource appraisal; geochemistry of metallic deposits

TSCHUDY, ROBERT H. **Bernadine**
Botanist
Ward Star Route
Jamestown, Colorado 80455
Br. of Paleontology & Stratigraphy
Denver **(303) 234-5862**
U. of Washington-PhD
Cretaceous palynology; megaspores

TUCKER, JUDITH A. **Richard**
Editorial Assistant
8464 W. Arizona Drive
Lakewood, Colorado 80226
Br. of Special Projects
Denver **(303) 234-2371**

TUFTIN, STEVEN E.
PST
Br. of Special Projects
Denver **(303) 234-2365**
Winona State U.-BS
Nevada test site gravity survey
Gravity interpretation; invertebrate paleontology; stratigraphy

TURKINGTON, DANIEL E.
Cartographic Aid **(415) 261-5386**
3226 Star Avenue
Oakland, California 94619
Office of Scientific Publications
Menlo Park **(415) 323-8111**
Sacramento State U.-BA

TURNER, MONICA L.
Contract Monitoring Clerk **(415) 365-4463**
3431 Page Street
Redwood City, California 94063
Office of Earthquake Studies
Menlo Park **(415) 323-8111 x2764**
U. of Cal. (Santa Cruz)-BA
Earthquakes; environmental planning

TURNER, ROBERT L. **Aileen**
Chemist
Br. of Exploration Research
Denver **(303) 234-6187**
U. of New Mexico-BA

TURNER, ROBERT M. **Toby**
PST **(703) 430-3877**
403 W. Beech Road
Sterling Park, Virginia 22170
Office of Resource Analysis
Reston **(703) 860-6455**
George Washington U.-BS
Computer applications to geology
Computer applications; geomorphology; remote sensing

TURNER-PETERSON, CHRISTINE E. **Fred**
Geologist **(303) 697-8454**
P.O. Box 471
Morrison, Colorado 80465
Br. of Uranium-Thorium Resources
Denver **(303) 234-5665**
College of William & Mary-BA; Northern Arizona U.-MS
Sedimentology & uranium geology, San Juan Basin, New Mexico
Sedimentology; uranium geology

TUTTLE, MICHELE L.
Chemist
Br. of Regional Geochemistry
Denver **(303) 234-3453**
U. of Colorado-BS; Colorado Schl. of Mines-MS
Geochemistry of plant material
Geochemical baselines

TWENHOFEL, WILLIAM S. **Beth**
Geologist **(303) 237-0780**
820 Estes Street
Lakewood, Colorado 80215
Br. of Special Projects
Denver **(303) 234-2112**
U. of Wisconsin-BA, PhD
Underground nuclear explosions; disposal of radioactive waste

TWETO, OGDEN L. **Nikki**
Geologist **(303) 237-6548**
1995 Taft Drive
Lakewood, Colorado 80215
Br. of Central Mineral Resources
Denver **(303) 234-3368**
U. of Montana-BA, MA; U. of Michigan-PhD
Geologic maps of Colorado
Colorado geology and mineral resources

TWICHELL, DAVID C., JR. **Nancy**
Oceanographer **(617) 540-3550**
90 Comanche Drive
Falmouth, Massachusetts 02540
Br. of Atlantic-Gulf of Mex. Geology
Woods Hole **(617) 548-8700**
Middlebury Coll.-BA; U. of Rhode Island-BS
Sedimentology; sedimentary processes; high-resolution geophysics

TYNER, RICHARD L. **Rita**
Cartographer **(602) 779-1275**
305 W. Navajo Road
Flagstaff, Arizona 86001
Br. of Astrogeologic Studies
Flagstaff **(602) 779-3311 x1534**
Northern Arizona U.-BS
Planetary mapping
Photo-cartography; remote sensing; digital image processing

TYSDAL, RUSSELL G. **Connie**
Geologist
Golden, Colorado 80401
Br. of Central Mineral Resources
Denver **(303) 234-6883**
Montana State U.-BS, MS; U. of Montana-PhD
Madison Range wilderness, Montana
Stratigraphy; structure

ULIBARRI, LORETTA J.
Cartographic Technician
Office of Scientific Publications
Denver **(303) 234-3648**
Community Coll. of Denver; Metropolitan State Coll.

ULRICH, GEORGE E. **Sally**
Geologist **(602) 774-3992**
1100 W. Thelma Way
Flagstaff, Arizona 86001
Br. of Central Environmental Geology
Flagstaff **(602) 779-3311 x1554**
Brown U.-AB; U. of Colorado-PhD
West Clear Creek wilderness area
Volcanic & structural geology

UNRUH, DANIEL M. **Becky**
Chemist **(303) 986-3647**
12296 W. Mississippi Avenue
Lakewood, Colorado 80228
Br. of Isotope Geology
Denver **(303) 234-3876**
Colorado Schl. of Mines-BS
Pb isotopes; U-Pb geochronology

URICK, MARY ALICE **Frederick**
Computer Systems Analyist **(703) 476-5409**
10731 Cross School Road
Reston, Virginia 22091
Office of the Chief Geologist
Reston **(703) 860-6544**
Gannon U.-BA
Administrative computer systems

UTTER, CHRISTOPHER G. **Pamela**
Photographer **(408) 269-0452**
2205 Quinn Avenue
Santa Clara, California 95051
Office of the Chief Geologist
Menlo Park **(415) 323-2214**
San Francisco City Coll.-AA

UTTER, PAMELA A. **Christopher**
PST **(415) 296-0452**
2205 Quinn Avenue
Santa Clara, California 95051
Br. of Pacific-Arctic Geology
Menlo Park **(415) 856-7101**
San Jose State U.-BA

UTTERBACK, PRISCILLA G. **Bob**
Secretary
358 Fort Evans Drive, #2
Leesburg, Virginia 22075
Office of Scientific Publications
Reston **(703) 860-6575**

VALDOVINO, YOLANDA L. **Ernest**
Travel-Clerk
3343 North King Street
Flagstaff, Arizona 86001
Br. of Astrogeologic Studies
Flagstaff **(602) 779-3311 x1538**

VALENTINE, PAGE C. **Lanci**
Geologist
Br. of Paleontology & Stratigraphy
Woods Hole **(617) 548-8700**
George Washington U.-BS, MS; U. of Cal. (Davis)-PhD
Atlantic & Gulf continental margins; post-Paleozoic biostratigraphy & depositional environments; calcareous nannofossils & ostracodes

VALLIER, TRACY L. **Trudy**
Geologist **(408) 732-4291**
142 Carlisle Way
Sunnyvale, California 94087
Br. of Pacific-Arctic Geology
Menlo Park **(415) 856-7048**
Iowa State U.-BS; Oregon State U.-PhD
Geology of Aleutian Ridge & North Pacific
Marine geology; igneous petrology; sedimentology

VAN ALSTINE, RALPH E. **Nancy**
Geologist **(301) 530-7582**
8907 Ewing Drive
Bethesda, Maryland 20034
Br. of Eastern Mineral Resources
Reston **(703) 860-6531**
Hamilton Coll.-BA; Northwestern U.-MS; Princeton U.-PhD
Fluorine resources of the world
Economic geology; nonmetals; fluorspar

VAN DRIEL, J. NICHOLAS **Dee Anne**
Geologist **(703) 430-6155**
RFD 2, Box 297
Sterling, Virginia
Br. of Eastern Environmental Geology
Reston **(703) 860-6421**
Iowa State U.-BS, MS
Computer applications; environmental geology

VAN HORN, RICHARD **Alice**
Geologist **(303) 526-0056**
Route 5, Box 875
Golden, Colorado 80401
Br. of Engineering Geology
Denver **(303) 234-3385**
Colorado Schl. of Mines-GE
Geology of Salt Lake City, Utah
Engineering geology; environmental geology; geomorphology

VAN LOENEN, RICHARD E. **Sharon**
Geologist **(303) 697-4881**
9288 Dirt Road
Morrison, Colorado 80465
Br. of Central Mineral Resources
Denver **(303) 234-6295**
Kansas State U.-BS; Colorado Schl. of Mines
Gallatin & Madison Range wilderness area, Montana
Mineral deposits; x-ray mineralogy

VAN LOENEN, SHARON D. **Richard**
PST **(303) 697-4881**
9288 Old Dirt Road
Morrison, Colorado 80465
Br. of Paleontology & Stratigraphy
Denver **(303) 234-5851**
U. of Colorado-BA
Palynology

VAN SCHAACK, JOHN R. **Donna**
Geophysicist
10376 Tonita Way
Cupertino, California 95014
Br. of Network Operations
Menlo Park **(415) 323-8111 x2584**
San Jose State U.-BS
Field experiment operations
Seismic instrumentation design; VHF radio telemetry

VAN TRUMP, GEORGE JR.
Mathematician **(303) 985-3508**
P.O. Box 26523
Lakewood, Colorado 80226
Br. of Exploration Research
Denver **(303) 234-2758**
Missouri Jr. Coll.-AS; U. of Kansas; U. of Wichita
Computer applications to geoscience
Computer science

VARNES, DAVID J. **Katharine**
Geologist **(303) 237-2639**
40 South Holland Street
Lakewood, Colorado 80226
Br. of Engineering Geology
Denver **(303) 234-3346**
Cal Tech-BS; Northwestern U.
Regional landslide studies
Landslides; soil & rock mechanics; engineering geological mapping

VARNES, KATHARINE L. **David**
Geologist **(303) 237-2639**
40 South Holland Street
Lakewood, Colorado 80226
Br. of Oil & Gas Resources
Denver **(303) 234-5235**
Bryn Mawr Coll.-AB, MS
Domestic oil & gas resource appraisal
Oil & gas resource appraisal; map editing

VAUGHAN, PATRICK R.
PST **(415) 328-4825**
1001 Fulton Street
Palo Alto, California 94301
Br. of Ground Motion & Faulting
Menlo Park **(415) 323-8111 x2064**
U. of California (Santa Cruz)-BS
Quaternary tectonics of southern California
Desert geomorphology; aerial photography interpretation

VAUGHN, ALVIN J. **Barbara**
PST **(408) 257-9975**
10720 Morengo Drive
Cupertino, California 95014
Br. of Network Operations
Menlo Park **(415) 323-8111 x2481**
Land permit & telemetry specialist

VEDDER, JOHN G. **Diane**
Geologist
285 Golden Oak Drive
Portola Valley, California 94025
Br. of Pacific-Arctic Geology
Menlo Park **(415) 323-2214**
Occidental Coll.; Pomona Coll.-BA; Claremont Coll.-MA
Tertiary & Quaternary marine molluscan faunas & their geophysical significance

VENNUM, WALTER R. **Barbara**
Geologist **(707) 823-2259**
3925 Kim Court
Sebastopol, California 95472
Br. of Central Environmental Geology
Denver **(303) 234-3624**
U. of Montana-BA; Stanford U.-PhD
Geologic mapping Orville Coast, Antarctica
Mineralogy; igneous & metamorphic petrology

VERBEEK, EARL R.
Geologist **(303) 278-1574**
63 S. Devinney Street
Golden, Colorado 80401
Br. of Central Environmental Geology
Denver **(303) 234-5259**
Penn State U.-BS, PhD
Mechanics of jointing, Piceance Creek basin
Structure of deformed rocks; mechanics of joint formation; fluorescence spectroscopy of minerals

VERCOUTERE, THOMAS L.
Geologist
837 Cambridge Avenue
Menlo Park, California 94025
Br. of Sedimentary Mineral Resources
Menlo Park **(415) 323-8111 x2306**
U. of Cal. (Santa Cruz)-BA; Cal State; San Jose
California phosphate
Stratigraphy; sedimentation

VIETOR, GLADYS L. **LeRoy**
Administrative Technician **(303) 238-9647**
10585 W. 9th Place
Lakewood, Colorado 80215
Br. of Paleontology & Stratigraphy
Denver **(303) 234-2741**

VIETS, JOHN G. **Marcia**
Chemist **(303) 278-3422**
13700 Berry Road
Golden, Colorado 80401
Br. of Exploration Research
Denver **(303) 234-6147**
Colorado State U.-BS
Geochemical analysis

VINE, JAMES D. **Blanche**
Geologist **(303) 526-9198**
21736 Panorama Drive
Golden, Colorado 80401
Br. of Sedimentary Mineral Resour.
Denver **(303) 234-3494**
U. of Michigan-BS; Harvard U.
Lithium resources
Mineral deposits & geochemistry of sedimentary rocks

VITALIANO, DOROTHY B. **Charles**
Geologist **(812) 332-2545**
1114 Brooks Drive
Bloomington, Indiana 47401
Library
Bloomington **(812) 337-7597**
Barnard Coll.-BA; Columbia U.-MA, Mph.
Translations, mainly from Russian & German
Archeological geology; geomythology; tephrochronology

VON ESSEN, JAREL C. **Mary Ellen**
Geologist **(415) 494-8490**
795 Cereza Drive
Palo Alto, California 94306
Br. of Isotope Geology
Menlo Park **(415) 323-8111 x2455**
San Jose State U.-BA

VULETICH, APRIL K.
PST
Br. of Isotope Geology
Denver **(303) 234-3876**
Metropolitan State Coll.-BS
Mass spectrometry

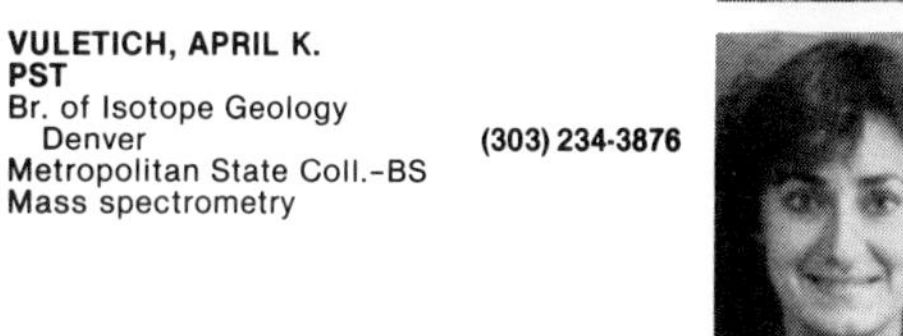

WAGNER, HOLLY C. Leslie
Geologist (415) 967-4470
1135 Lisa Lane
Los Altos, California 94022
Br. of Pacific-Arctic Geology
Menlo Park (415) 856-7013
UCLA-BA, MA; U. of Kansas
Geologic hazards, northern Puget Sound offshore
Marine geology; structural geology; stratigraphy

WAHLBERG, JAMES S. Delilah
Chemist (303) 567-2353
Box 87
Idaho Springs, Colorado 80452
Br. of Analytical Laboratories
Denver (303) 234-2521
U. of Colorado-BA
X-ray spectroscopy
Development of x-ray methods

WAITT, RICHARD B., JR. Cynthia
Geologist
Br. of Western Environmental Geology
Menlo Park (415) 323-2902
U. of Texas-BS, MA; U. of Washington-PhD
Surficial geology; Wenatchee 1x2° sheet, Washington
Glacial geology; surficial processes; Quaternary stratigraphy

WALCOTT, CHARLES D. Mary
Geologist-Paleontologist
Office of the Director
Reston (703) 860-6531
Utica Academy
Biostratigraphy of North American Cambrian; trilobites, Burgess Shale
Scientist, administrator; presidential advisor; traveler

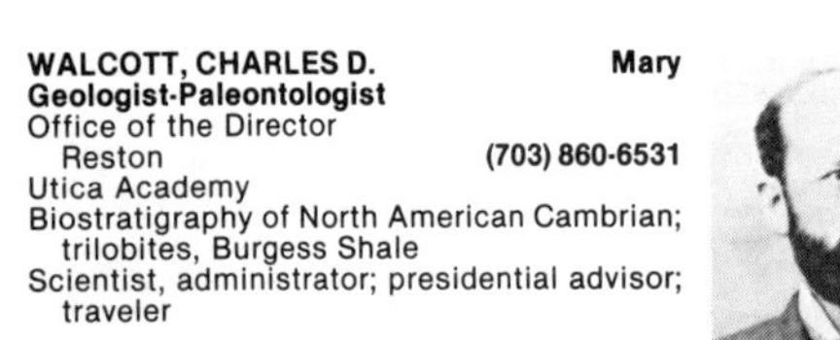

WALKER, GEORGE W. Barbara
Geologist (415) 948-5200
190 Osage Avenue
Los Altos, California 94022
Br. of Western Mineral Resources
Menlo Park (415) 323-8111 x2285
Stanford-AB, MS
Mineral resources of Oregon
Volcanic & economic geology; petrology

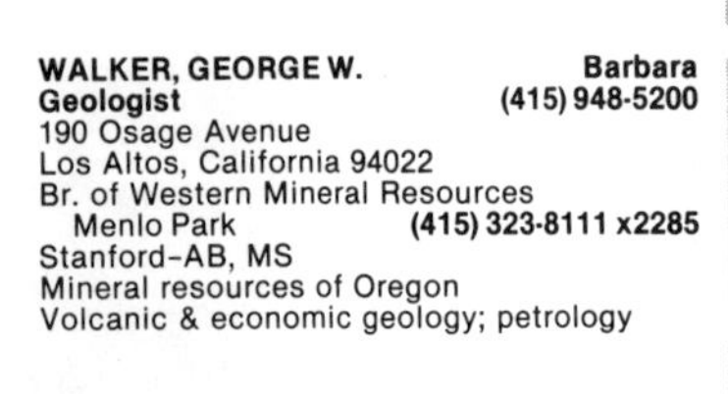

WALKER, WILLIAM M. Malissia
PST (703) 437-8325
300 East Tazewell Road
Sterling Park, Virginia 22170
Br. of Atlantic-Gulf of Mex. Geology
Reston (703) 860-7164
West Virginia State Coll.-BS; American U.; Howard U.
Marine organic geochemistry

WALLACE, CHESTER A.
Geologist
1023 S. Beech Drive
Lakewood, Colorado 80228
Br. of Central Environmental Geology
Denver (303) 234-5257
Antioch Coll.-BA; U. of Cal. (Santa Barbara)-PhD
Butte 2° quadrangle, Montana
Stratigraphy; sedimentology; structure

WALLACE, LAURE G. Don
Geologist (703) 327-4623
P.O. Box 201
Aldie, Virginia 22001
Br. of Oil & Gas Resources
Reston (703) 860-6595
U. of Maryland-BS
Appalachian basin Devonian shales program
Stratigraphy

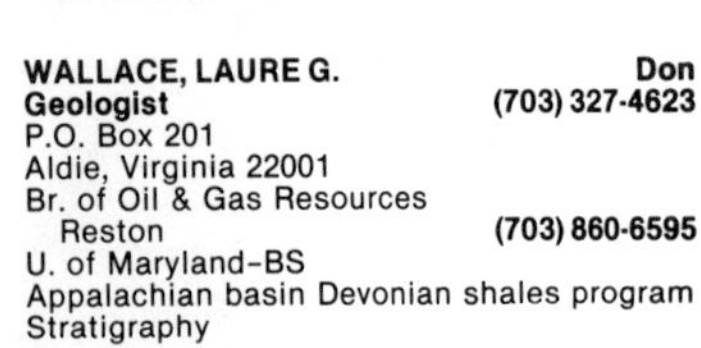

WALLACE, ROBERT E. Trudy
Geologist
240 Cervantes Road
Portola Valley, California 94025
Office of Earthquake Studies
Menlo Park (415) 323-8111 x2751
Northwestern U.-BS; Cal Tech-MS, PhD
Tectonics of active faults
Earthquake geology; geomorphology; mineral resources

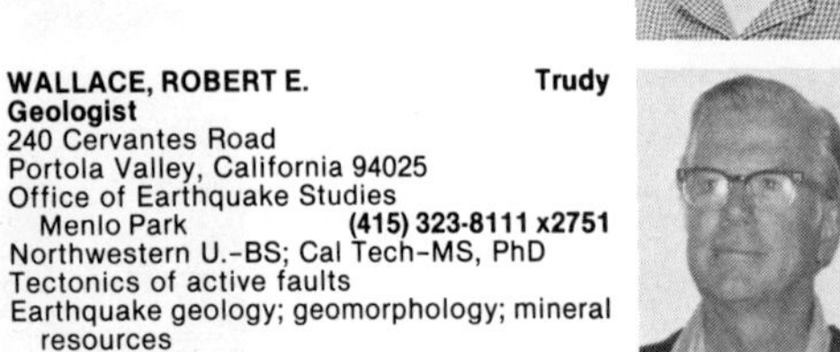

WALTHALL, FRANK G. Peggy
Chemist (703) 759-3036
9435 Brian Jac Lane
Great Falls, Virginia 22066
Br. of Analytical Laboratories
Reston (703) 860-7248
Roanoke Coll.-BS; U. of North Carolina
Geochemical analysis
Emission spectroscopy; wavelength deter ation

WALTHER, LINDA M.
Accounting Technician
Office of International Geology
Reston (703) 860-6522
Virginia Polytech. Inst.-BS

WALTON, RONALD J. Judith
Computer Systems Admin. (303) 674-1314
24767 Giant Gulch Road
Evergreen, Colorado 80439
Office of the Chief Geologist
Denver (303) 234-6488
Indiana U.-BS; American U.-MS
ADP coordinator
Geologic & geophysical computer applications; marine geology

WANDLESS, GREGORY A. Mary-Virginia
Chemist (703) 476-5719
12056 Greywing Square
Reston, Virginia 22091
Br. of Analytical Laboratories
Reston (703) 860-6852
George Mason U.-BS
Neutron activation analysis
Trace element analysis; radiochemical analysis

WANDLESS, MARY-VIRGINIA Gregory
Chemist
12056 Greywing Square
Reston, Virginia 22091
Br. of Exper. Geochem. & Mineralogy
Reston (703) 860-7780
George Mason U.-BA
Transmission electron microscopy
Lunar highland rocks; pyroxenes; sample preparation

WANG, FRANK H. Sandra
Geologist
758 Mayview Avenue
Palo Alto, California 94303
Br. of Pacific-Arctic Geology
Menlo Park (415) 323-2214
National Southwestern U. (China)-BS; U. of Washington-PhD
Circum-Pacific mapping project

WARD, A. WESLEY
GEOLOGIST (602) 526-5188
2515 N. Oakmont Drive
Flagstaff, Arizona 86001
Br. of Central Environmental Geology
Flagstaff (602) 779-3311 x1302
Washington State U.-BS, MS; U. of Washington-PhD
Arizona desert winds; Martian surficial geology
Geomorphology; petrology

WARD, DWIGHT E.
Geologist (303) 444-2871
1700 Balsam Avenue
Boulder, Colorado 80302
Br. of Coal Resources
Denver (303) 234-3689
U. of Colorado-BA, PhD
Coal geology; stratigraphy

WARD, KAREN M. Peter
Geophysicist
Office of Earthquake Studies
Menlo Park (415) 323-8111 x2823
CUNY-BS; Cal. State U.-MPA
Seismicity of Imperial Valley; dam-induced seismicity

WARD, LAUCK W. Theresa
Geologist (703) 476-4124
12835 Framingham Court
Herndon, Virginia 22070
Br. of Paleontology & Stratigraphy
Reston (703) 860-7745
Frederick Coll.-BS; U. of South Carolina-MS, PhD
Tertiary molluscan paleontology; Atlantic & Gulf Coastal Plain biostratigraphy

WARD, PETER L. Karen
Geophysicist (415) 322-3183
380 Nova Lane
Menlo Park, California 94025
Br. of Seismology
Menlo Park (415) 323-8111 x2838
Dartmouth-BA; Columbia-MA, PhD
Developing computer methods for handling earthquake data and research
Seismology; volcanology; computer science

WARDLAW, BRUCE R. Jeanne
Geologist (303) 935-2800
1415 S. Eaton Court
Lakewood, Colorado 80226
Br. of Paleontology & Stratigraphy
Denver (303) 234-5932
U. of Cal. (Riverside)-BA; Case Western Reserve U.-PhD
Biostratigraphy & organic metamorphism of the overthrust belt
Brachiopods, conodonts; stratigraphy

WARLOW, RALPH C.
Geologist (703) 860-1269
12601 Pinecrest Road
Herndon, Virginia 22070
Br. of Coal Resources
Reston (703) 860-6684
Southern Methodist U.-BS; George Washington U.
Coal resources, Wind River Basin, Wyoming
Stratigraphy; structure; energy resource evaluation

WARREN, CHARLES R. Joyce
Geologist (202) FE3-1952
3606 Whitehaven Parkway NW
Washington, D.C. 20007
Br. of Eastern Environmental Geology
Reston (703) 860-6595
Yale-BS, PhD
Massachusetts Cooperative
Glacial geology; geomorphology

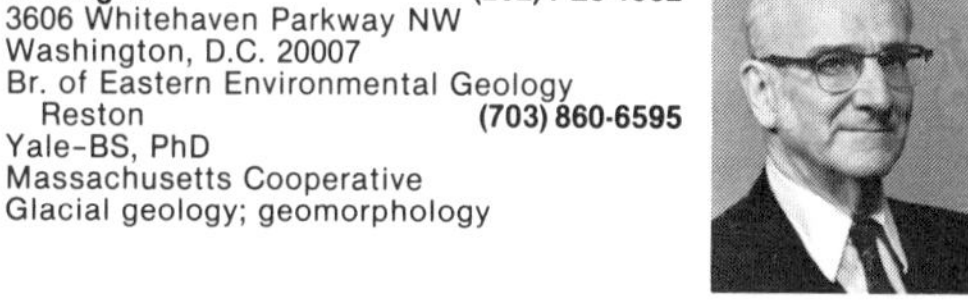

WARREN, DAVID H.
Geophysicist (415) 493-4108
275 Ventura Avenue, Apt. 12
Palo Alto, California 94306
Br. of Seismology
Menlo Park (415) 323-8111 x2531
Renssalaer Polytech. Inst.-BS; Columbia-MA
Aftershocks of Willits, Cal., earthquake
Seismic crustal structure; studies of small earthquakes

WARRICK, RICHARD E. Gudrun
Geophysicist (415) 326-5967
1808 Mark Twain
Palo Alto, California 94303
Br. of Ground Motion & Faulting
Menlo Park (415) 323-8111 x2757
Stanford-BS
Ground motion instrumentation
Shear wave generation & measurement

WARSHAW, CHARLOTTE M. Israel
Geologist (301) 652-4173
3703 Stewart Driveway
Chevy Chase, Maryland 20015
Br. of Field Geochemistry & Petrology
Reston (703) 860-7451
Smith Coll.-BA; Bryn Mawr-MA; Penn State-PhD
Petrology of Jemez Mountains, N.M.
Igneous petrology; clay mineralogy

WATKINS, CAROLINE A.
Geologist (303) 238-2860
35 South Carr Street
Lakewood, Colorado 80226
Office of International Geology
Denver (303) 234-3708
U. of Rochester-BA
International geology, participant programs

WATSON, DONALD E. Paula
Geologist (303) 238-7975
1850 Alkire Court
Golden, Colorado 80401
Office of the Chief Geologist
Denver (303) 234-2910
U. of Pittsburgh-BS
Magnetics, paleomagnetism

WATSON, KENNETH Norma
Geophysicist (303) 986-2185
2054 S. Moore Court
Lakewood, Colorado 80227
Br. of Petrophysics & Remote Sens.
Denver (303) 234-2349
U. of Toronto-BA; Cal Tech-MS, PhD
Thermal infrared geophysics
Remote sensing

WATT, ARTHUR D.
PST
Br. of Paleontology & Stratigraphy
Washington, D.C. (202) 343-2490
Bethel (Kansas); Texas Coll. of Mines-BS; U. of Kansas
Paleobotany collections & library
Bibliography of American paleobotany; index of generic names of fossil plants

WATTERSON, JOHN R. Lynne
Chemist (303) 979-3260
9300 Coal Mine Road
Littleton, Colorado 80123
Br. of Exploration Research
Denver (303) 234-3237
Arizona State U.-BA
Biogeochemical research
Biogeochemistry; geomicrobiology; selenium-tellurium commodity

WATTS, RAYMOND D.
Geophysicist (303) 447-1136
505 Aurora Avenue
Boulder, Colorado 80302
Br. of Electromag. & Geomagnetism
Denver (303) 234-3493
Pomona Coll.-BA; MIT; U. of Toronto-PhD
Radioactive waste disposal
Radio/radar methods; electromagnetic theory

WEANT, SHARI L. Vic
Administrative Assistant
Office of Environmental Geology
Reston (703) 860-6417

WEATHERHEAD, SHARON K. Chuck
Budget & Fiscal Asst. (703) 860-4291
11222 Lagoon Lane
Reston, Virginia 22091
Office of the Chief Geologist
Reston (703) 860-6537
Madison Coll.

WEAVER, JEAN NOE
PST (303) 237-7440
815 Lee Street
Lakewood, Colorado 80215
Br. of Coal Resources
Denver (303) 234-3560
Vassar Coll.-BA

WEBER, FLORENCE R. Albert
Geologist (907) 479-6446
P.O. Box 80745
Fairbanks, Alaska 99708
Br. of Alaskan Geology
Fairbanks (907) 479-7245
U. of Chicago-MS
Geology, Yukon-Tanana Upland, Alaska
Structure; stratigraphy; glacial geology

WEBRING, MICHAEL W.
PST
P.O. Box 382
Golden, Colorado 80401
Br. of Regional Geophysics
Denver **(303) 234-5473**
Colorado Schl. of Mines-BS
Development programming in gravity & magnetics
Computer algorithms

WEDDLE, MARY I.
Secretary
2193 Allison Street
Lakewood, Colorado 80215
Office of Earthquake Studies
Denver **(303) 234-4029**
Red Rocks College; Millikin U.

WEED, ELAINE G.A. **Charles**
Geologist **(703) 356-9339**
7121 Thrasher Road
McLean, Virginia 22101
Br. of Oil & Gas Resources
Reston **(703) 860-7538**
Mount Holyoke Coll.-BA
Eastern wilderness resources
Atlantic margin geology; environmental & marine geology

WEEDMAN, ROSALIE E. **Caroll**
Editorial Assistant **(303) 985-7173**
13792 West Dakota Avenue
Lakewood, Colorado 80228
Br. of Engineering Geology
Denver **(303) 234-3819**

WEEMS, ROBERT E.
Geologist **(703) 860-2136**
12-C, 2319 Freetown Court
Reston, Virginia 22091
Br. of Eastern Environmental Geology
Reston **(703) 860-6503**
Randolph Macon-BS; VPI-MS; George Washington U.-PhD
Charleston earthquake project
Stratigraphy; vertebrate paleontology

WEGENER, STEVEN S. **Laurel**
Physical Scientist **(415) 854-5887**
2024 Santa Cruz Avenue
Menlo Park, California 94025
Br. of Tectonophysics
Menlo Park **(415) 323-8111 x2930**
U. of Cal. (Davis)-BA
Fault zone properties

WEHRLE, FREDERIC P., JR.
PST **(303) 757-3746**
3830 South Narcissus Way
Denver, Colorado 80237
Br. of Exploration Research
Denver **(303) 234-6165**
Colorado Coll.-BA
Electronics; computer sciences

WEIDE, DAVID L.
Geologist **(702) 646-1535**
1300 Shadow Mountain Place
Las Vegas, Nevada 89108
Br. of Central Environmental Geology
Denver **(303) 234-2650**
UCLA-PhD
Surface mining reclamation, San Juan Basin, New Mexico
Geomorphology; Quaternary geology

WEIR, DORIS B. **Gordon**
Geologist **(602) 526-3207**
2016 Fox Hill Road
Flagstaff, Arizona 86001
Br. of Astrogeologic Studies
Flagstaff **(602) 261-1455**
Bryn Mawr Coll.-AB
Map & manuscript editing

WEIR, GORDON W. **Doris**
Geologist **(602) 526-3207**
2016 Fox Hill Road
Flagstaff, Arizona 86001
Br. of Central Environmental Geology
Flagstaff **(602) 261-1455**
UCLA-BA
Holbrook, Arizona 1° x 2° quadrangle
Stratigraphy; geologic mapping

WEIS, PAUL L. **Alice**
Geologist **(509) 448-0828**
S. 5106 Sunward Drive
Spokane, Washington 99203
Br. of Central Mineral Resources
Spokane **(509) 456-4677**
U. of Wisconsin-BS, PhD
Challis 2° sheet & Garden Valley-Boiling Springs quadrangles
Economic geology; graphite; regional geology of northwest U.S.

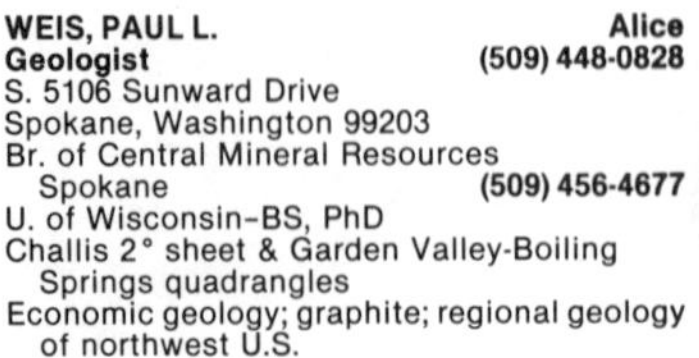

WEISS, MALCOLM P.
Geologist **(815) 753-1944**
127 Ilehamwood Drive
DeKalb, Illinois 60115
Br. of Central Environmental Geology
Denver **(303) 234-3624**
U. of Minnesota-BBA, MS, PhD
Price, Utah 2° sheet
Stratigraphy; sedimentary petrology

WEISSENBOM, ALBERT E. **Frances**
Geologist **(509) 624-6261**
W519 24th Avenue
Spokane, Washington 99203
Office of Mineral Resources
Spokane **(509) 456-4677**
Lehigh U.-EM; Washington State U.; E. Wash. State U.
Economic geology; mineral resources

WELD, BETSY A.
Geologic Reports Specialist
510 21st Street, N.W.
Washington, D.C. 20006
Office of Scientific Publications
Reston **(703) 860-6579**
U. of Washington-BA

WELLS, JOHN D. **Ruth**
Geologist **(303) 232-0475**
12635 W. 15th Place
Lakewood, Colorado 80215
Br. of Central Mineral Resources
Denver **(303) 234-6724**
Kansas State U.-BS, MS; U. of Colorado
Nonfuel minerals of Montana
Geology and geochemistry of mineral deposits

WELLS, RAY E.
Geologist
Br. of Western Environmental Geology
Menlo Park **(415) 323-8111 x2223**
Penn State-BS; U. of Oregon-MS; U. of California (Santa Cruz)
Geology & tectonics of S.W. Washington Coast Range
Cenozoic geology, Pacific NW; paleomagnetism & plate tectonics; igneous petrology

WENRICH-VERBEEK, KAREN J.
Geologist **(303) 278-1574**
63 S. Devinney Street
Golden, Colorado 80401
Br. of Uranium-Thorium Resources
Denver **(303) 234-5552**
Penn State U.-BS, MS, PhD
Exploration guides for uranium in silicic volcanic rocks
Volcanology; geochemical exploration for uranium

WENTWORTH, CARL M. **Carol**
Geologist **(415) 854-6750**
280 Dedalera Drive
Menlo Park, California 94025
Br. of Western Environmental Geology
Menlo Park **(415) 323-2474**
Dartmouth Coll.-BA; Stanford-PhD
U.S. tectonics & earthquakes
Tectonics; environmental geology; sedimentology

WERRE, RAYMOND W. **Charlotte**
Engineering Technician **(703) 471-7215**
1308 Dulles Place
Herndon, Virginia 22070
Br. of Analytical Laboratories
Reston **(703) 860-7443**
Instrument development

WESLEY, JANNETTE SHARP
Technical Information Specialist
Br. of Oil & Gas Resources
Denver **(303) 234-5235**
U. of Colorado-BA; U. of Oklahoma-MLS

WESSON, ROBERT L. **Cornelia**
Geophysicist **(703) 860-9608**
2402 Bugle Lane
Reston, Virginia 22091
Office of Earthquake Studies
Reston **(703) 860-7488**
MIT-SB; Stanford-MS, PhD
Research administration
Earthquake prediction; seismology; tectonophysics

WESTERLUND, ROBERT E.
PST
937 5th Avenue
Redwood City, California 94063
Br. of Tectonophysics
Menlo Park **(415) 323-8111 x2937**
California State U.-BA
Rock physics laboratory; air gun experiments

WETLAUFER, PAMELA H. **Gerald**
Geologist
Br. of Exper. Geochem. & Mineralogy
Reston **(703) 860-6795**
Vassar-BA; George Washington U.-MS
Environmental of ore deposition
Economic geology; remote sensing; geothermal studies

WEVER, THERESA L.
Administrative Assistant
276 Manassas Drive
Manassas, Virginia 22110
Office of Mineral Resources
Reston **(703) 860-6571**

WHEELER, RUSSELL L. **Peggy A. Lentz**
Geologist
4223 King Street
Denver, Colorado 80211
Br. of Earthquake Tectonics & Risk
Denver **(303) 234-5087**
Yale-BS; Princeton-PhD
Seismogenic structures in southeastern U.S.
Structural geology; applied statistics; seismic hazards

WHELAN, JOSEPH F.
Geologist
10081 W. 9th Drive
Lakewood, Colorado 80215
Br. of Isotope Geology
Denver **(303) 234-3876**
Ball State-BS; Penn State-MS
Isotope geochemistry of sediments & sediment-hosted ore deposits

WHIPPLE, JAMES W. **Susan**
Geologist
Br. of Central Mineral Resources
Denver **(303) 234-6295**
Western Washington U.-BA; U. of Colorado-MS
Sedimentology of Precambrian rocks in Glacier National Park
Sedimentology & genesis of Precambrian stratabound ore deposits

WHITCOMB, HARRY S. **Nell**
Geophysicist
315 Oneida Street
Boulder, Colorado 80303
Br. of Global Seismology
Denver **(303) 234-5083**
Montana School of Mines-BS
Seismic Observatories; branch projects management

WHITE, DONALD E. **Helen**
Geologist **(415) 325-3966**
222 Blackbury Avenue
Menlo Park, California 94025
Br. of Field Geochemistry & Petrology
Menlo Park **(415) 323-8111 x2367**
Stanford U.-BA; Princeton-PhD
Thermal waters
Active geothermal systems; hydrothermal ore deposits

WHITE, DOROTHY C.
Librarian
Library
Reston **(703) 860-6613**
South Carolina State Coll.-BS; U. of Denver-MLS

WHITE, ELLEN R.
Earth Science Information Technician
Br. of Alaskan Geology
Menlo Park **(415) 323-8111 x2342**
U. of Colorado-BA; U. of Denver-MLS
Province technical information on earth science for Alaska

WHITEBREAD, DONALD H. **Virginia**
Geologist **(408) 252-2741**
1077 Arlington Lane
San Jose, California 95129
Br. of Western Mineral Resources
Menlo Park **(415) 323-8111 x2356**
U. of Colorado-BA, MS
Massive sulfides of Nevada
Structural geology; metallic mineral deposits

WHITLOW, AGNES W. **Jesse**
Administrative Officer
1200 E. Lee Road
Laurel, Maryland 22092
Br. of Eastern Environmental Geology
Reston **(703) 860-6403**

WHITLOW, JESSE W. **Agnes**
Geologist **(703) 437-4061**
1200 E. Lee Road
Sterling, Virginia 22170
Br. of Eastern Mineral Resources
Reston **(703) 860-7356**
Virginia Polytechnic Inst.-BS; Cornell U.
Charlotte 2° quadrangle, N.C.-S.C.
Element distribution as determined by geochemical methods

WHITMORE, FRANK C., JR. **Martha**
Geologist **(301) 593-6618**
20 Woodmoor Drive
Silver Spring, Maryland 20901
Br. of Paleontology & Stratigraphy
Washington, D.C. **(202) 343-2333**
Amherst Coll.-BA; Penn State-MS; Harvard-PhD
Neogene vertebrate biostratigraphy, Atlantic & Gulf Coastal Plain
Neogene cetacea; tertiary stratigraphy

WHITNEY, C. GENE **Sue**
Geologist
Br. of Sedimentary Mineral Resour.
Denver **(303) 234-3624**
Western Washington U.-BS, MS; U. of Illinois-PhD
Bentonites
Petrology of clay minerals

WHITTINGTON, CHARLES L. — Alice
Geologist (303) 744-0687
1080 South Fillmore Way
Denver, Colorado 80209
Br. of Exploration Research
Denver (303) 234-6152
DePauw U.-AB; Johns Hopkins
Geochemical exploration of Medford (Oregon) 2° quadrangle
Geochemical prospecting

WIECZOREK, GERALD F. — Gayle
Civil Engineer (415) 366-2619
452 6th Avenue
Menlo Park, California 94025
Br. of Engineering Geology
Menlo Park (415) 856-7113
U. of Cal. (Berkeley)-PhD
Earthquake, induced landslides
Geological engineering; soil mechanics

WIEGAND, GAIL P.
Secretary (301) 568-4826
3807 Swann Road, Apt. 102
Suitland, Maryland 20023
Office of the Chief Geologist
Reston (703) 860-6631

WIER, KAREN E.
Geologist (703) 361-1297
11591 Purse Drive
Manassas, Virginia 22110
Br. of Eastern Environmental Geology
Reston (703) 860-6503
U. of Washington-BS; Bryn Mawr Coll.-PhD
Metamorphic map, Iron River, Mich. 2°
Petrology of metamorphic rocks

WIER, KENNETH L. — Elizabeth
Geologist (303) 279-6415
14100 Crabapple Road
Golden, Colorado 80401
Br. of Central Mineral Resources
Denver (303) 234-3830
U. of Alaska-BS
Sedimentary iron deposits; Precambrian geology

WIGGINS, LOVELL B.
Geologist
Br. of Exper. Geochem. & Mineralogy
Reston (703) 860-7619
U. of Virginia-BA; Virginia Polytech. Inst.-MS
Microprobe
Phase equilibria studies

WILBER, PATRICIA L.
Cartographic Technician
Office of Scientific Publications
Denver (303) 234-3624

WILCOX, RAY E. — Mary
Geologist
Br. of Field Geochemistry & Petrology
Denver (303) 234-2836
U. of Wisconsin-PhB, PhM, PhD
Volcanic ashbeds correlation
Igneous petrology; petrographic methods

WILDE, AMANDA M.
Administrative Technician (512) 855-1124
921 Miramar Place
Corpus Christi, Texas 78411
Br. of Atlantic-Gulf of Mex. Geology
Corpus Christi (512) 888-3294

WILEY, M. CLAIRE
PST
2702 Ramos Court
Mountain View, California 94040
Br. of Pacific-Arctic Geology
Menlo Park (415) 856-7013
San Jose State U.-BS
Geologic hazards; offshore northern Puget Sound, Strait of Juan de Fuca
Marine geology

WILKINS, ELEANORE E.
Librarian (415) 322-8604
1347 American Way
Menlo Park, California 94025
Library
Menlo Park (415) 323-8111 x2364
Washington U. (St. Louis)-BA; Carnegie Inst. Tech.-BSLS

WILLIAMS, DAVID L.
Geophysicist (303) 277-1696
3935 Douglas Mtn. Drive
Golden, Colorado 80401
Br. of Regional Geophysics
Denver (303) 234-5160
U. of Texas-BA; MIT-PhD
Geophysics of young volcanic systems
Heat flow; hydrothermal circulation; aeromagnetic interpretation

WILLIAMS, EDNA L. — Joe
Computer Technician
1717 Waldron Road
Corpus Christi, Texas 78420
Br. of Atlantic-Gulf of Mex. Geology
Corpus Christi (512) 888-3294

WILLIAMS, JACKIE M.
PST
Br. of Regional Geophysics
Denver (303) 234-5508
Langston U.-BS; U. of Colorado
East coast magnetics

WILLIAMS, JOAN B. — Richard
Clerk-Typist (617) 428-2975
80 Lakeshore Drive
Marstons Mills, Massachusetts 02648
Br. of Atlantic-Gulf of Mex. Geology
Woods Hole (617) 548-8700 x182

WILLIAMS, JOHN R.
Geologist
2905 Bryant Street
Palo Alto, California 94306
Br. of Alaskan Geology
Menlo Park (415) 323-8111 x2924
Harvard-AB, AM
Northwest Alaskan gas pipeline project; Copper River basin, AK
Glacial geomorphology; ground water; engineering geology

WILLIAMS, KATHLEEN M.
Geologist
Br. of Western Environmental Geology
Menlo Park (415) 323-8111 x2066
San Jose State U.-BS
Western Transverse Ranges, California
Petrographic analysis; earthquake & fault correlation

WILLIAMS, PAUL L. — Jean
Geologist (303) 526-9213
Route 5, Box 737-A
Golden, Colorado 80401
Br. of Central Environmental Geology
Denver (303) 234-3624
U. of Washington (Seattle)-BS, MS, PhD
Geothermal resources, Snake River Plain
Geology related to geothermal areas; volcanic stratigraphy; geology of Antarctica

WILLIAMS, VAN S. — **Lhakpa Sherpa**
Geologist — **(303) 989-3461**
483 S. Hoyt Street
Lakewood, Colorado 80226
Br. of Central Environmental Geology
Denver — **(303) 234-5169**
Colorado Schl. of Mines-GE; U. of Washington-MS, PhD
Surficial geologic mapping of coal lands in southern Utah
Quaternary geology; engineering geology

WILLINGHAM, BETTY L.
PST
2858½ Topeka
Corpus Christi, Texas 78404
Br. of Atlantic-Gulf of Mex. Geology
Corpus Christi — **(512) 888-3241**
Corpus Christi State U.-BS
Sedimentology laboratory
Sedimentological methods; clay mineralogy

WILLINGHAM, HELEN P. — **Bill**
Administrative Officer — **(703) 430-2581**
1053 South Ironwood Road
Sterling, Virginia 22092
Office of International Geology
Reston — **(703) 860-6348**

WILLINGHAM, JAMES W.
PST — **(512) 853-6673**
1516 Lazy Lane
Corpus Christi, Texas 78415
Br. of Atlantic-Gulf of Mex. Geology
Corpus Christi — **(512) 888-3297**
Corpus Christi State U.-BS
Marine electronics; chemical analysis

WILLIS, CAROLYN A. — **James**
Secretary — **(703) 281-1544**
2765 Hill Road
Vienna, Virginia 22180
Office of Marine Geology
Reston — **(703) 860-7241**

WILSHIRE, HOWARD G.
Geologist
Br. of Western Environmental Geology
Menlo Park — **(415) 323-8111 x2091**
U. of Oklahoma-BA; U. of Cal. (Berkeley)-PhD
Surface processes in arid lands
Environmental geology; upper mantle processes

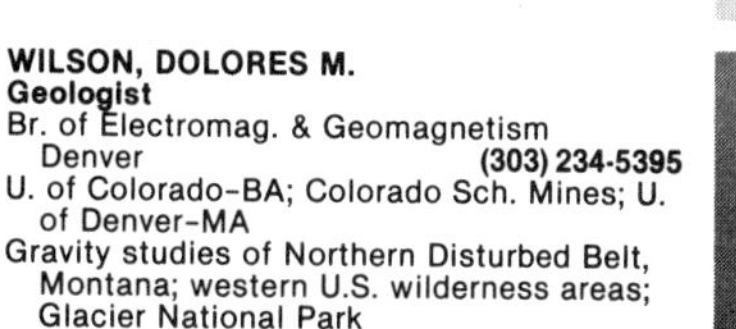

WILSON, DOLORES M.
Geologist
Br. of Electromag. & Geomagnetism
Denver — **(303) 234-5395**
U. of Colorado-BA; Colorado Sch. Mines; U. of Denver-MA
Gravity studies of Northern Disturbed Belt, Montana; western U.S. wilderness areas; Glacier National Park

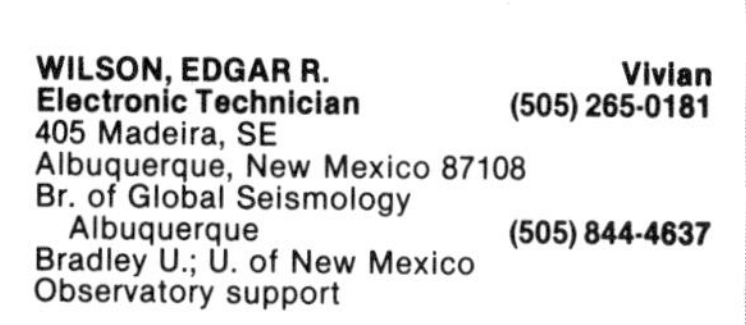

WILSON, EDGAR R. — **Vivian**
Electronic Technician — **(505) 265-0181**
405 Madeira, SE
Albuquerque, New Mexico 87108
Br. of Global Seismology
Albuquerque — **(505) 844-4637**
Bradley U.; U. of New Mexico
Observatory support

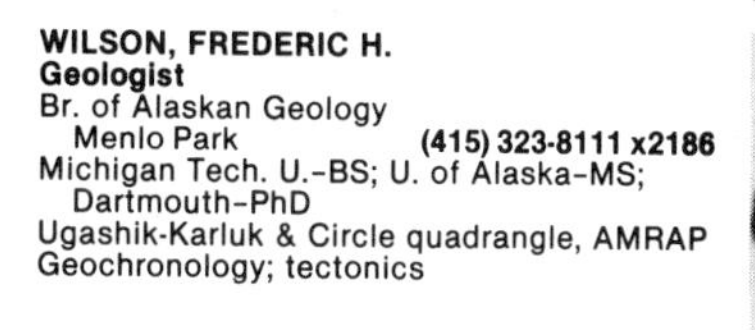

WILSON, FREDERIC H.
Geologist
Br. of Alaskan Geology
Menlo Park — **(415) 323-8111 x2186**
Michigan Tech. U.-BS; U. of Alaska-MS; Dartmouth-PhD
Ugashik-Karluk & Circle quadrangle, AMRAP
Geochronology; tectonics

WILSON, FREDERICK A.
Geologist
Oxon Hill, Maryland 20021
Br. of Eastern Environmental Geology
Reston — **(703) 860-6503**
Brooklyn Coll. (CUNY)-BS, MA; George Washington U.
Geophysics & geology, Charlotte 2° sheet
Geophysics & geology; vulcanology; geothermal studies

WILSON, LANNY R. — **Bonnie**
Geophysicist — **(303) 499-1807**
420 Oneida Street
Boulder, Colorado 80303
Br. of Electromag. & Geomagnetism
Denver — **(303) 234-5178**
Kansas State U.-BA
Data processing; magnetic observatories

WILSON, LEONARD A. — **Joanne**
Geologist — **(303) 985-4015**
12021 W. Center Place
Lakewood, Colorado 80228
Br. of Paleontology & Stratigraphy
Denver — **(303) 234-5852**
Arizona State U.-BS

WILSON, RAYMOND C. — **Marcia**
Geologist
Br. of Engineering Geology
Menlo Park — **(415) 856-7126**
Rice U.-BA, MA; Texas A&M-PhD
Seismic-induced ground failures
Engineering geology; geomechanics; numerical modeling

WILSON, STEPHEN A.
Chemist
1917 S. Shields, B-4
Fort Collins, Colorado 80526
Br. of Analytical Laboratories
Denver — **(303) 234-6404**
U. of New Hampshire-PhD
Ion chromatography
Metal speciation, humic acids

WILT, LINDA S.
Administrative Clerk
150 Laviel Way Drive, Apt. 1A
Herndon, Virginia 22070
Office of Geochemistry & Geophysics
Reston — **(703) 860-6585**

WINDOLPH, JOHN F., JR. — **Catherine**
Geologist — **(301) 926-6293**
16901 White Grounds Road
Boyds, Maryland 20720
Br. of Coal Resources
Reston — **(703) 860-7734**
U. of Notre Dame-BS; U. of Maryland; George Washington U.; MIT
Coal resources of the Wind River Indian Reservation
LANDSAT; stratigraphy; geotectonics

WINGET, ELIZABETH A.
Editor
Br. of Atlantic-Gulf of Mex. Geology
Woods Hole — **(617) 548-8700**
Southeastern Mass. U.-BA

WINHAM, DENA J.
Secretary
6006 Norvel Drive
Corpus Christi, Texas 78412
Br. of Atlantic-Gulf of Mex. Geology
Corpus Christi — **(512) 888-3294**

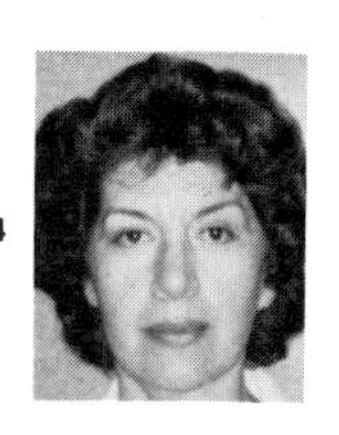

WINKLER, GARY R. **Susan Bartsch-Winkler**
Geologist
Br. of Alaskan Geology
Anchorage **(907) 271-4150**
U. of Utah-BS, MS; Stanford U.
Regional geology, southern Alaska
Physical stratigraphy; tectonics

WINSOR, HENRY C.
PST **(415) 493-0560**
Br. of Field Geochemistry & Petrology
Menlo Park **(415) 323-8111 x2194**
U. of the Pacific-BS; San Jose State U.-MS
CUSMAP

WITHERS, HELEN P.
Geologist
Office of Scientific Publications
Reston **(703) 860-6517**
U. of Texas (Austin)-BA; George Washington
Geologic inquiries

WITKIND, IRVING J. **Zelda**
Geologist **(303) 233-6065**
30 Ammons Street
Lakewood, Colorado 80226
Br. of Central Environmental Geology
Denver **(303) 234-3624**
Brooklyn Coll.-BA; Columbia U.-MA; U. of Colorado-PhD
Central Utah energy lands
Structural geology

WOBUS, REINHARD A. (BUD) **Sherry**
Geologist **(413) 458-8639**
20 Grandview Drive
Williamstown, Massachusetts 01267
Br. of Central Environmental Geology
Williams Coll., Mass. **(413) 597-2470**
Washington U.-BA; Harvard-MA; Stanford-PhD
Aztec 1° x 2° quadrangle, New Mexico
Precambrian of southern Rocky Mountains; igneous petrology

WOLD, RICHARD J.
Geophysicist **(303) 674-1233**
3910 S. Mossy Rock Lane
Evergreen, Colorado 80439
Br. of Electromag. & Geomagnetism
Denver **(303) 234-3388**
U. of Wisconsin-BS, PhD
Marine geophysics; seismology; potential field methods

WOLFE, EDWARD W.
Geologist **(602) 526-4920**
3007 E. Pine Drive
Flagstaff, Arizona 86001
Br. of Central Environmental Geology
Flagstaff **(602) 779-3311 x1556**
Coll. of Wooster-BA; Ohio State U.-PhD
San Francisco volcanic field; wilderness studies in central Arizona
Areal geology; volcanology; petrology

WOLFE, RUTH F.
Mathematician **(602) 774-8983**
603 N. San Francisco Street
Flagstaff, Arizona 86001
Br. of Astrogeologic Studies
Flagstaff **(602) 779-3311 x1465**
Northern Arizona U.-BS, MS
Computer modeling of the solar system
Mathematics; computer programming

WONG, HELEN C.
Radio Telemetry Technician
1540 Middle Avenue
Menlo Park, California 94025
Br. of Network Operations
Menlo Park **(415) 323-8111 x2585**

WONG, WILLIAM B.
Electronics Technician
Br. of Ground Motion & Faulting
Menlo Park **(415) 323-8111 x2514**
U. of Cal. (Berkeley)-BSEE
Alaska project
Filters

WOO, CHING CHANG
Geologist
36 Milton Street
Teaticket, Massachusetts 02543
Br. of Atlantic-Gulf of Mex. Geology
Woods Hole **(617) 548-8700**
National U. of Beijing-BS; U. of Chicago-MS, PhD

WOOD, DEWEY L. **Linda**
Cartographic Technician
126 North Johnson Road
Sterling, Virginia 22170
Office of Scientific Publications
Reston **(703) 860-6495**
Columbia Technical Inst.

WOOD, GORDON H., JR. **Eleanor**
Geologist **(301) 229-3247**
6115 Ramsgate Road
Bethesda, Maryland 20016
Br. of Coal Resources
Reston **(703) 860-7734**
U. of New Mexico-BS
Coal resource assessment group
Resource assessment; coal geology; structure of coal fields

WOOD, JANET M.
Administrative Clerk
Office of International Geology
Reston **(703) 860-6348**

WOOD, JOHN D. **Afaf**
Geophysicist **(303) 666-9608**
9014 Tahoe Lane
Boulder, Colorado 80301
Br. of Electromag. & Geomagnetism
Denver **(303) 234-5458**
Mississippi State U.-BS
Magnetic observatories & field surveys
Geomagnetism

WOODS, TERRI L. **John**
Geologist **(703) 435-2856**
543 Florida Avenue
Herndon, Virginia 22070
Br. of Exper. Geochem. & Mineralogy
Reston **(703) 860-6795**
U. of Delaware-BS; U. of Arizona-MS
Environment of ore deposition; Creede fossil geothermal fluid inclusion; mineral chemistry

WORL, PATRICIA B. **Ronald**
Secretary **(303) 989-3265**
12624 West Hawaii Avenue
Lakewood, Colorado 80228
Br. of Oil & Gas Resources
Denver **(303) 234-5008**
Utah State U.; U. of Wyoming; CSU

WORL, RONALD G. **Pat**
Geologist **(303) 989-3265**
12624 West Hawaii Avenue
Lakewood, Colorado 80228
Br. of Central Mineral Resources
Denver **(303) 234-6811 x3830**
Utah State U.-BS; U. of Wyoming-MA, PhD
Bridger wilderness, Wyoming
Mineral deposits; metamorphic structures; Precambrian geology

WORLEY, DAVID M.
Electronics Technician **(303) 422-8316**
6401 Quail, #143
Arvada, Colorado 80004
Br. of Engineering Geology
Denver **(303) 234-4477**
Tri-State U.
Instrumentation

WRATHER, WILLIAM E. **Alice**
Geologist
Office of the Director
Reston **(703) 860-6531**
U. of Chicago-PhB
Petroleum geology of U.S.; mineral resources; Permian stratigraphy
Scientist; administrator; consultant; teacher

WRIGHT, DAVID L.
Electronics Engineer **(303) 989-3905**
1024 S. Alkire Street
Lakewood, Colorado 80228
Br. of Electromag. & Geomagnetism
Denver **(303) 234-5462**
Wheaton Coll.-BS; U. of Illinios-MS; U. of New Mexico-PhD
Nuclear waste program
Electromag. methods; digital computation; prototype instrument design & develop.

WRIGHT, EVELYN L. **Charles**
Mathematician **(617) 564-4250'**
58 Bay Road
North Falmouth, Massachusetts 02556
Br. of Atlantic-Gulf of Mex. Geology
Woods Hole **(617) 548-8700**
U. of Michigan-BS
Digital facilities project
Numerical analysis; computer programming

WRIGHT, NANCY A.
Geologist
Office of Resource Analysis
Reston **(703) 860-6451**
Vassar Coll.-BA
Computer applications to resource appraisal

WRIGHT, THOMAS L.
Geologist **(202) 882-8316**
7520 Morningside Drive, NW
Washington, D.C. 20012
Br. of Exper. Geochem. & Mineralogy
Reston **(703) 860-6694**
Pomona Coll.-BA; Johns Hopkins-PhD
Geology of Columbia River basalt
Igneous petrology; volcanology

WRIGHT, WILLIAM H., JR. **Nancy**
Geologist
744 Dranesville Road
Herndon, Virginia 22070
Office of the Chief Geologist
Reston **(703) 860-6544**
Villanova-BS; Siena Coll.-MBA
Management information systems
Program planning; budget management

WRIGHT, WILNA B.
Geologist
615 South Fairfax Street
Alexandria, Virginia 22314
Office of Scientific Publications
Reston **(703) 860-6511**
U. of North Carolina; George Washington U.-BS
Stratigraphy & stratigraphic nomenclature
Iron ore; energy; marine geology

WRUCKE, CHESTER T. **Ruth Ann**
Geologist **(415) 851-8435**
30 Cima Way
Portola Valley, California 94025
Br. of Western Mineral Resources
Menlo Park **(415) 323-8111 x2347**
Stanford-BS, MS, PhD
Mazatzal Mountains, Ariz., Shoshone Range, Nevada
Structural geology; igneous & metamorphic rocks

WU, SHERMAN S.C. **Ankee**
Physical Scientist **(602) 774-3987**
Br. of Astrogeologic Studies
Flagstaff **(602) 779-3311**
Chinese Survey Coll.-BS; Syracuse U.-MS; U. of Arizona-PhD.
Planetary exploration—U.S. Space Program
Planetary mapping; photogrammetry; statistics

WUOLLET, GERALDINE M. **James**
Geophysicist **(303) 232-6398**
12820 Willow Lane, #18
Golden, Colorado 80401
Br. of Ground Motion & Faulting
Denver **(303) 234-5604**
U. of Arizona
Southern Great Basin seismicity study
Earthquake tectonics

WYANT, BARBARA D. **Dorman**
Secretary
2435 Alsop Court
Reston, Virginia 22091
Office of Geochemistry & Geophysics
Reston **(703) 860-6584**
Madison Coll.

WYATT, NANCY J.
Editorial Assistant
5435 Eldridge
Arvada, Colorado 80002
Br. of Central Environmental Geology
Denver **(303) 234-4391**
Community College-AA

WYNN, JEFFREY C. **Louise**
Geophysicist **(703) 430-8345**
16 Silo Mill Court
Sterling, Virginia 22170
Office of Geochemistry & Geophysics
Reston **(703) 860-6582**
U. of California (Berkeley)-BA; U. of Illinois-MS; U. of Arizona-PhD
Strategic minerals geophysics
Mining geophysics; geoelectrical studies; archaeophysics

YAMASHITA, KENNETH M.
PST
Br. of Field Geochemistry & Petrology
Hawaiian Volcano Observ. **(808) 967-7328**

YAMAMOTO, AKIRA **Nobue**
PSA **(808) 966-9331**
P.O. Box 5
Volcano, Hawaii 96785
Br. of Field Geochemistry & Petrology
Hawaiian Volcano Observ. **(808) 967-7328**
Hilo College

YEDNOCK, CATHERINE M.
Accounting Technician
Office of International Geology
Reston **(703) 860-6522**

YEEND, WARREN E.
Geologist **(415) 948-4968**
12725 La Cresta Drive
Los Altos Hills, California 94022
Br. of Alaskan Geology
Menlo Park **(415) 323-8111 x2541**
WSU-BS; U. of Colorado-MS; U. of Wisconsin-PhD
Placers of Alaska
Surficial geology; field mapping; environmental studies

YEHLE, LYNN A. Fran
Geologist (303) 237-0839
1235 Everett Court
Lakewood, Colorado 80215
Br. of Engineering Geology
Denver (303) 234-2999
U. of Wisconsin-BS, MS
Surficial geology; geomorphology; glacial geology

YENOWINE, JAMES H.
PST
Br. of Oil & Gas Resources
Denver (303) 234-5235
Indiana State U.-BS
Domestic oil & gas resource appraisal
Petroleum geology

YEOMAN, ROSS A.
PST (303) 447-9115
971 Pleasant
Boulder, Colorado 80302
Br. of Central Mineral Resources
Denver (303) 234-2773
Brown U.-BS; UCLA
Ore deposits; igneous petrology

YERKES, ROBERT F.
Geologist
Br. of Engineering Geology
Menlo Park (415) 323-8111 x2350
Claremont Coll.-MA
Seismotectonics of western Transverse Ranges
Tectonic geology of southern California

YETTER, THOMAS J.
PST
174 Vista Verde Way
Portola Valley, California 94025
Br. of Western Environmental Geology
Menlo Park (415) 323-8111 x2939
Humboldt State U.-BS
Quaternary geology; environmental geology

YOCHELSON, ELLIS L. Sally
Geologist (301) 262-1563
12303 Stafford Lane
Bowie, Maryland 20715
Br. of Paleontology & Stratigraphy
Washington, D.C. (202) 343-3232
U. of Kansas-BS, MS; Columbia-PhD
Paleozoic biostratigraphy of mollusks; paleozoic stratigraphy; paleoecology

YORK, JUDY K.
PSA (415) 368-1703
635 Acacia Lane
Redwood City, California 94062
Br. of Seismology
Menlo Park (415) 323-8111 x2192
San Francisco State U.

YOSHIFUJI, MARY N. Sakaye
Administrative Clerk
339 San Antonio Avenue
San Mateo, California 94401
Office of the Chief Geologist
Menlo Park (415) 323-8111 x2219

YOUD, T. LESLIE Denice
Civil Engineer (415) 792-6636
4614 Richmond Avenue
Fremont, California 94536
Br. of Engineering Geology
Menlo Park (415) 856-7117
Brigham Young U.-BES; Iowa State U.-PhD
Mapping of liquefaction potential
Geotechnical engineering; earthquake engineering; liquefaction of sands

YOUNG, EDWARD J. Amy
Geologist (303) 237-8580
3000 Union Street
Lakewood, Colorado 80215
Br. of Uranium & Thorium Resources
Denver (303) 234-3427
Rutgers U.-BS; MIT-MS, PhD
Mineralogy; petrology; geochemistry

YOUNG, GARNET S. Robert
Administrative Officer
Br. of Engineering Geology
Denver (303) 234-2431

YOUNG, ROBERT J.
PST (303) 936-1695
4151 W. Greenwood
Denver, Colorado
Br. of Analytical Laboratories
Denver (303) 234-6401
Colorado Schl. of Mines-ME
XRF-continuing

YOUNG, ROBERT L. Margarit
Electronics Technician (505) 296-7267
12637 Elyse Place SE
Albuquerque, New Mexico 87123
Br. of Global Seismology
Albuquerque (505) 474-3815
Microprocessor-based seismic systems
Fault monitor; digital record systems

YOUNSE, GARY A. Kathy
Geologist (415) 253-2450
889 Cottonwood Drive
Cupertino, California 95014
Office of International Geology
Menlo Park (415) 323-8111 x2926
San Jose State U.-BS, MS
Phosphates; hydrocarbons

YOUNT, JAMES C. Sheryl
Geologist (408) 726-5675
600 Terrace Avenue
Half Moon Bay, California 94019
Br. of Western Environmental Geology
Menlo Park (415) 323-8111 x2905
U. of Washington-BS; U. of Colorado-MS
Seismic ground response; Puget Sound, Washington
Sedimentology, tectonics of Pacific Northwest; glaciation in western U.S.

ZABLOCKI, CHARLES J. Carole
Geophysicist
Office of Geochemistry & Geophysics
Reston (703) 860-6583
Syracuse U.-BS; U. of Colorado
Volcanic minerals; geophysical methods;' electrical magnetic

ZALE, ELIZABETH A.
Technical Editor
2955 Carnegie
Boulder, Colorado 80302
Office of Scientific Publications
Denver (303) 234-3283
Michigan State U.; U. of Colorado (Boulder)-BA

ZARTMAN, ROBERT E. Ruth
Geologist (303) 973-0405
10077 W. Tufts Place
Littleton, Colorado 80123
Br. of Isotope Geology
Denver (303) 234-3876
Penn State-BS; Cal Tech-MS, PhD
Geochronology
Radioactive dating; isotope geology; igneous petrology

ZBIGNEWICH, EDWARD W. Mary
Engineering Technician (301) 933-3629
12202 Grandview Avenue
Wheaton, Maryland 20902
Br. of Analytical Laboratories
Reston (703) 860-7442
Instrument development

ZEHNER, RICHARD E.
PST
Br. of Alaskan Geology
Menlo Park (415) 323-8111 x2471
U. of Cal. (Santa Cruz)-BS
Mt. Hayes quadrangle (AMRAP)

ZELIBOR, JOSEPH L. Brenda Kay
Microbiologist (703) 591-6072
3149 Bayswater Court
Fairfax, Virginia 22030
Br. of Atlantic-Gulf of Mex. Geology
Reston (703) 860-7662
Purdue U.-BS
Geomicrobiology; marine geochemistry
Marine microbiology; biogeochemistry

ZELLER, KARL A. Dorothy
Photographer (602) 526-0849
4510 N. Hamblim
Flagstaff, Arizona 86001
Br. of Astrogeologic Studies
Flagstaff (602) 779-3311 x1318

ZEN, E-AN
Geologist (703) 860-0845
11923 Escalante Court
Reston, Virginia 22091
Br. of Exper. Geochem. & Mineralogy
Reston (703) 860-6621
Cornell U.-BA; Harvard-MA, PhD
Petrology & tectonics of Pioneer Mtns., Montana; bedrock geologic map of Mass.
Petrology; structural & regional geology

ZERMANE, ALBERT J. Leona
Electronic Technician (703) 435-1625
312 Missouri Avenue
Herndon, Virginia 22070
Br. of Analytical Laboratories
Reston (703) 860-6778
DeVry Technical Institute-AAS
Laboratory & instrument automation; computer hardware & software design

ZIELINSKI, ROBERT A. Kathryn
Geochemist
690 Estes Street
Lakewood, Colorado 80215
Br. of Uranium-Thorium Resources
Denver (303) 234-4201
Rutgers U.-BA; MIT-PhD
Uranium source rocks-volcanic
Trace element geochemistry

ZIEMAN, MARGARET A. Richard
Librarian
11229 Leatherwood Drive
Reston, Virginia 22091
Library
Reston (703) 860-6613
Vanderbilt U.-BA; George Peabody Coll.-MLS

ZIETZ, ISIDORE Frances
Geophysicist (703) 790-1343
8340 Greensboro Drive
McLean, Virginia 22102
Br. of Regional Geophysics
Reston (703) 860-7236
CCNY-BS, MS; Catholic U.-PhD
Compilation of National Aeromagnetic Map
Magnetics; gravity-potential theory

ZIHLMAN, FREDERICK N.
PST (301) 283-6262
#1 Green Meadows Court
Indian Head, Maryland 20640
Br. of Oil & Gas Resources
Reston (703) 860-7258
U. of Maryland-BS
Computer service

ZIMMERMAN, PATRICIA A. Vernon
Administrative Technician
1075 S. Cody Street
Lakewood, Colorado 80226
Office of the Chief Geologist
Denver (303) 234-3622

ZIMMERMANN, ROBERT A.
Geologist
Route 1, Box 551-B
Conifer, Colorado 80433
Br. of Isotope Geology
Denver (303) 234-4201
U. of Pennsylvania-PhD
Fission track dating studies
Regional uplift; tectonics; geochronology

ZIONY, JOSEPH I. Denise
Geologist (415) 328-4218
1640 Escobita Avenue
Palo Alto, California 94306
Office of the Chief Geologist
Menlo Park (415) 323-8111 x2214
UCLA-BA, MA, PhD
Earthquake hazards; regional tectonics & structure

ZOBACK, MARK D. Mary Lou
Geophysicist (415) 322-2776
1175 Forest Avenue
Palo Alto, California 94301
Br. of Tectonophysics
Menlo Park (415) 323-8111 x2034
U. of Arizona-BS; Stanford-MS, PhD
Fault zone properties; in-situ stress measurement
Earthquake mechanics

ZOBACK, MARY LOU Mark
Geophysicist (415) 322-2776
1175 Forest Avenue
Palo Alto, California 94301
Br. of Earthquake Tectonics & Risk
Menlo Park (415) 323-8111 x2944
Stanford-BS, MS, PhD
Earthquake hazards-Wasatch fault
Basin & Range tectonics; state of stress in conterminous U.S.

ZOHDY, ADEL A.R. Sohair
Geophysicist (303) 985-9752
2088 S. Zephyr Court
Lakewood, Colorado 80227
Br. of Regional Geophysics
Denver (303) 234-5465
U. of Alexandria (Egypt)-BS; U. of California-MS; Stanford-PhD
Resistivity interpretation
Applied research-electrical resistivity for groundwater & geothermal exploration

ZOLLER, HENRY H. Marion
Librarian
11601 Sourwood Lane
Reston, Virginia 22091
Library
Reston (703) 860-6613
Wayne State U.-BA, MLS

ZUBOVIC, PETER Elizabeth C.
Research Chemist (301) 831-7154
1259 Ridge Road
Mt. Airy, Maryland 21771
Br. of Coal Resources
Reston (703) 860-6044
Shippensburg State Coll.-BS; American U.-BS, MS
Geochemistry of coal of eastern U.S.
Coal geochemistry; coal utilization effects—environment & conversion processes

ZUCCA, JOHN J.
Geophysicist
Br. of Seismology
Menlo Park **(415) 323-8111 x2823**
U. of Cal. (Santa Barbara)-BA; Stanford-MS
Velocity structure of Hawaiian volcanoes
Seismic refraction and gravity

OFFICE AND BRANCH ROSTERS

OFFICE OF THE CHIEF GEOLOGIST

Blum, John E.
Boyer, Robert W.
Drake, Avery A., Jr.
Eaton, Gordon P.
Eichler, Helen E.
Fichtner, Wayne E.
Gall, R. Michael
Glaister, Carole S.
Goldberg, Jerald M.
Griggs, M. Louise
Hanshaw, Penelope M.
Hoffmann, Susan T.
Johnson, Maureen G.
Keefer, William R.
Kettell, Nan L.
Kirschmer, Beverly A.
Koteff, Carl
Like, Linda L.
Mankinen, Jeanne C.
Marranzino, Albert P.
May, Irving
McGaha, Ana C.
Mercilliott, Beverly A.
Miller, Harley P.
Miller, Hope D.
Miller, Janet L.
Moore, Sharon T.
Myles, Percy R.
O'Donnell, Claudine S.
Oscarson, Robert L.
Peck, Dallas L.
Pillera, Joseph S.
Pitts, Susan E.
Rambo, William L.
Reynolds, Mitchell W.
Richards, Earl, Jr.
Rusling, Donald H.
Schray, Carole A.
Seyler, David A.
Sharkey, Beatrice D.
Smoot, Nancy M.
Sumida, Theodore T.
Swann, Gordon A.
Tedder, Sandi
Thomas, Virginia L.
Urick, Mary Alice
Utter, Christopher G.
Walton, Ronald J.
Watson, Donald E.
Weatherhead, Sharon K.
Wiegand, Gail P.
Wright, William H., Jr.
Yoshifugi, Mary N.
Zimmerman, Patricia A.
Ziony, Joseph I.

OFFICE OF SCIENTIFIC PUBLICATIONS

Aaron, John M.
Alden, Andrew L.
Barnhart, Merilee A.
Boore, Sara A.
Brigida, Miriam J.
Brokaw, James A.
Buffa, Elizabeth A.
Carter, Lorna M.
Chapman, Nancy L.
Chopko, Glenda I.
Cravotta, Charles A., III
D'Agostino, Catherine M.
Davis, Robert E.
Derr, John S.
Detterman, Janis S.
Eister, Margaret F.
Estabrook, James R.
Fuller, H. Kit
Goldsmith, June W.
Gonsalves, Janet S.
Good, Elizabeth E.
Hansen, Ellen E.
Havach, George A.
Hawkins, Bernard W.
Hillier, Barbara C.
Hodgen, Laurie D.
Hodgson, Helen E.
Hoss, Karen L.
Hubert, Marilyn L.
Humphreys, D. Darlene
Iannacite, Sherri E.
Jones, Diane N.
Jussen, Virginia M.
Kinney, Douglas M.
Kook, Ann E.
Kopf, Rudolph W.
Kropschot, Susan J.
Langenheim, Virginia A.
Love, Eleanor M.
Luttrell, Gwendolyn W.
MacLachlan, Marjorie E.
Major, Virginia L.
Martin, Jane H.
Martin, Peter L.
Mayfield, Susan E.
Morgan, Ida M.
Mowinckel, Penelope K.
Nelson, Clifford M.
Newman, William L.
Owen, Mary J.
Parcel, Alice A.
Penman, Martha K.
Pinkerton, James B.
Poppe, Barbara B.
Porter, Pearl B.

Powell, Yvonne J.
Przedpelski, Mary Ann
Quigley, Margaret R.
Ratliff, Mary L.
Russell, Jerry M.
Sable, Vera H.
Sakss, Yula E.
Sangree, Anne C.
Schmidt, Paul W.
Schnabel, Diane C.
Schray, Karen A.
Selander, Laurie L.
Spall, Henry
Sparks, Delores M.
Swanson, Roger W.
Tamamian, Nancy J.
Tarbert, Anita W.
Tarr, Arthur C.
Thomas, Julia A.
Thompson, Marian O.
Timmins, Jane D.
Turkington, Daniel E.
Ulibarri, Loretta J.
Utterback, Priscilla G.
Weld, Betsy A.
Wilbur, Patricia L.
Withers, Helen P.
Wood, Dewey L.
Wright, Wilna B.
Zale, Elizabeth A.

Library

Albright, Verna T.
Baker, Kay M.
Behrendt, Elizabeth C.
Bier, Robert A., Jr.
Blair, Nancy L.
Borders, Glenn H.
Branch, Benjamin H., Jr.
Brodes, Betty B.
Caldwell, Delia M.
Chappell, Barbara A.
Dalechek, Marjorie E.
Dennis, Phyllis R.
Freeberg, Jacquelyn H.
Glammeyer, Roy A.
Goodwin, George H., Jr.
Havener, Alice F.
Hayes, Jeannette M.
Horan, Carol L.
Imlay, Wilma G.
Karakov, Yelena M.
Ketch, Jane K.
King, Helen M.
Lewis, Diane M.
Liszewski, Edward H.
Martna, Maret H.
Macqueen, Laura M.
McGregor, Joseph K., Jr.
Perry, Jane W.
Powers, M. Susann W.
Reed, Bobby M.
Roberts, John K., III
Rogge, Betty M.
Sanders, William
Sasscer, Richard S.
Scaun, Anatole
Sellin, Jon B.
Shields, Caryl L.
Sinnott, Trudy M.
Stover, Sarah A.
Swann, Inez L.
Vitaliano, Dorothy B.
White, Dorothy C.
Wilkins, Eleanore E.
Zieman, Margaret A.
Zoller, Henry H.

OFFICE OF MINERAL RESOURCES

Andreani, Veronica A.
Barton, Paul B., Jr.
Beikman, Helen M.
Feeney, Diane D.
Fridrich, Douglas K.
Goudarzi, Gus H.
Grybeck, Donald J.
Kiilsgaard, Thor H.
Montague, Nancy L.
Mulhausen, Lorraine E.
Ovenshine, A. Thomas
Pilgrim, Margaret
Wever, Theresa L.
Weissenborn, Albert E.

Branch of Alaskan Geology

Albert, Nairn R.D.
Balin, Donna F.
Bartsch-Winkler, Susan
Berg, Henry C.
Brew, David A.
Brosgé, William P.
Carlson, Christine
Carter, Claire
Carter L. David
Chapman, Robert M.
Churkin, Michael, Jr.
Cobb, Edward H.
Collins, Irene L.
Coonrad, Warren L.
Csejtey, Bela, Jr.
Cushing, Grant W.
Decker, John E.
Detterman, Robert L.
Diggles, Michael F.
Dumoulin, Julie A.
Dusel-Bacon, Cynthia
Eberlein, G. Donald
Ellersieck, Inyo F.
Elliott, Raymond L.
Ferrians, Oscar J., Jr.
Ford, Arthur B.
Foster, Helen L.
Galloway, John Paul
Grantz, Arthur
Green, Marsha R.
Greene, RoxAnn E.
Hamilton, Thomas D.
Hartz, Roger William
Herzon, Paige Leigh
Hoare, Joseph M.
Hopkins, David M.
Huie, Carl
Hunt, Susan J.
Johnson, Bruce R.

Johnson, Kathleen M.
Kachadoorian, Reuben
Karl, Susan M.
Koch, Richard D.
Lamb, Beth M.
Le Compte, James R.
Loney, Robert A.
Malloy, Mary J.
Mayfield, Charles F.
Meehan, Richard H.
Miller-Hoare, Marti L.
Miller, Thomas P.
Miyaoka, Ronny T.
Moll, Elizabeth J.
Moore, Thomas E.
Morgenstern, Karen M.
Mull, Charles G.
Nelson, Willis H.
Nokleborg, Warren J.
Obi, Curtis M.
Patton, William W., Jr.
Plafker, George
Reed, Bruce L.
Reiser, Hillard N.'
Schwab, Carl E.
Shew, Nora B.
Silberman, Miles L.
Smith, Peggy A.
Sonnevil, Ronald A.
Trexler, James H., Jr.
Trollman, Winifred M.
Weber, Florence R.
White, Ellen R.
Williams, John R.
Wilson, Frederic H.
Winkler, Gary R.
Yeend, Warren E.
Zehner, Richard E.

Office of Resource Analysis

Attanasi, Emil D.
Barari, Rachel A.
Bawiec, Walter J.
Bliss, James D.
Bowen, Roger W.
Bridges, Nancy J.
Calkins, James A.
Cargill, Simon M.
Chesson, Sharon Ann
Cookro, Theresa M.
DeYoung, John H., Jr.
Drew, Lawrence J.
Fulton, Patricia A.
Grundy, Wilbur D.
Guild, Philip W.
Hanley, J. Thomas
Huber, Donald F.
Keefer, Eleanor K.
Kesecker, Patricia A.
Kork, John O.
Long, Keith R.
Manley, Dorothy J.
Mason, George T., Jr.
McCammon, Richard B.
McQueen, David R.
Medlin, Antoinette L.
Menzie, W. David
Meyer, Richard F.
Michelitch, Arleen V.
Mosier, Dan L.
Olson, Jane Ciener
Orris, Greta J.
Paidakovich, Matthew E.
Root, David H.
Salem, Bruce B.
Schruben, Paul G.
Scott, William A.
Shaffer, Glenn L.
Singer, Donald A.
Starritt, Bruce C.
Stoltz, Melissa L.
Swanson, James R.
Turner, Robert M.
Wright, Nancy A.

Branch of Exploration Research

Allcott, Glenn H.
Antweiler, John C.
Barton, Harlan N.
Berger, Byron R.
Botinelly, Theodore
Campbell, Wesley L.
Canney, Frank C.
Carlson, Robert R.
Cathrall, John B.
Chaffee, Maurice A.
Chao, Tsun Tien
Church, Stanley E.
Cooley, Elmo F.
Crim, William D.
Dietrich, John A.
Domenico, James A.
Duncan, Karen M.
Erickson, M. Suzanne
Erickson, Ralph L.
Ficklin, Walter H.
Friskin, James G.
Griffitts, Wallace R.
Grimes, David J.
Hinkle, Margaret E.
Hopkins, Roy T.
Hubert, Arthur E.
King, Harley D.
Leach, David L.
Learned, Robert E.
Leinz, Reinhard W.
Marsh, Sherman P.
McCarthy, J. Howard, Jr.
McDanal, Steven K.
McHugh, John B.
Meier, Allen L.
Miller, William R.
Mosier, Elwin L.
Nakagawa, Harry M.
Negri, John C.
Neuerburg, George J.
Nowlan, Gary A.
Odland, Sarah K.
O'Leary Richard M.
Rosenblum, Sam
Sanzolone, Frank V.
Sanzolone, Richard F.
Siems, David F.
Smith, David B.
Theobald, Paul K.

Tripp, Richard B.
Turner, Robert L.
VanTrump, George, Jr.
Viets, John G.
Watterson, John R.
Wehrle, Frederic P., Jr.
Whittington, Charles L.

Branch of Eastern Mineral Resources

Bell, Henry, III
Brobst, Donald A.
Brown, C. Ervin
Cameron, Cornelia C.
Cannon, William F.
Clark, Sandra H.B.
D'Agostino, John P.
Ericksen, George E.
Foose, Michael P.
Force, Eric R.
Gair, Jacob E.
Grosz, Andrew E.
Hosterman, John W.
Kirkemo, Harold
Lesure, Frank G.
Lipin, Bruce R.
Loferski, Patricia J.
Luce, Robert W.
Neeley, Cathy L.
Patterson, Sam H.
Prinz, William C.
Schmidt, Robert Gordon
Slack, John F.
Stricker, Mary F.
Van Alstine, Ralph E.
Whitlow, Jesse W.

Branch of Central Mineral Resources

Back, Judith M.
Blackmon, Paul D.
Bohannon, Robert G.
Bush, Alfred L.
Christian, Ralph P.
Clémensen, Margaret A.
Cox, Leslie J.
Coxe, Berton W.
Cunningham, Charles G.
Desborough, George A.
Doyle, Brien F.
Duffy, Mary T.
Eckel, Edwin B.
Ekren, E. Bartlett
Elliott, James E.
English, Kathy S.
Finnell, Tommy L.
Fisher, Frederick S.
Foord, Eugene E.
Foord, Suzann C.
Hall, Robert B.
Harrison, Glenda K.
Hasler, J. William
Hedricks, Louise S.
Heidel, Robert H.
Heyl, Allen V.
Houser, Brenda B.
King, Robert U.
Koesterer, Mary Ellen
Landis, Gary P.
Leonard, Benjamin F., III
Lindsey, David A.
Ludington, Stephen D.
Maxwell, Charles H.
McIntyre, David H.
Miller, Mary H.
Modreski, Peter J.
Moench, Robert H.
Moore, Samuel L.
Nelson, Karen A.
Nishi, James M.
Norton, James J.
Pearson, Robert C.
Pinckney, Darrell M.
Poole, Forrest G.
Pratt, Walden P.
Raymond, William H.
Schmidt, Arlene C.
Schoenfeld, Richard E.
Segerstrom, Kenneth
Sekulich, Michael J.
Sharp, William N.
Shawe, Daniel R.
Sheridan, Douglas M.
Simmons, George C.
Simons, Frank S.
Sims, Paul K.
Smith, J. Fred, Jr.
Stairs, Joseph G.
Starkey, Harry C.
Steven, Thomas A.
Taylor, Richard B.
Thorman, Charles H.
Tschanz, Charles M.
Tweto, Odgen L.
Tysdal, Russell G.
Van Loenen, Richard E.
Weis, Paul L.
Wells, John D.
Wier, Kenneth L.
Worl, Ronald G.
Whipple, James W.
Yeoman, Ross A.

Branch of Western Mineral Resources

Albers, John P.
Alsop, Karen S.
Amamoto, Nobuko
Armin, Richard A.
Ashley, Roger P.
Bailey, Edgar H.
Barnard, James B.
Barnes, Carol S.
Barnhart, Rebecca K.
Bergquist, Joel R.
Böhlke, John Karl, F.P.
Bradley, Robin
Briskey, Joseph A., Jr.
Coats, Robert R.
Cornwall, Henry R.
Cox, Dennis P.
Dohrenwend, John C.
Evarts, Russell C.
Harner, Joy L.
Haxel, Gordon B.
James, Harold L.
Johannesen, Dann C.
Johnson, Maureen G.

Keith, William J.
Kleinhampl, Frank J.
Macleod, Norman S.
Madrid, Raul J.
Morris, Hal T.
Moore, William J.
Page, Norman J.
Peterson, Donald W.
Peterson, Jocelyn A.
Rinehart, C. Dean
Roberts, Ralph J.
Seitz, James F.
Shock, Everett L.
Sorensen, Martin L.
Stager, Harold K.
Stewart, John H.
Tanti, Dorothy V.
Theodore, Ted G.
Tooker, Edwin W.
Walker, George W.
Whitebread, Donald H.
Wrucke, Chester T.

OFFICE OF INTERNATIONAL GEOLOGY

Addicott, Warren O.
Bergin, Marion J.
Bergquist, Wenonah E.
Brown, Glen F.
Buck, Carolyn E.
Cadigan, Geraldine C.
Chidester, Alfred H.
Clark, Allen L.
Cole, James C.
Cook, Jennifer L.
Davidson, David F.
Dematteo, Ronald E.
duBray, Edward A.
Fleming, Hershell L.
Fary, Raymond W., Jr.
Gawarecki, Stephen J.
Gray, Helen M.
Greene, Robert C.
Hadley, Donald G.
Holzle, Alvin F.
Kellogg, Karl S.
Long, Shirley L.
Maberry, Andrea L.
Marinenko, Olga H.
Mikuni, Diane E.
Miller, Ralph L.
Mills, Francis R.
Morgan, Joseph O.
Olive, Wilds W.
Phipps, Carolyn G.
Reinemund, John A.
Roberts, Albert E.
Roen, Jeanne H.
Rooney, Lawrence F.
Rosario, Henry
Rosenblum, Lenore
Sable, Edward G.
Smedley, Ocie V.
Spike, Barbara P.
Stanin, S. Anthony
Tatlock, Donald B.
Terman, Maurice J.
Tolbert, Gene E.
Walther, Linda M.
Watkins, Caroline A.
Willingham, Helen P.
Wood, Janet M.
Yednock, Catherine M.
Younse, Gary A.

OFFICE OF GEOCHEMISTRY & GEOPHYSICS

DeDontney, Dorothy M.
Dietz, Barbara D.
Ellick, Dona M.
Ferrier, Susan C.
Goerlitz, Patricia A.
Gramm, Alice E.
Hartley, Debra L.
Hemley, J. Julian
Jenkins, Rosemary
Keith, John R.
Klick, Donald W.
Kover, Allan N.
Line, Mildred W.
McKeown, Helen L.
Preston, Margaret W.
Roberts, Carol E.
Scott, Dorothy L.
Seitsinger, Katheryn M.
Stonehocker, Leland K.
Tilling, Robert I.
Wilt, Linda S.
Wyant, Barbara D.
Wynn, Jeffrey C.
Zablocki, Charles J.

Branch of Experimental Geochemistry & Mineralogy

Belkin, Harvey E.
Bethke, Philip M.
Chou, I-Ming
Clynne, Michael A.
Consagra, Kathleen M.
Czamanske, Gerald K.
Dick, Patricia A.
Erd, Richard C.
Evans, Howard T., Jr.
Fournier, Robert O.
Gartner, Anne E.
Harris, David M.
Helz, Rosalind T.
Huebner, J. Stephen
James, Odette B.
Kieffer, Susan W.
Konnert, Judith A.
McGee, Kenneth A.
Mrose, Mary E.
Nord, Gordon L., Jr.
Robie, Richard A.
Robinson, Gilpin R., Jr.
Roedder, Edwin W.
Roseboom, Eugene H., Jr.
Ross, Malcolm
Sato, Motoaki
Schafer, Constance M.
Shaw, Herbert R.
Stewart, David B.
Thompson, J. Michael
Thornber, Carl R.
Thurnau, Ellen M.
Toulmin, Priestley, III

Truesdell, Alfred H.
Wandless, Mary-Virginia
Wetlaufer, Pamela H.
Wiggins, Lovell B.
Woods, Terri L.
Wright, Thomas L.
Zen, E-an

Branch of Field Geochemistry & Petrology

Bacon, Charles R.
Bailey, Roy A.
Bargar, Keith E.
Barker, Fred
Beeson, Melvin H.
Calk, Lewis C.
Christiansen, Robert L.
Coleman, Robert G.
Cook, Amy E.
Dickson, Jane J.
Dodge, Franklin C.W.
Donato, Mary M.
Donnelly-Nolan, Julie M.
Duffield, Wendell A.
Dwornik, Deborah
Ernst, Wallace G.
Guffanti, Marianne C.
Gottfried, David
Gray, Karen J.
Hall, Wayne E.
Hearn, B. Carter
Hildreth, Edward W.
Holcomb, Robin T.
Kays, M. Allan
Keith, Terry E.C.
Koeppen, Robert P.
Krauskopf, Konrad B.
Luedke, Robert G.
Mallis, Robert R.
Moore, James G.
Morgan, Benjamin A., III
Morgenstern, Joseph C.
Muffler, L.J. Patrick
Nathenson, Manuel
Ota, Yoshiko Pat
Pallister, John S.
Pflaum, Bernard H.
Phair, George
Polovtzoff, Oleg C.
Reed, Marshall J.
Russell-Robinson, Susan L.
Smith, Robert L.
Swanson, Donald A.
Tilbury, Carol A.
Tobisch, Othmar T.
Warshaw, Charlotte M.
White, Donald E.
Wilcox, Ray E.
Winsor, Henry C.

Branch of Field Geochemistry & Petrology: Hawaiian Volcano Observatory

Banks, Norman G.
Casadevall, Tom J.
Decker, Robert W.
Dzurisin, Daniel
English, Thomas T.
Forbes, John C.
Greenland, L. Paul
Honma, Kenneth T.
Jackson, Dallas B.
Klein, Fred W.
Kojima, George
Koyanagi, Robert Y.
Lockwood, John P.
Moore, Richard B.
Nakata, Jennifer S.
Okamura, Arnold T.
Okamura, Reginald T.
Puniwai, Gary S.
Sako, Maurice K.
Yamashita, Kenneth M.
Yamamoto, Akira

Branch of Regional Geophysics

Ackermann, Hans D.
Barnes, David F.
Becker, Keir
Behrendt, John C.
Bisdorf, Robert J.
Blakely, Richard J.
Bond, Kevin R.
Campbell, Wallace H.
Case, James E.
Daniels, David L.
Dansereau, Danny A.
Finn, Carol A.
Godson, Richard H.
Goodman, Minnie H.
Grim, Muriel S.
Griscom, Andrew
Hamilton, Warren B.
Hanna, William F.
Hassemer, Jerry H.
Hildenbrand, Thomas G.
Hill, Patricia L.
Jones, William J.
Kane, Martin F.
Kaufmann, Harold E.
King, Elizabeth R.
Kucks, Robert P.
Linton, Jimmie R.
Mabey, Don R.
Michalski, Daniel C.
Pankratz, Leroy W.
Petrafeso, Frank A.
Phillips, Jeffrey D.
Plesha, Joseph L.
Roberts, Carter W.
Saltus, Richard W.
Seginak, Emil P.
Sikora, Robert F.
Snyder, David B.
Snyder, Stephen L.
Sorensen, Scott B.
Spydell, D. Randall
Sweeney, Ronald E.
Tang, Roger W.
Theel, Marjorie A.
Webring, Michael W.
Williams, David L.
Williams, Jackie M.
Zietz, Isidore
Zohdy, Adel A.R.

Branch of Isotope Geology

Abeyta, Vona V.
Afra, Donna W.
Aleinikoff, John N.
Arth, Joseph G.
Berry, Anne L.
Billings, Patty
Briggs, Nancy D.
Bunker, Carl M.
Burr, George S.
Bush, Charles A.
Dalrymple, G. Brent
Day, John H.
Delevaux, Maryse H.
Dobson, Steven W.
Doe, Bruce R.
Doering, Willis P.
Dooley, John R., Jr.
Dube, Marcel J.
Emsing, Sandra L.
Fenicle, Karen A.
Fischer, Lynn B.
Friedman, Irving
Futa, Kiyoto
Heard, Irvin, Jr.
Hedge, Carl E.
Holmstrom, Carl A.
Huebner, Mark
Huestis, Gary M.
Kelley, M. Lea
Kennedy, Mark W.
Kistler, Ronald W.
Kovach, Jack
Kwak, Loretta M.
Kyser, T. Kurtis
Lanphere, Marvin A.
LeLange, John E.
Luetscher, John D.
Mangum, John H.
Mann, Dorothy J.
Marvin, Richard F.
May, Rodd J.
McNair, Donald W.
Mehnert, Harald H.
Mikesell, Jon L.
Mosely, Roderick C.
Naeser, Charles W.
Newell, Marcia F.
Obradovich, John D.
O'Neil, James R.
Peterman, Zell E.
Robinson, Allen C.
Robinson, Stephen W.
Rosholt, John N.
Rubin, Meyer
Rye, Robert O.
Senftle, Frank E.
Sherrill, Nathaniel D.
Spiker, Elliott C.
Stacey, John S.
Stern, Thomas W.
Szabo, Barney J.
Tanner, Allan B.
Tatsumoto, Mitsunobu
Trimble, Deborah A.
Unruh, Daniel M.
Von Essen, Jarel C.
Vuletich, April K.
Whelan, Joseph F.
Zartman, Robert E.
Zimmermann, Robert A.

Branch of Regional Geochemistry

Anderson, Barbara M.
Boerngen, Josephine G.
Connor, Jon J.
Coubrough, Mary A.
Dean, Walter E.
Ebens, Richard J.
Erdman, James A.
Fleischer, Michael
Gough, Larry P.
Harms, Thelma F.
Harrach, George H.
Herring, James R.
Hinkley, Todd K.
Jachens, Robert C.
Lee, Donald E.
McNeal, James M.
Mendes, Roy V.
Miesch, Alfred T.
Morin, Robert L.
Oliver, Howard W.
Papp, Clara S.E.
Plouff, Donald
Ponce, David A.
Reif, Louise M.
Schultz, Leonard G.
Severson, Ronald C.
Smith, Kathleen S.
Tidball, Ronald R.
Tompkins, Mary L.
Tourtelot, Harry A.
Tuttle, Michele L.

Branch of Electromagnetism & Geomagnetism

Abrams, Gerda A.
Alldredge, Leroy R.
Bankey, Viki L.
Berarducci, Alan M.
Bradley, Jerry A.
Bramsoe, Erik
Broker, Michael M.
Cain, Joseph C.
Caldwell, Jill E.
Campbell, Donald M.
Christopherson, Karen R.
Cooke, James E.
Davis, Leo E.
DeJonge, Cheryl D.
Fabiano, Eugene B.
Fitterman, David V.
Flanigan, Vincent J.
Frischknecht, Frank C.
Green, Richard G.
Greenhaus, Michael R.
Henry, Charles P.
Heran, William D.
Herzog, Donald C.
Hoover, Donald B.
Klein, Douglas P.
Kleinkopf, Dean
Krizman, Robert W.

Kuberry, Richard W.
Long, Carl L.
Martin, Ronny A.
Mitchell, Charles M.
Nelson, Maurice M.
Nervick, Kevin H.
O'Donnell, Jim E.
Osbakken, Willis E.
Papp, John E.
Paul, Lawrence E.
Peddie, Norman W.
Rohret, Donald H.
Roubique, Charles J.
Sauter, Edward A.
Senterfit, Robert M.
Sherrard, Mark S.
Sneddon, Richard A.
Stearns, Charles O.
Tilton, Sandra P.
Tippens, Charles L.
Towle, James N.
Townshend, John B.
Travis, Allen H.
Watts, Raymond D.
Wilson, Dolores M.
Wilson, Lanny R.
Wold, Richard J.
Wood, John D.
Wright, David L.

Branch of Petrophysics & Remote Sensing

Barth, Joseph J.
Beck, Myrl E., Jr.
Bressler, Stephen L.
Campbell, David L.
Cady, John W.
Ceder, Herbert W.
Champion, Duane E.
Cole, David
Crownover, Linda M.
Daniels, Jeffrey J.
Duval, Joseph S.
Elston, Donald P.
Friedman, Jules D.
Gassaway, Judith S.
Grommé, Sherman
Hagstrum, Jonathan T.
Hasbrouck, Wilfred P.
Heller, Joan S.
Hillhouse, John W.
Huff, William E.
Hughes, Eric M.
Hunt, Graham R.
Johnson, Gordon R.
Kelley, Edmund
Kingston, Marguerite J.
Kipfinger, Roy P., Jr.
Knepper, Daniel H., Jr.
Krohn, M. Dennis
Lewis, Elizabeth M.
Mankinen, Edward A.
McGee, Linda C.
Miller, Susanne H.
Milton, Nancy
Mohr, Pamela J.
Nelms, Charles A.
Olhoeft, Gary R.
Peters, Douglas C.
Pitkin, James A.
Podwysocki, Melvin H.
Pohn, Howard A.
Purdy, Terri L.
Raines, Gary L.
Rowan, Lawrence C.
Sampson, Jay A.
Sawatzky, Don L.
Scott, James H.
Seeley, Robert L.
Segal, Donald B.
Shoemaker, Eugene M.
Simpson, Shirley L.
Stehle, John P.
Watson, Kenneth

Branch of Analytical Laboratories

Alban, Eugene H.
Allingham, James M.
Anderson, Kenneth E.
Aruscavage, Philip J.
Baedecker, Philip A.
Baker, James W.
Bartel, Ardith J.
Bell, Edith E.
Berman, Sol
Bradley, Leon A.
Brandt, Elaine L.
Briggs, Paul H.
Brown, Floyd W.
Budahn, James R.
Burrow, George T.
Callahan, Betty L.
Campbell, Esma
Carr, Joseph F.
Childress, Anne E.
Conklin, Nancy M.
Crandell, William B.
Cremer, Marcelyn J.
Crock, James G.
d'Angelo, William M.
Dawson, Margaret P.
Dorrzapf, Anthony F., Jr.
Dwornik, Edward J.
Elsheimer, H. Neil
Engleman, Edythe E.
Fabbi, Brent P.
Finkelman, Robert B.
Flanagan, Francis J.
Fleming, Stanley L., II
Fletcher, Janet D.
Fries, Terry L.
Gandy, Gerald D.
Gent, Carol A.
Golightly, Danold W.
Gwyn, Mary E.
Haffty, Joseph
Hamilton, John C.
Harmon, Forrest L.
Harris, Joseph L.
Hatfield, D. Brooke
Haubert, Adolph W.
Hearn, Paul P.
Heropoulos, Chris
Jackson, Larry L.
Jenkins, Lillie B.
Johnson, Robert G.
Kane, Jean S.

Kawakita, Gary M.
Keaten, Barbara A.
Kelsey, James
King, Bi-shia
Kirchenbaum, Herbert
Klock, Paul R.
Knight, Roy J.
Lamothe, Paul J.
Larson, Richard R.
Layman, Lawrence R.
Leister, John W., Jr.
Lichte, Frederick E.
Lindsay, James R.
Malcolm, Mollie J.
Marinenko, John W.
Massoni, Camillo J.
Mays, Robert E.
McCall, Benjamin A.
McDade, Johnnie
McGregor, Robert E.
McKown, David M.
Mei, Leung
Merritt, Violet M.
Millard, Hugh T., Jr.
Morgan, John W.
Mossotti, Victor G.
Myers, Alfred T.
Neil, Sarah T.
Neiman, Harriet G.
Neuville, Arnett K.
Norton, Daniel R.
Pickering, Michael J.
Rait, Norma
Raspet, Rudolph
Riddle, George O.
Rose, Harry J.
Schnepfe, Marian M.
Schwarz, Louis J.
Sellers, George A.
Seeley, James L.
Shaw, Van E.
Shipley, Gaylord D.
Silk, Eleana S.
Skeen, Carol J.
Smith, Brian L.
Smith, Hezekiah
Solt, Merlyn W.
Sutton, Art L.
Taggart, Joseph E., Jr.
Thaxton, Charlene V.
Thomas, James A.
Tranter, Mary C.
Wahlberg, James S.
Walthall, Frank G.
Wandless, Gregory A.
Werre, Raymond W.
Wilson, Stephen A.
Young, Robert J.
Zbignewich, Edward W.
Zermane, Albert J.

OFFICE OF ENERGY RESOURCES

Benton, Lee L.
Butler, William C.
Chase, Barbara R.
Donnelly, Cyril A.
Girard, Oswald W., Jr.
Hoover, Linn
Lane, Marilyn K.
Lantz, Robert J.
Masters, Charles D.
McKelvey, Vincent E.
Murphy, Jack F.
Noble, E.A.
Pastrana, Lourdes N.

Branch of Oil & Gas Resources

Alley, Lonnie B.
Ambeau, Elsie Pinnace
Anders, Donald E.
Anderson, Robert C.
Arnal, Robert E.
Arthur, Michael A.
Balch, Alfred H.
Ball, Mahlon M.
Barker, Charles E.
Bayer, Kenneth C.
Becke, Donna M.
Becker, David G.
Beyer, Larry A.
Blank, Jeanne N.
Bostick, Neely H.
Bruns, Terry R.
Butler, Laurie L.
Carlson, Kurt H.
Carr, David R.
Cassidy, Robert J.
Charpentier, Ronald R.
Claypool, George E.
Coit, Theresa A.
Cook, Harry E.
Coury, Anny B.
Crovelli, Robert A.
Dalziel, Mary C.
DeMarinis, Susan K.
deWitt, Wallace, Jr.
Dolton, Gordon L.
Donovan, Terrence J.
Dyman, Thaddeus S.
Fisher, Michael A.
Fouch, Thomas D.
Fox, D. Dianne
Fox, James E.
Frezon, Sherwood E.
Gautier, Donald L.
Ging, Tom G.
Hahn, Deborah A.
Haley, Boyd R.
Halley, Robert B.
Harris, Leonard D.
Hendricks, John D.
Henry, Mitch E.
Howell, David G.
Hudson, J. Harold
Isaacs, Caroline M.
Keller, Margaret A.
Kepferle, Roy C.
Khan, Abdul S.
Krivoy, Harold L.
Lane, Brent L.
Law, Ben E.
LeFeber, Josephine A.

Lidz, Barbara H.
Linenberger, William
Lister, Jean H.
Long, Allan T.
Love, Alonza H.
Markochick, Dennis J.
Martinez, Roberto J.
Mast, Richard F.
Mattick, Robert E.
Maughan, Edwin K.
McKee, Edwin D.
McLean, Hugh
Megeath, Joe D.
Merewether, E. Allen
Michalski, Thomas C.
Molenaar, Cornelius M.
Nichols, Kathryn M.
Oman, Joanne K.
Palacas, James G.
Pawlewicz, Mark J.
Perry, William J., Jr.
Peterson, James A.
Pike, Robert S.
Pollack, Barry M.
Pollastro, Richard M.
Powers, Richard B.
Rhodehamel, Edward C.
Rice, Dudley D.
Robbin, Daniel M.
Robbins, Stephen L.
Roberts, Alan A.
Roen, John B.
Ryder, Robert T.
Schenk, Christopher J.
Scholle, Peter A.
Scott, Edward W.
Shinn, Eugene A.
Shurr, George W.
Spencer, Charles W.
Taylor, David J.
Thompson, Charles R.
Varnes, Katharine L.
Wallace, Laure G.
Weed, Elaine G.A.
Wesley, Jannette Sharp
Worl, Patricia B.
Yenowine, James H.
Zihlman, Frederick N.

Branch of Coal Resources

Affolter, Ronald H.
Altschuler, Zalman S.
Arndt, Harold H.
Babcock, Richard N.
Barnes, Connie A.
Berlage, Linda J.
Blake, Dorsey
Bodenlos, Alfred J.
Bragg, Linda J.
Breger, Irving A.
Carey, Mary Alice
Carter, M. Devereux
Cathcart, J. Daniel
Cecil, C. Blaine
Chao, Edward C.T.
Coleman, S. Lynn
Connor, Carol Waite
Correia, George A.
Crowley, Sharon S.
Culbertson, William C.
Dulong, Frank T.
Englund, Kenneth J.
Fitch, Sherry L.
Foose, Peggy L.
Freeman, Val L.
Flores, Romeo M.
Gordon, David W.
Griffith, Jean K.
Gualtieri, James L.
Hackman, Bettie S.
Hansen, Dan E.
Hardie, John K.
Haschke, Laura R.
Hatcher, Patrick G.
Henry, Thomas W.
Hickling, Nelson L.
Hildebrand, Ricky T.
Hobbs, Robert G.
Hunter, Judy F.
Jones, Vicky L.
Kehn, Thomas M.
Kent, Bion H.
Kerr, Patricia T.
Kirschbaum, Mark A.
Krasnow, Marta R.
Krohn, Kathleen Kozey
Landis, Edwin R.
Lemaster, Mary E.
Lyons, Paul C.
Mapel, William J.
McLellan, Marguerite W.
McPhillips, Maureen
Medlin, Jack H.
Meissner, Charles R., Jr.
M'Gonigle, John W.
Minkin, Jean A.
Molnia, Carol L.
Montijo, Elouise V.
Mytton, James W.
Needham, Russell E.
O'Neal, Rebecca K.
Olson, Annabel B.
Oman, Charles L.
Pastore, Eileen R.
Pitts, JoAnn K.
Pugh, Young-Ja
Rega, Noreen H.
Rice, Charles L.
Richmond, Deanne L.
Robbins, Eleanora I.
Roehler, Henry W.
Schneider, Gary B.
Schweinfurth, Stanley P.
Sigleo, Wayne R.
Smith, Ricke J.
Stamm, Margaret E.
Stricker, Gary D.
Teasdale, Warren E.
Thompson, Carolyn L.
Trent, Virgil A.
Ward, Dwight E.
Warlow, Ralph C.
Weaver, Jean Noe
Windolph, John F., Jr.
Wood, Gordon H., Jr.
Zubovic, Peter

Branch of Sedimentary Mineral Resources

Andrews, J. Stacey
Asher-Bolinder, Sigrid
Bodine, Marc W., Jr.
Brumley, Edith B.
Buntenbah, Carole J.
Cashion, William B.
Cathcart, James B.
Dallin, Madelyn
Donnell, John R.
Durham, David L.
Gude, Arthur J., III
Gulbrandsen, Robert A.
Hail, William J.
Hite, Robert J.
Humphreys, Richard D.
Johnson, Ronald C.
Jones, Charles L.
Kasmen, Nancy G.
Keighin, C. William
Ketner, Keith B.
Madsen, Beth M.
Mahrt, Louis R., Jr.
Milton, Charles
Pantea, Michael P.
Raup, Omer B.
Schack, Artis M.
Sheldon, Richard P.
Sheppard, Richard A.
Simmons, Florence M.
Smith, George I.
Smith, Marjorie C.
Vercoutere, Thomas L.
Vine, James D.
Whitney, C. Gene

Branch of Uranium & Thorium Resources

Armbrustmacher, Theodore J.
Armstrong, Frank C.
Been, Josh M.
Boudette, Eugene L.
Brownfield, Isabelle K.
Byers, Virginia P.
Cadigan, Robert A.
Campbell, John A.
Condon, Steven M.
Craig, Lawrence C.
Denson, Norman M.
Dickinson, Kendell A.
Dodge, Harry W., Jr.
Dubiel, Russell F.
Felmlee, J. Karen
Finch, Warren I.
Franczyk, Karen J.
Goldhaber, Martin B.
Granger, Harry C.
Grauch, Richard I.
Green, Morris W.
Hammond, David J.
Hills, F. Allan
Huffman, A. Curtis
Kirk, Allan R.
Leventhal, Joel S.
Lupe, Robert D.
Macke, David L.
Murrey, Donald G.
Nash, J. Thomas
Offield, Terry W.
Otton, James K.
Peterson, Fred
Phillips, Carol A.
Reimer, G. Michael
Reynolds, Richard L.
Ridgley, Jennie L.
Robertson, Jacques F.
Robinson, Keith
Santos, Elmer S.
Seeland, David A.
Sikkink, Pamela G.
Simmons, Kathleen R.
Spirakis, Charles S.
Staatz, Mortimer H.
Steele-Mallory, Brenda A.
Stuckless, John S.
Suits, Vivian J.
Thaden, Robert E.
Tapia-Frisken, Marie L.
Turner-Peterson, Christine E.
Wenrich-verbeek, Karen J.
Young, Edward J.
Zielinski, Robert A.

OFFICE OF ENVIRONMENTAL GEOLOGY

Atherton, Nancy C.
Beeson, Fern E.
Bennett, Glenn E.
Cronin, Thomas M.
Fatz, Lorrie P.
Harris, Linda D.
Krushensky, Richard D.
Maberry, John O.
McGuire, Virginia A.
Morris, Robert H.
Morton, Douglas M.
Nichols, Lindsay R.
Olson, Rita J.
Tilley, Scott E.
Weant, Shari L.

Branch of Eastern Environmental Geology

Aquilino, Linda M.
Beardsley, Mary A.
Black, Douglas F.B.
Bonham, Selma M.
Clarke, James W.
Cranford, S. Linda
DeCillis, Maria C.
Dempsey, William J.
Duty, Dennis W.
Epstein, Jack B.
Froelich, Albert J.
Force, Lucy M.
Gohn, Gregory S.
Goldsmith, Richard
Hackman, Robert J.
Hatch, Norman L., Jr.
Higgins, Michael W.
Horton, J. Wright
Larsen, Frederick D.
Leavy, Brian D.
Lee, Kwang Yuan
Leo, Gerhard W.

London, Elizabeth B.H.
Lyttle, Peter T.
Markewich, Helaine Walsh
McDowell, Robert C.
Mills, Hugh H.
Milton, Daniel J.
Mixon, Robert B.
Newell, Wayne L.
Niccolls, Linda J.
Novak, Steven W.
Osberg, Philip H.
Outerbridge, William F.
Owens, James P.
Pavich, Milan J.
Pavlides, Louis
Peper, John D.
Peterson, Warren L.
Pomeroy, John S.
Prowell, David C.
Queen, Donald G.
Rachlin, Jack
Rankin, Douglas W.
Ratcliffe, Nicholas M.
Reinhardt, Juergen
Rodgers, John
Ruane, Paul J.
Schafer, John Phillip
Schneider, Ray R.
Stanley, Rolfe S.
Stone, Byron D.
Stone, Janet R.
Sutton, Robin C.
Taylor, Alfred R.
Thomas, Roger E.
Trask, Newell J.
Van Driel, J. Nicholas
Warren, Charles R.
Weems, Robert E.
Whitlow, Agnes W.
Wier, Karen E.
Wilson, Frederick A.

Branch of Central Environmental Geology

Amos, Dewey H.
Baltz, Elmer H.
Bryant, Bruce H.
Cady, Wallace M.
Carrara, Paul E.
Christiansen, Ann Coe
Coates, Donald A.
Colman, Steven M.
Colton, Roger B.
Condit, Christopher D.
Cressman, Earle R.
Derick, James L.
Dixon, H. Roberta
Eggleton, Richard E.
Fullerton, David S.
Gable, Dolores J.
Gibbons, Anthony B.
Glick, Ernest E.
Hansen, Wallace R.
Harrison, Jack E.
Hawkins, Fred F.
Heggem, Flora A.
Hereford, Richard
Hobbs, S. Warren
Hoggan, Roger D.
Izett, Glen Arthur
Johnson, Darline E.
Johnson, William D., Jr.
Karlstrom, Thor N.V.
Kuntz, Mel A.
Lidke, David J.
Lipman, Peter W.
Lopez, David A.
Love, J. David
Lucchitta, Ivo
Luft, Stanley J.
Machette, Michael N.
Madole, Richard F.
Malde, Harold E.
Manley, Kim
McBroome, Lisa A.
Moore, David W.
Myers, Donald A.
Nealey, L. David
O'Connell, Everett M.
O'Neill, J. Michael
Oriel, Steven S.
Pierce, Kenneth L.
Pillmore, Charles L.
Platt, Lucian B.
Prichard, George E.
Raup, Robert B., Jr.
Reed, John C., Jr.
Richmond, Gerald M.
Rowley, Peter D.
Ruppel, Edward T.
Sargent, Kenneth A.
Schmidt, Dwight L.
Schmidt, Robert George
Scott, Glenn R.
Sharps, Joseph A.
Shawe, Fred R.
Shride, Andrew F.
Shroba, Ralph R.
Skipp, Betty A.
Snyder, George L.
Stravers, Jay A.
Sutton, Robert L.
Ulrich, George E.
Vennum, Walter R.
Verbeek, Earl R.
Wallace, Chester A.
Weide, David L.
Weir, Gordon W.
Weiss, Malcolm P.
Williams, Paul L.
Williams, Van S.
Witkind, Irving J.
Wobus, Reinhard A.
Wolfe, Edward W.
Wyatt, Nancy J.

Branch of Western Environmental Geology

Adam, David P.
Allison, Annie L.
Atwater, Brian F.
Barker, Judy A.
Barnett, Catherine A.
Bartow, J. Alan

Blake, Milton C., Jr.
Brabb, Earl E.
Campbell, Catherine C.
Carr, Michael D.
Carson, Scott E.
Clark, Joseph C.
Coleman, Annette L.
Crittenden, Max D., Jr.
Doukas, Michael P.
Dupré, William R.
Ellen, Stephenson D.
Fischer, Charlene R.
Fox, Kenneth F., Jr.
Frizzell, Virgil A., Jr.
Griffin, Elizabeth A.
Harden, Jennifer W.
Harwood, David S.
Heller, Paul L.
Helley, Edward J.
Herd, Darrell G.
Hietanen-Makela, Anna M.
Hoggatt, Wendy C.
Howard, Keith A.
Huber, N. King
John, Barbara E.
Jones, Libby L.
King, Philip B.
Krodel, Lida L.
Lettis, William R.
Marchand, Denis E.
Mark, Robert K.
Matti, Jonathan C.
McLaughlin, Robert J.
McMasters, Catherine R.
Meyer, Charles E.
Milan, Mary P.
Miller, David M.
Miller, Fred K.
Minard, James P.
Moore, Susan L.
Newman, Evelyn B.
Nilsen, Tor H.
Norman, Meade B., II
Pernokas, Martha A.
Peterson, David M.
Pierce, William G.
Pike, Jane E.
Robinson, Gershon D.
Rodriguez, Eduardo A.
Rogers, Bruce W.
Rusmore, Margaret E.
Russell, Paul C.
Sarna-Wojcicki, Andrei M.
Schlocker, Julius
Seiders, Victor M.
Sorg, Dennis H.
Stone, Paul
Stuart-Alexander, Desiree E.
Swint, Theresa R.
Tabor, Rowland W.
Taylor, Fred A.
Throckmorton, Constance K.
Tinsley, John C., III
Waitt, Richard B., Jr.
Wells, Ray E.
Wentworth, Carl M.
Williams, Kathleen M.
Wilshire, Howard G.
Yetter, Thomas J.
Yount, James C.

Branch of Engineering Geology

Agard, Sherry S.
Barker, Rachel M.
Baskerville, Charles A.
Bennett, Michael J.
Bennetti, John B., Jr.
Bostwick, Larry G.
Buchanan-Banks, Jane M.
Butterfield, Willard
Campbell, Russell H.
Chen, Albert T.F.
Chleborad, Alan F.
Collins, Donley S.
Crandell, Dwight R.
Davies, William E.
Dickey, Dayton D.
Diehl, Sharon F.
Dobrovolny, Ernest E.
Dow, Virginia L.
Dunrud, C. Richard
Ege, John R.
Erickson, George S.
Farrow, Richard A.
Fleming, Robert W.
Gardner, Cynthia A.
Hait, Mortimer H., Jr.
Harp, Edwin L.
Hays, William H.
Hinrichs, E. Neal
Hoblitt, Richard P.
Holzer, Thomas L.
Hoose, Seena N.
Houser, Fred N.
Jackson, Charlotte E.
Jimenez, Irene S.
Johnson, Michelina J.
Kanizay, Stephen P.
Keefer, David K.
Lee, Fitzhugh T.
Lee, Homa J.
Lindvall, Robert M.
McGill, John T.
McGregor, Edward E.
McLaughlin, Philip V.
Miller, C. Dan
Miller, Carter H.
Miller, Danny R.
Miller, Robert D.
Miller, Theron E.
Mullineaux, Donal R.
Nichols, Donald R.
Nichols, Roger W.
Nichols, Thomas C., Jr.
Obermeier, Stephen F.
Odum, Jack K.
Ohlmacher, Gregory C.
Olsen, Harold W.
Osterwald, Frank W.
Pampeyan, Earl H.
Pippin, Marjorie F.
Poole, Patricia C.
Powers, Philip S.
Radbruch-Hall, Dorothy H.
Reheis, Marith C.
Robertson, Eugene C.
Savage, William Z.
Schmoll, Henry R.
Schuster, Robert L.
Shaler, Sam
Shifflett, Carol M.

Simpson, Howard E.
Smith, William K.
Swolfs, Henri S.
Tannaci, Nancy E.
Van Horn, Richard
Varnes, David J.
Weedman, Rosalie E.
Wieczorek, Gerald F.
Wilson, Raymond C.
Worley, David M.
Yehle, Lynn A.
Yerkes, Robert F.
Youd, T. Leslie
Young, Garnet S.

Branch of Special Projects

Bath, Gordon D.
Byers, Frank M., Jr.
Carr, Wilfred J.
Carroll, Roderick D.
Corchary, George S.
Cunningham, David R.
Dixon, Gary L.
Ellis, William L.
Fernald, Arthur T.
Gard, Leonard M., Jr.
Glanzman, Virginia M.
Hamblin, Irene F.
Healey, Don L.
Hoover, David L.
Jenkins, Evan C.
Kibler, John D.
Kibler, Joyce E.
Magner, Jerry E.
Maldonado, Florian
Moser, Jeanette M.
Muller, Douglas C.
Orkild, Paul P.
Rivers, Willie C.
Rosenbaum, Joseph G.
Snyder, Richard P.
Spengler, Richard W.
Stead, Frank W.
Starkey, Ruth W.
Swadley, W. C.
Tucker, Judith A.
Tuftin, Steven E.
Twenhofel, William S.

Branch of Astrogeologic Studies

Arthur, David W.G.
Bachstein, Elizabeth
Barcus, Loretta A.
Batson, Raymond M.
Boudreau, Ramona L.
Breed, Carol S.
Bridges, Patricia M.
Briggs, Melrose M.
Carr, Michael H.
Carroll, Linda J.
Clow, Gary D.
Conley, Edward
Davis, Susan L.
Dial, Arthur L., Jr.
Edwards, Kathleen L.
Ferguson, Holly M.
Grolier, Maurice J.
Hall, Virginia M.
Harper, Billie J.
Hodges, Carroll Ann
Holt, Henry E.
Inge, Jay L.
Jordan, Raymond
Kadish, Ruth L.
Kieffer, Hugh H.
Lucchitta, Baerbel K.
Lugn, Richard V.
Masursky, Harold
McCauley, John F.
Moore, Henry J.
Morris, Elliot C.
Pike, Richard J.
Roddy, David J.
Roeming, Susan S.
Schaber, Gerald G.
Schafer, Francis J.
Scott, David H.
Soderblom, Laurence A.
Strobell, Mary E.
Swann, Jody
Tyner, Richard L.
Valdovino, Yolanda L.
Ward, A. Wesky
Weir, Doris B.
Wolfe, Ruth F.
Wu, Sherman S.C.
Zeller, Karl A.

Branch of Paleontology & Stratigraphy

Ager, Thomas A.
Andrews, George W.
Armstrong, Augustus K.
Balanc, Marija
Baldauf, Jack G.
Barron, John A.
Bedette, Barbara A.
Berdan, Jean M.
Blackwelder, Blake W.
Bown, Thomas M.
Bradbury, J. Platt
Burkholder, Robert E.
Burnham, Robyn J.
Bybell, Laurel M.
Christopher, Raymond A.
Cobban, William A.
Cole, Grace L.
Compton, Ellen E.
Crawford, Janie M.
Cronan, Mary
Dieterich, Kathryn V.
Doher, L. Imogene
Douglass, Raymond C.
Dutro, J. Thomas, Jr.
Edwards, Lucy E.
Flot, Terry R.
Frederiksen, Norman O.

Gibbs, JoAnn
Gibson, Thomas G.
Gordon, Mackenzie, Jr.
Hanley, John H.
Harris, Anita G.
Hazel, Joseph E.
Hershiser, Robert W.
Hunter, Patricia A.
Imlay, Ralph W.
Karklins, Olgerts L.
Karlson, Kathryn H.
Kosanke, Robert M.
Leonard, Susan E.
Low, Doris L.
MacDonald, Edward F.
Mamay, Sergius H.
Margerum, Richard
Marincovich, Louie N., Jr.
Martin, Wayne E.
Mason, Charles E.
May, Fred E.
McDougall, Kristin
McKinney, Robert H.
Miller, John W.
Miller, Laurie D.
Mochizuki, Haruo E.
Moore, Ellen J.
Moore, Keith R.
Neuman, Robert B.
Newton, Sally K.
Nichols, Douglas J.
O'Donnell, Robert W.
Oftedahl, Orrin G.
Oliver, William A., Jr.
Paulson, William C.
Pinckney, Wiliam C., Jr.
Poag, C. Wylie
Pojeta, John, Jr.
Poore, Richard Z.
Repenning, Charles A.
Repetski, John E.
Ross, Reuben James, Jr.
Ross, William O.
Rouse, Mary A.
Russell, Dale M.
Sakamoto, Kenji
Saldukas, R. Birute
Sando, William J.
Saunders, Harold I.
Schneider, Jonathan N.
Scholten, John R.
Shaw, Effie G.
Silberling, Norman J.
Sliter, William V.
Smith, Charles C.
Sohl, Norman F.
Sohn, Israel Gregory
Suess, Hans E.
Trombley, Rose M.
Tschudy, Robert H.
Valentine, Page C.
Van Loenen, Sharon D.
Vietor, Gladys L.
Ward, Lauck W.
Wardlaw, Bruce R.
Watt, Arthur D.
Whitmore, Frank C., Jr.
Wilson, Leonard A.
Yochelson, Ellis L.

OFFICE OF EARTHQUAKE STUDIES

Andrews, Dudley J.
Bachert, Les
Brown, Robert D.
Costello, Joyce A.
Devine, James F.
Filson, John R.
Frazeur, W. Scott
Greene, Gordon W.
Jones-Cecil, Meridee
Mead, Jennifer R.
Miles, Cheryl
Nelson, Glenn D.
O'Neill, Mary E.
Orsini, Nicholas A.
Phelps, William E., Jr.
Potts, Rosemarie
Raleigh, C. Barry
Reeves, Jessie F.
Seiders, Wanda H.
Simirenko, Marie J.
Stewart, Roger M.
Turner, Monica L.
Wallace, Robert E.
Ward, Karen M.
Weddle, Mary I.
Wesson, Robert L.

Branch of Global Seismology

Arnold, Edouard P.
Baldwin, Francis W.
Bolchunos, Marilyn C.
Bowman, Ethel G.
Britton, O.J.
Butler, Howell M.
Choy, George L.
Clark, Harold E., Jr.
Dewey, James W.
Dunphy, Gerald J.
Echert, Elise D.
Engdahl, Eric R.
Evans, David H.
Green, William N.
Hoffman, John P.
Holcomb, L. Gary
Howell, Trevor H.
Hunter, Roger N.
Hutt, Charles R.
Jacobs, Willis S.
Jaksha, Lawrence H.
Jordan, James N.
Kangas, Reino
Kerry, Leonard E.
Kimball, Barbara
Koontz, Allen H.
Medina, Edward S.
McCarthy, Robert P.
Mendoza, Carlos
Minsch, John H.
Morris, Lynn G.
Murdock, James N.
Nevin, Jan M.
Odell, Lyndon A.
Peterson, Jon R.

Reagor, Bobby G.
Reynolds, Robert D.
Schmieder, William H.
Senecal, Joseph M.
Sipkin, Stuart A.
Sloan, Barbara J.
Smith, Paula K.
Spence, William J.
Stover, Carl W.
Taggart, James N.
Thompson, John B.
Whitcomb, Harry S.
Wilson, Edgar R.
Young, Robert L.

Branch of Seismology

Bakun, William H.
Bennett, Hugh F.
Bufe, Charles G.
Buhr, Grover S.
Ellsworth, William L.
Evans, John R.
Fischer, Frederick G.
Fluty-Harvey, LuAnn
Fuis, Gary S.
Gunn, Mari L.
Hall, Philip C.
Hill, David P.
Hirscher, Elsie H.
Iyer, H. Mahadeva
Koesterer, Charles L.
Kohler, William M.
Lee, William H.K.
Leone, Laurel E.
Mavko, Barbara B.
Pelton, John R.
Pfluke, John H.
Pitt, Andrew M.
Rapport, Amy L.
Spieth, Mary Ann
Stewart, Samuel W.
Ward, Peter L.
Warren, David H.
York, Judy K.
Zucca, John J.

Branch of Seismic Engineering

Porcella, Ronald L.
Silverstein, Barry L.

Branch of Ground Motion & Faulting

Archuleta, Ralph J.
Baker, Lawrence M.
Bonilla, Manuel G.
Boore, David M.
Borcherdt, Roger D.
Burke, Dennis B.
Clark, Malcolm M.
Condrotte, Charles G.
Covington, Pamela A.
Cranswick, Edward
Fogleman, Kent A.
Gibbs, James F.
Harmsen, Stephen C.
Hazlewood, Robert M.
Helton, Suzanne M.
Joyner, William B.
Kennedy, George L.
Lahr, John C.
Lajoie, Kenneth R.
Mathieson, Scott A.
Nason, Robert D.
Navarro, Richard
Olsen, Alice L.
Page, Robert A.
Park, Robert B.
Ponti, Daniel J.
Raugh, Mike R.
Rogers, Al M.
Rogers, John A.
Roth, Edward F.
Stephens, Christopher D.
Vaughan, Patrick R.
Warrick, Richard E.
Wuollet, Geraldine M.
Wong, William B.

Branch of Earthquake Tectonics & Risk

Algermissen, Sylvester T.
Anderson, R. Ernest
Barnhard, Lynn M.
Bice, Tom
Bogdon, Marianne
Brockman, Stanley R.
Bucknam, Robert C.
Burford, Robert O.
Carver, David L.
Crone, Anthony J.
Dart, Richard L.
Diment, William H.
Espinosa, Alvaro F.
Hamilton, Robert M.
Harding, Samuel T.
Hopper, Margaret G.
Irwin, William P.
Keefer, Marcia Mergner
Langer, Charley J.
McCune, Dorothy G.
McKeown, Frank A.
Overturf, Dee E.
Perkins, David M.
Rhea, B. Susan
Ricotta, Ralph J.
Ross, Donald C.
Russ, David P.
Rymer, Michael J.
Shedlock, Kaye M.
Sims, John D.
Thomasson, Carol J.
Tibbetts, Reva F.
Wheeler, Russell L.
Zoback, Mary Lou

Branch of Tectonophysics

Byerlee, James D.
Daul, William B.
Dieterich, James H.
Galanis, S. Peter, Jr.
Gallanthine, Steven K.

Grubb, Fredd V.
Harper, Kenneth R.
Harsh, Philip W.
Herriot, James W.
Husk, Robert H., Jr.
Iwatsubo, Eugene Y.
Johnston, Malcolm J.
Jones, Alan C.
Keller, Vincent G.
Kennelly, John P., Jr.
King, Chi-Yu
King, Nancy E.
Lachenbruch, Arthur H.
Langbein, John O.
Liu, Hsi-Ping
Lisowski, Michael
Lockner, David A.
Lutter, William J.
Marshall, B. Vaughn
Mase, Charles W.
Mavko, Gerald M.
McNutt, Marcia K.
Miranda, Sandra L.
Mortensen, Carl E.
Moses, Thomas H.
Mueller, Robert J., Jr.
Munroe, Robert J.
Murray, Thomas L.
Myren, Glenn D.
Peselnick, Louis
Pinkston, John C.
Prescott, William H.
Rodriguez, Thelma R.
Sandoval, Nancy C.
Sass, John H.
Savage, James C.
Schulz, Sandra S.
Silverman, Stan A.
Smith, Eugene P.
Svitek, Joseph F.
Thatcher, Wayne
Wegener, Steven S.
Westerlund, Robert E.
Zoback, Mark D.

Branch of Network Operations

Allen, Rex V.
Butler, Frederick L.
Coakley, John M.
Eaton, Jerry P.
Ellis, James O.
Hall, Wesley D.
Jackson, Thomas C.
Jensen, E. Gray
Jensen, Richard E.
Kaderabek, Ronald M.
Lee, Clearthur
Locquiao, Alejo M.
McClearn, Robert W.
Somera, Robert G.
Stevenson, Peter R.
Taylor, David E.
Taylor, Jeanne M.
Tomey, Jack W.
Van Schaack, John R.
Vaughn, Alvin J.
Wong, Helen C.

OFFICE OF MARINE GEOLOGY

Corwin, Gilbert
Edgar, N. Terence
Kempema, Edward W.
O'Leary, Dennis W.
Rowland, Robert W.
Taylor, Mary E.
Teleki, Paul G.
Tracey, Joshua I., Jr.
Willis, Carolyn A.

Branch of Pacific-Arctic Geology

Allan, Mary A.
Alpha, Tau R.
Atwood, Thomas J.
Bailey, Shirley A.
Barnes, Peter W.
Baskett, Sharon L.
Basler, James R.
Biagi, Carlo H.
Bingham, Michael P.
Bischoff, James L.
Blunt, David J.
Boucher, Gary W.
Buffington, Edwin C.
Bukry, John D.
Caras, Geoffrey B.
Carlson, Paul R.
Carpenter, Carrie E.
Chase, Thomas E.
Chin, John L.
Clague, David A.
Clifton, H. Edward
Cochrane, Guy R.
Cooper, Alan K.
Crockett, Reuben D.
Dadisman, Shawn V.
Dates, Merid D.
Dingler, John R.
Dinter, David A.
Eittreim, Stephen L.
Field, Michael E.
Fischer, Jeffrey M.
Fletcher, Charles H.
Friesen, Walter B.
Garlow, Richard A.
Goodfellow, Robert W.
Graham, Marjorie A.
Greenberg, Jonathan J.
Gutmacher, Christina E.
Hart, Patrick E.
Hein, James R.
Hess, Gordon R.
Hill, Gary W.
Hirozawa, Carol A.
Howd, Peter A.
Hunter, Ralph E.
Hutchison, La Vernné W.
Jones, David M.
Jones, D.R.
Knipe, Diane E.
Kvenvolden, Keith A.
Kollmann, Auriel C.
Koski, Randolph A.
Lander, Diane L.
Larsen, Matthew C.

Lee, Jack W.
Leong, Kam Wo
Lichtman, Grant S.
Luepke, Gretchen
Mann, Dennis M.
Marlow, Michael S.
Mascardo, Juanita S.
Mathews, Susan K.
May, Steven D.
McClellan, Patrick H.
McCulloh, Thane H.
McHendrie, A. Graig
Miller, M. Meghan
Minkler, Peter W.
Molnia, Bruce F.
Moore, George W.
Morley, James M.
Moyer, Frances V.
Nelson, C. Hans
Nelson, Martha J.
Nichols, Frederic H.
Normark, William R.
Ogle, Helen M.
Orlando, Robert C.
Peake, Loren G.
Pearl, James E.
Pierce, Charlotte J.
Piper, David Z.
Quinterno, Paula J.
Rapp, John B.
Rappeport, Mel
Rearic, Douglas M.
Reimnitz, Erk
Richmond, Bruce M.
Rosenbauer, Robert J.
Rubin, David M.
Rubin, Jason S.
Ruder, Vic C.
Sallenger, Asbury H.
Sanders, Rex
Sicard, Dorothy M.
Sinnock, Shannon K.
Snavely, Parke D., Jr.
Steele, William C.
Swenson, Doris R.
Thompson, Janet K.
Utter, Pamela A.
Vallier, Tracy L.
Vedder, John G.
Wagner, Holly C.
Wang, Frank H.
Wiley, M. Claire

Branch of Atlantic-Gulf of Mexico Geology

Aldrich, Thomas C.
Ambuter, Bruce P.
Bailey, Norman G.
Barta, Jimmie L.
Berryhill, Henry L.
Blackwood, Dann S.
Booth, James S.
Botbol, Joseph M.
Bothner, Michael H.
Bouma, Arnold H.
Bowker, Paul C.
Bowles, Robert M.
Bryden, Cynthia G.
Burke, Janet B.
Butman, Bradford
Cashman, Katharine V.
Circe, Ron C.
Commeau, Judith A.
Commeau, Robert F.
Conley, Sandra J.
Cousins, Carolyn G.
Coward, Elizabeth L.
Dahl, Alfred G.
Davis, Raymond E.
Deadmon, Charles E.
Dillon, John J.
Dillon, William P.
Dodd, James E.
Dorey, Susan E.
Egelson, David C.
Eisner, Sara B.
Escowitz, Edward C.
Evenden, Gerald I.
Ferrebee, Wayne M.
Folger, David W.
Foote, Richard Q.
Forrestel, Patricia L.
Fugate, James K.
Garrison, Louis E.
Gelinas, Janet L.
Green, Arthur W., Jr.
Grow, John A.
Hall, Raymond E.
Hampson, John C., Jr.
Harrison, D. George
Hathaway, John C.
Hempenius, Margaret W.
Holmes, Charles W.
Hutchinson, Deborah R.
Kent, Kathleen M.
Kindinger, Jack L.
Kindinger, Mary E.
Kirby, John R.
Klitgord, Kim D.
Knebel, Harley J.
Larson, John C.
Lerch, Harry E., III
Ligon, Dennis T., Jr.
Lillard, Neal M.
Lombard, Charles W.
Manheim, Frank T.
Martin, E. Ann
Martin, Ray G.
Mason, David H.
Massingill, Linda M.
McCarthy, Gerard V.
McDaniel, Susan E.
McGregor, Bonnie A.
Mihalyi, Dale L.
Miller, Robert E.
Miller, Ronald J.
Mons-wengler, Margaret Clare
Musialowski, Frank R.
Needell, Sally W.
Noble, Marlene A.
North, Lester D.
O'Hara, Charles J.
Oldale, Robert N.
O'Leary, Dennis W.
Owen, Douglass E.
Parmenter, Carol M.

Parolski, Kenneth F.
Paull, Charles K.
Popenoe, Peter
Poppe, Lawrence J.
Purdy, Susan J.
Ray, James D.
Reed, Betty S.
Reed, Bobbi J.
Rendigs, Richard R.
Rice, Cynthia A.
Rice, Thomas L.
Robb, James M.
Scanlon, Kathryn M.
Schlee, John S.
Schultz, David M.
Shideler, Gerald L.
Shoukimas, Mary E.
Shurtleff, Joan L.
Soderberg, Nancy K.
Stelting, Charles F.
Swift, B. Ann
Sylwester, Richard W.
Trippet, Anita L.
Trumbull, James V.A.
Twichell, David C., Jr.
Walker, William M.
Wilde, Amanda M.
Williams, Edna L.
Williams, Joan B.
Willingham, Betty L.
Willingham, James W.
Winget, Elizabeth A.
Winham, Dena J.
Woo, Ching Chang
Wright, Evelyn L.
Zelibor, Joseph L.

NOTES

NOTES

NOTES

NOTES

NOTES

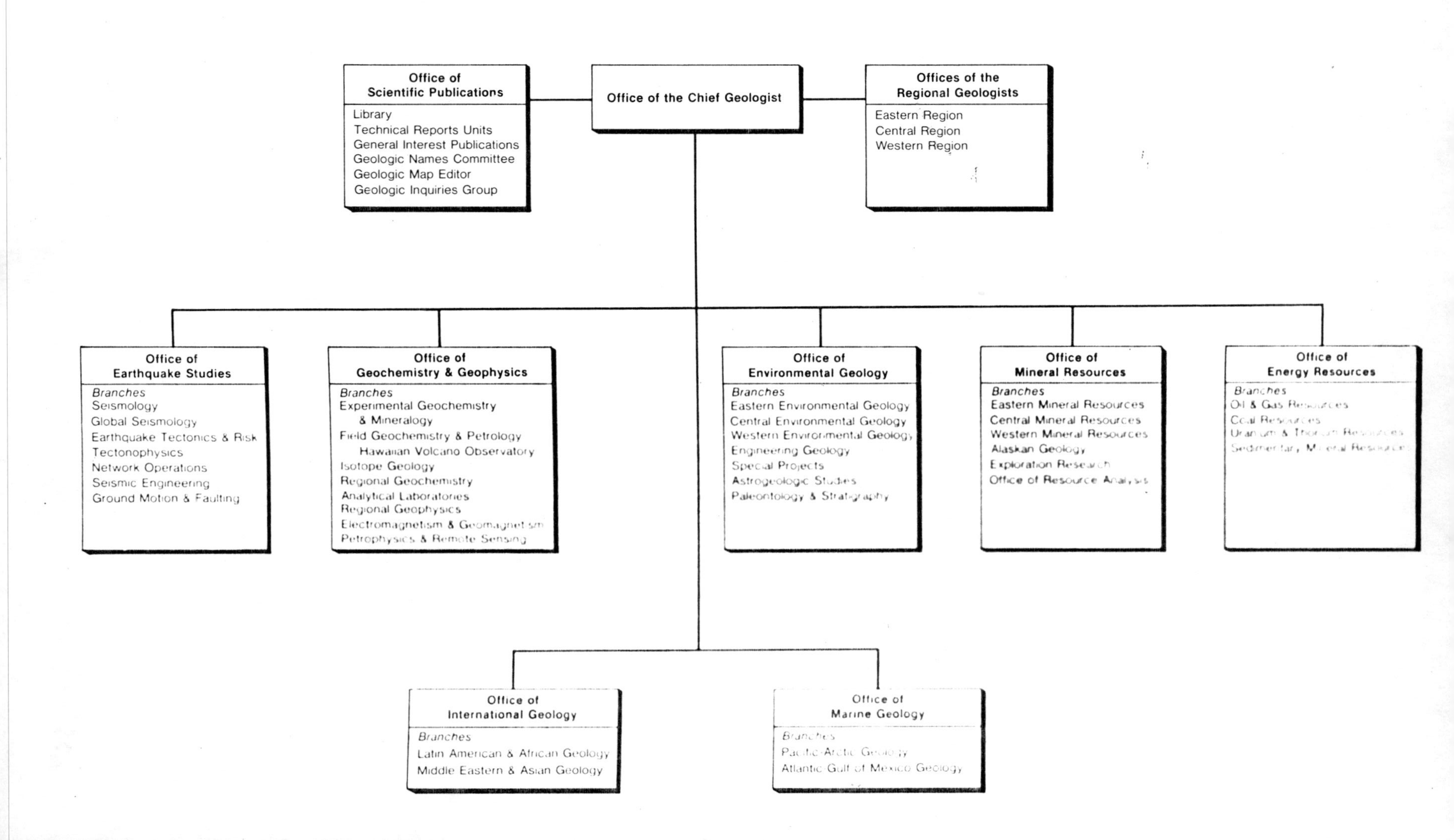
GEOLOGIC DIVISION ORGANIZATION
Office of the Chief Geologist
Office of Scientific Publications
Library
Technical Reports Units
General Interest Publications
Geologic Names Committee
Geologic Map Editor
Geologic Inquiries Group
Offices of the Regional Geologists
Eastern Region
Central Region
Western Region
Office of Earthquake Studies
Branches
Seismology
Global Seismology
Earthquake Tectonics & Risk
Tectonophysics
Network Operations
Seismic Engineering
Ground Motion & Faulting
Office of Geochemistry & Geophysics
Branches
Experimental Geochemistry & Mineralogy
Field Geochemistry & Petrology
Hawaiian Volcano Observatory
Isotope Geology
Regional Geochemistry
Analytical Laboratories
Regional Geophysics
Electromagnetism & Geomagnetism
Petrophysics & Remote Sensing
Office of Environmental Geology
Branches
Eastern Environmental Geology
Central Environmental Geology
Western Environmental Geology
Engineering Geology
Special Projects
Astrogeologic Studies
Paleontology & Stratigraphy
Office of Mineral Resources
Branches
Eastern Mineral Resources
Central Mineral Resources
Western Mineral Resources
Alaskan Geology
Exploration Research
Office of Resource Analysis
Office of Energy Resources
Branches
Oil & Gas Resources
Coal Resources
Uranium & Thorium Resources
Sedimentary Mineral Resources
Office of International Geology
Branches
Latin American & African Geology
Middle Eastern & Asian Geology
Office of Marine Geology
Branches
Pacific-Arctic Geology
Atlantic-Gulf of Mexico Geology

1896